Quantum Gravity Phenomenology

Quantum Gravity Phenomenology

Editors

Arundhati Dasgupta
Alfredo Iorio

Basel • Beijing • Wuhan • Barcelona • Belgrade • Novi Sad • Cluj • Manchester

Editors
Arundhati Dasgupta
University of Lethbridge
Lethbridge, AB
Canada

Alfredo Iorio
Charles University
Prague
Czech Republic

Editorial Office
MDPI AG
Grosspeteranlage 5
4052 Basel, Switzerland

This is a reprint of articles from the Special Issue published online in the open access journal *Universe* (ISSN 2218-1997) (available at: https://www.mdpi.com/journal/universe/special_issues/Quantum_Gravity_Phenomenology).

For citation purposes, cite each article independently as indicated on the article page online and as indicated below:

Lastname, A.A.; Lastname, B.B. Article Title. *Journal Name* **Year**, *Volume Number*, Page Range.

ISBN 978-3-7258-2373-4 (Hbk)
ISBN 978-3-7258-2374-1 (PDF)
doi.org/10.3390/books978-3-7258-2374-1

© 2024 by the authors. Articles in this book are Open Access and distributed under the Creative Commons Attribution (CC BY) license. The book as a whole is distributed by MDPI under the terms and conditions of the Creative Commons Attribution-NonCommercial-NoDerivs (CC BY-NC-ND) license.

Contents

About the Editors vii

Preface ix

Carlo Rovelli
Considerations on Quantum Gravity Phenomenology
Reprinted from: *Universe* **2021**, *7*, 439, doi:10.3390/universe7110439 1

Nick E. Mavromatos, Pablo Pais and Alfredo Iorio
Torsion at Different Scales: From Materials to the Universe
Reprinted from: *Universe* **2023**, *9*, 516, doi:10.3390/universe9120516 5

Igor I. Smolyaninov and Vera N. Smolyaninova
Analogue Quantum Gravity in Hyperbolic Metamaterials
Reprinted from: *Universe* **2022**, *8*, 242, doi:10.3390/universe8040242 50

V. V. Sreedhar and Amitabh Virmani
Maximal Kinematical Invariance Group of Fluid Dynamics and Applications
Reprinted from: *Universe* **2022**, *8*, 319, doi:10.3390/universe8060319 58

Mikhail O. Katanaev
Spin Distribution for the 't Hooft–Polyakov Monopole in the Geometric Theory of Defects
Reprinted from: *Universe* **2021**, *7*, 256, doi:10.3390/universe7080256 76

Paolo Castorina, Alfredo Iorio and Helmut Satz
Hunting Quantum Gravity with Analogs: The Case of High-Energy Particle Physics
Reprinted from: *Universe* **2022**, *8*, 482, doi:10.3390/universe8090482 83

Giovanni Acquaviva, Alfredo Iorio, Pablo Pais and Luca Smaldone
Hunting Quantum Gravity with Analogs: The Case of Graphene
Reprinted from: *Universe* **2022**, *8*, 455, doi:10.3390/universe8090455 105

Ednardo Paulo Spaniol, R. G. G. Amorim and Sergio Costa Ulhoa
On the Hilbert Space in Quantum Gravity
Reprinted from: *Universe* **2022**, *8*, 413, doi:10.3390/universe8080413 128

Hugo García-Compeán and Daniel Mata-Pacheco
Lorentzian Vacuum Transitions in Hořava–Lifshitz Gravity
Reprinted from: *Universe* **2022**, *8*, 237, doi:10.3390/universe8040237 138

Claus Gerhardt
A Unified Quantization of Gravity and Other Fundamental Forces of Nature
Reprinted from: *Universe* **2022**, *8*, 404, doi:10.3390/universe8080404 157

About the Editors

Arundhati Dasgupta

Arundhati Dasgupta is an associate professor at the University of Lethbridge, Canada. Her research is in theoretical physics, gravitational physics, quantum gravity, and cosmology. She is currently working on studying interactions of gravitational waves with matter, which will also provide clues to the existence of the quanta of gravity, the graviton. She is very active in international efforts to change the under-representation of women in physics. She is currently an elected member of the International Union of Pure and Applied Physics Mathematical Commission and Regional Councilor for Alberta, Nunavut and Northwest Territories, Canadian Association of Physicists.

Alfredo Iorio

Alfredo Iorio is a full professor of theoretical physics at Charles University, Prague, Czech Republic. His research is in quantum field theories, both at high and low energy, gravitational physics, and mathematical modeling of various phenomena, even beyond high energy theory, such as virus structures and DNA condensation. In the last decade, he has focused on analog gravity, proposing a prominent role of graphene, and related materials, for the reproduction of black hole phenomena. He created an innovative elementary school in Prague and founded a scientific consulting company where theoretical physicists face the big challenges of climate change.

Preface

1. Introduction

In 1915, Albert Einstein presented his original work on General Relativity to the Prussian Academy of Sciences. It was a new theory of nature, changing previous paradigms about gravity. Spacetime curvature and gravitation became synonymous in this theory and the new paradigm has survived direct experimental tests. Yet, a hundred years later, General Relativity cannot be theoretically quantized using standard methods, and we are still searching for the quanta of gravity.

A decade earlier than the discovery of General Relativity, Max Planck postulated that light was a stream of photons, which led to the discovery of the quantum world, and Einstein's paper on photoelectric effect had used that result. The theoretical discovery of gravitational waves had followed the formulation of General Relativity as early as 1916, and they were observed in LIGO, a sensitive laser interferometer, in 2015. However, the quanta of the waves, the gravitons, the counterpart of the photons for gravity, have not been observed, or even theoretically formulated.

The reason that quantum gravity has not been discovered is connected to what is known as the hierarchy problem of nature. Gravitational force is 10^{-38} orders weaker than electrodynamics. One has to probe length scales of 10^{-35} m and obtain energies of 10^{19} GeV in colliders to see experimental evidence of quantum gravity. Theoretically too, the standard quantization schemes require new techniques to be invented due to the non-polynomial and non-linearity of the Einstein action. However, Hawking radiation, area quantization, and black hole thermodynamics were predicted theoretically, which suggested quantum origins, but these require experimental confirmation. The search for primordial black holes is an effort in that direction. The theoretical computation of graviton interactions is non-renormalizable, which makes the particles more mysterious. There are various studies including very recent ones that argue that gravity is perhaps classical all the way through. This places gravity on a pedestal, different from the other interactions of nature.

With the discovery of very weak gravitational waves, in 2016, the search for gravitons received new impetus, and 'Quantum Gravity Phenomenology' could have a future. Precision experiments have given us access to the gravitational waves emitted from distant events. Gravitational waves are measured up to an amplitude of 10^{-21} using interferometers. As the precision measurements of nature head towards the quantum gravity scale, there might be indirect verification of the quantum gravity physics of the microscopic spacetime. Quantum phenomena might have significant effects at macroscopic and large scales, as these are emergent from the microscopic scale. The very metric which measures the curvature of spacetime is the classical limit of a quantum operator, and therefore semi-classical fluctuations should be visible at some length scales larger than the Planck scale. At this time, there are several competing theoretical formulations of quantum gravity, some extensions of standard versions, and some new ones. These include loop quantum gravity, path-integral quantization, discrete models, causal dynamical triangulation, asymptotically safe gravity, and causal set theory. Various formulations have matured to discuss observational predictions. In this Special Issue of *Universe*, we discuss some of these, and provide new insights into the future direction of this research area.

Of the various ways to verify the theoretical predictions of quantum gravity, two broad categories exist. These are the following: (i) Using analog models which simulate gravitational systems and the quantum phenomena associated with these systems. (ii) Direct and indirect evidence of quantum gravity predictions in natural experiments.

There has been considerable research in both of the above avenues for exploring the existence of new quantum physics for gravity. A volume dedicated to papers in this field is the need of the hour, and *Universe*'s topical collections serve that purpose.

When we were asked to suggest topics for a Special Issue for the journal *Universe*, the topic of 'Quantum Gravity Phenomenology' was a natural choice. There have been a number of papers addressing both ways of obtaining experimental evidence on the nature of quantum gravitational physics in the past few decades. When we embarked on this venture of editing a Special Issue on quantum gravity phenomenology, we wanted a clear direction to emerge in this field. We have collected 22 paper contributions, now published in two volumes. We would like to thank profusely our Assistant Editor from *Universe*, Ms. Cici Xia, who has made the two volumes of paper contributions happen. Initially our aim was to seek papers only on the above two approaches (i and ii) for obtaining observational physics in quantum gravity. However, over the course of the time we took to finalize the volumes, the focus has diversified. We are very thankful to the authors for publishing their papers in *Universe*. Although we are still inconclusive about the true nature of spacetime at quantum length scales, we have some highlights on the current status of research in this field. In the following, we offer some general considerations and briefly discuss on the papers in the two volumes, with Volume I being primarily focused on analog models and Volume II on astrophysics.

2. Probing Quantum Gravity through Analogs (Volume I)

Although direct experiments are to be preferred, given that all phenomena insist on the same fundamental dynamics, we can extract information on quantum gravity from systems other than quantum astrophysical systems. Volume I is primarily dedicated to this indirect search.

We shall now briefly describe the contributions to Volume I of this Special Issue, where the general frame just discussed does not enter yet in its full declination, but some aspects of it do, one way or the other.

It is our pleasure to open this collection of articles with the contribution of Carlo Rovelli, who has shaped an important part of the theoretical search for a consistent theory of Quantum Gravity over the last two decades. His contribution points to the importance of the experimental search, and is a highlight coming from a master of this highly theoretical field. We then present a review paper by a leading expert of the phenomenology of Quantum Gravity, Nick Mavromatos, coauthored with Pablo Pais and Alfredo Iorio. In this review, both the approaches of the direct and the analog are merged together in the quest for the role of torsion in cosmology as well as in condensed matter.

The following five papers are then fully focused on the analog approach: The contribution of Smolyaninov and Smolyaninova deals with Hyperbolic Metamaterials, which are fascinating condensed mater systems, recently discovered, whose potential for Quantum Gravity is still to be fully exploited. The work of Sreedhar and Virmani reviews the group theoretical structure common to supernova explosions and laboratory plasma implosion. This research was initiated by Lochlainn O'Raifeartaigh. Michail Katanaev, in his original contribution, describes the 't Hooft–Polyakov monopole within the elastic theory analog approach to geometry/gravity, which he pioneered. The series of papers on the analog approach is closed by two reviews, fully dedicated to the use of two different systems, high-energy scattering (by Castorina, Iorio, and Satz) and graphene (by Acquaviva, Iorio, Pais, and Smaldone), in the hunt for the reproduction of a variety of aspects of the high-energy theoretical research, from Quantum Gravity to Supersymmetry.

The last three papers are original contributions and are focused on the direct approach: the work of Spaniol, Amorim, and Ulhoa deals with the Hilbert space for various solutions of the Einstein equations, while the work of García-Compeán and Mata-Pacheco focuses on the Hořava–Lifshitz theory of gravity. We close this volume with the paper of Gerhardt that offers a suggestion for a unified way to quantize fundamental interactions, including gravity.

Arundhati Dasgupta and Alfredo Iorio
Editors

Considerations on Quantum Gravity Phenomenology

Carlo Rovelli [1,2,3]

1 Aix Marseille University, Université de Toulon, CNRS, CPT, 13288 Marseille, France; rovelli.carlo@gmail.com
2 Perimeter Institute, 31 Caroline Street North, Waterloo, ON N2L 2Y5, Canada
3 Institute of Philosophy, 1151 Richmond St. N, London, ON N6A 5B7, Canada

Abstract: I describe two phenomenological windows on quantum gravity that seem promising to me. I argue that we already have important empirical inputs that should orient research in quantum gravity.

Keywords: gravity entanglement; dark matter; primordial black holes

I do two things in this brief note. First, I describe the two directions towards quantum gravity phenomenology that seem more promising to me. Then, I list some considerable empirical information that we have obtained lately, which I think is relevant for understanding the quantum properties of gravity. This is also the opportunity for some general considerations on the topic.

1. Where Are We Going to See Quantum Gravity Effects

1.1. Gravity-Induced Entanglement

It is possible to probe a plausible and genuine quantum gravity effect in the laboratory with technology that is not far from the one available today. Surprisingly, nobody had realized that this was the case until a few years ago. The trick that makes this possible is that this is a (genuine, but) non-relativistic quantum gravitational effect.

Here is the main idea (for related ideas, see [1]). Two systems, A and B, each with mass m, are each put into the quantum superpositions of two different positions, say, L and R. This generates a state formed by four branches:

$$(|R\rangle_A + |L\rangle_A) \otimes (|R\rangle_B + |L\rangle_B) = |R,R\rangle + |R,L\rangle + |L,R\rangle + |L,L\rangle. \tag{1}$$

The systems are arranged in such a way that in one of these four branches, the two masses are at a small distance d from each other, and they are kept so for a time t. Then, the two components of each of the two systems are recombined.

The vicinity of the masses in one of the branches generates a gravitational interaction. This has the effect of altering the evolution of the phase of the branch. In a relativistic picture, this is because the gravitational field is different in each of the branches: the gravitational field is in a superposition of classical configurations; in the branch where the particles are close, each particle feels the time dilatation due to the vicinity of the other mass [2]. In the non-relativistic picture, the same effect is interpreted as due to gravitational potential energy $V = -Gm^2/d$. Since the phase evolves with the energy H as in $exp\{-iHt/\hbar\}$, the total change in the phase of the branch is then clearly

$$\delta\phi = \frac{Gm^2 t}{\hbar d} \tag{2}$$

This has the effect of entangling the two systems, which, as (1) shows, were not entangled to start with. The fact that they are entangled can then be tested in the lab.

The crucial observation is that today's technology is not far from the possibility of keeping nano-particles in a superposition and at a distance d from each other for a time t, such that $\delta\phi \sim \pi$ [3]. Hence, if the gravitational field can be in a superposition, the effect follows. Since we know from general relativity that the gravitational field is the same entity as the geometry of spacetime, the measurement of this effect amounts to detecting an effect that follows from the superposition of spacetime geometries.

The power of this setup is in fact even stronger. The reason is a well-known fact in quantum information: it is not possible to entangle two quantum systems by having them both interact with a third classical system. In this setup, the two systems are A and B, and the third system is the gravitational field. If we find A and B entangled by the gravitational interaction, then the gravitational field cannot be classical [4].

To be sure, the knowledge that gravity is mediated by a field (in fact, a relativistic field) is needed for the interpretation of the experiment. If gravity was an instantaneous action at a distance and not mediated by a field, then we could not conclude anything from the experiment itself. Hence, the subtlety at the basis of this experiment is that it can be performed in a non-relativistic regime, but its full implication requires the knowledge (that we have) that gravity is mediated by a relativistic field. In other words, a positive outcome of the experiment is not compatible with a description of gravity as the result of a classical field.

When successfully performed, the importance of this experiment will be major. It could well be the first clear manifestation of the fact that spacetime geometry is not classical.

Since it is a non-relativistic regime, this experiment would not differentiate current tentative theories of gravity (such as loop quantum gravity, string theory, asymptotic safety, or others). All current tentative theories predict it. It could, instead, rule out speculations such as those exploring the (unlikely) possibility that gravity is not quantized, or that there is a gravitationally induced physical collapse of the wave function. (A variant of the experiment has been proposed that might actually access the relativistic regime and test the discreteness of proper time [5], but this would require a much higher experimental sensitivity.)

In the past, there have been numerous other ideas on testing the effects of hypothetical quantum gravity, but—as far as I could understand—none considered a plausible effect; namely, an effect predicted by the current credible quantum gravity theories. The gravity-induced entanglement experiment does so.

This, I believe, is a general point. I find that there is a common false impression that since quantum gravity is an open problem, then everything is possible and any wide speculation can be counted as a "possible" quantum gravity phenomenon. This is not good science, in my opinion. Quantum gravity is an open problem because no quantum gravity effect has been measured yet, because there are a few competing theories about what exactly happens at the Planck scale, and because we do not have a way of empirically probing them. But all these theories are expected to generically give the same indications about what does or does not happen at lower scales. As always in science, a priori everything is possible, but there is a profound difference between testing a wild speculation and testing the predictions of a plausible, coherent framework.

1.2. Dark Matter as Quantum Gravity Stabilized White Holes

The first black hole signal was detected long before any black hole signal was recognized as such. In fact, a strong radio signal from Sagittarius A*, the gigantic black hole at the center of our galaxy, has been detected by radio antennas since the dawn of radio astronomy, without people suspecting it could be due to a black hole.

It might be the same with quantum gravity. Dark matter is a major unclear phenomenon [6]. There are many candidate theories for explaining dark matter, virtually all of which require the hypothesis of new physics. But there is also a possibility that dark matter could be explained without any recourse to new physics (which makes this hypothesis more, not less, interesting). The possibility is that dark matter might be formed by long-living Planck-size remnants of evaporated black holes. The black holes could have been formed in the early universe, or alternatively, if the Big Bang was a Big Bounce, they might have crossed the bounce.

The idea of black hole remnants is an old one, recently revived by quantum gravity calculations that provide them with a realistic model: white holes with a large interior and a small horizon stabilized by quantum gravity [7]. A large body of theoretical research converge today, indicating that spacetime can be continued past the central singularity of a black hole and into an anti-trapped region, namely, a white hole. The singularity itself is replaced by a quantum region where the Einstein equations are briefly violated.

Macroscopic white holes are unstable because they can easily re-collapse into black holes, but Planck-size ones are stabilized by quantum theory [8]. White hole remnants need to be long-lived because the information they store needs a long time to exit, in the form of low-frequency radiation.

This scenario is attractive, difficult to falsify, but also hard to confirm. In this article, I do not cover the current work that explores its phenomenology [9]. What I intended to point out is that it might (well) be that we are already seeing a massive quantum gravity effect: dark matter.

2. What Do We Already Know about Quantum Gravity?

Allow me come to the second topic—results that are relevant to quantum gravity, which are already providing us with crucial information.

2.1. Lorentz Invariance

The breaking of the Lorentz invariance at the Planck scale may simplify the construction of a quantum theory of gravity [10]. This observation sparked a large theoretical enthusiasm for Lorentz-breaking theories some time ago, and rightly so. But that bubble of enthusiasm has been deflated by empirical observations. A large campaign of astrophysical observations has failed to reveal the Planck-scale breaking of the Lorentz invariance in situations where it would have been expected if this track for understanding quantum gravity had been the good one [11].

A methodological consideration is important at this point. Popperian falsifiability is an important demarcation criterium for scientific theories (that is, if a theory is not falsifiable, then we better not call it "science"); however, Popperian falsification is very rarely the way theories gain or lose credibility in science.

The way scientific theories gain or loose credibility in real science is rather through a Bayesian gradual increase or decrease of the positive or negative confirmation from empirical data. That is, when a theory predicts a novel phenomenon and we find that theory to be right, our confidence in the theory grows; when it predicts a novel phenomenon and we do not find it, our confidence in the theory decreases. Failed predictions rarely definitely kill a theory, because theoreticians are very good at patching up and adjusting. But failed predictions do make the success of a research program far less probable: we loose confidence in it.

Hence, this has been the effect of not finding Lorentz violations in astrophysics: tentative quantum gravity theories that break Lorentz invariance might perhaps still be viable in principle, but in practice, far fewer people bet on them.

2.2. Supersymmetry

What I wrote above is particularly relevant to the spectacular non-discovery of supersymmetry at the LHC [12]. While in the Popperian sense, the non-appearance of supersymmetric particles at the TeV scale does not rule out all the theories based on supersymmetry, including string theory, in practice, the strong disappointment of not finding what was expected counts heavily as a strong dis-confirmation, in the Bayesian sense, of all those theories.

People have written that the non-discovery of supersymmetry is a crisis for theoretical physics. This is nonsense, of course. It is only a crisis for those who bet on supersymmetry and string theory. For all the alternative theoretical quantum gravity programs that were never convinced by the arguments for low-energy supersymmetry, the non-discovery of supersymmetry is not a crisis: it is a victory.

Precisely for the same reason that the discovery of supersymmetry would have been a confirmation of the ideas supporting the string supersymmetry research direction, the non-discovery of supersymmetry at the LHC is a strong empirical indication against the search for quantum gravity in the direction of supersymmetric theories and strings.

Nature talks, and we better listen.

2.3. Cosmological Constant

A case similar to the one above but even stronger concerns the sign of the cosmological constant. The cosmological constant is a fundamental constant of nature, part of the Einstein equations (since 1917), whose value had not been measured until recently. An entire research community has long worked, and is still working, under general hypotheses that lead to the expectation for the sign of the cosmological constant to be negative. Even today, the vast majority of the theoretical work in that community assume it to be so.

Except that the sign of the cosmological constant is not negative. It is positive, as observation has convincingly shown [6].

Once again, this counts as a strong dis-confirmation of the hypotheses on which a large community has worked in the past, and is still working on today.

So far, we lack any direct evidence of a quantum gravitational phenomenon; however, the non-detection of Lorentz violations around the Planck scale, the non-discovery of super symmetric particles at the LHC, and the measurement of a positive cosmological constant are strong indications from Nature that disfavor the tentative quantum gravity theories that naturally imply these phenomena.

Funding: This research was funded by the JFT—QISS grant #61466.

Conflicts of Interest: The authors declare no conflict of interest.

References

1. Howl, R.; Vedral, V.; Naik, D.; Christodoulou, M.; Rovelli, C.; Iyer, A. Non-Gaussianity as a Signature of a Quantum Theory of Gravity. *arXiv* **2021**, arXiv:2004.01189.
2. Christodoulou, M.; Rovelli, C. On the possibility of laboratory evidence for quantum superposition of geometries. *Phys. Lett. B* **2018**, *792*, 64–68. [CrossRef]
3. Bose, S.; Mazumdar, A.; Morley, G.W.; Ulbricht, H.; Toroš, P.M.M.; Geraci, A.A.; Barker, P.F.; Kim, M.S.; Milburn, G. Spin Entanglement Witness for Quantum Gravity. *Phy. Rev. Lett.* **2017**, *119*, 240401. [CrossRef] [PubMed]
4. Marletto, C.; Vedral, V. Witness gravity's quantum side in the lab. *Nature* **2017**, *547*, 156–158. [CrossRef] [PubMed]
5. Christodoulou, M.; Rovelli, C. On the possibility of experimental detection of the discreteness of time. *arXiv* **2018**, arXiv:1812.01542v1.
6. Adam, R. Planck 2015 results: I. Overview of products and scientific results. *Astron. Astrophys.* **2016**, *594*, 1–38.
7. Bianchi, E.; Christodoulou, M.; Ambrosio, F.D.; Haggard, H.M.; Rovelli, C. White holes as remnants: A surprising scenario for the end of a black hole. *Class. Quantum Grav.* **2018**, *35*, 225003. [CrossRef]
8. Rovelli, C.; Vidotto, F. Small black/white hole stability and dark matter. *Universe* **2018**, *4*, 127. [CrossRef]
9. Vidotto, F. Quantum insights on Primordial Black Holes as Dark Matter. *arXiv* **2018**, arXiv:1811.08007.
10. Horava, P. Quantum gravity at a Lifshitz point. *Phys. Rev. D Part. Fields Grav. Cosmol.* **2009**, *79*, 084008. [CrossRef]
11. Liberati, S. Tests of Lorentz invariance: A 2013 update. *Class. Quantum Grav.* **2013**, *30*, 133001. [CrossRef]
12. Canepa, A. Searches for supersymmetry at the Large Hadron Collider. *Rev. Phys.* **2019**, *4*, 100033. [CrossRef]

Review

Torsion at Different Scales: From Materials to the Universe

Nick E. Mavromatos [1,2,*], Pablo Pais [3,4] and Alfredo Iorio [4]

1. Physics Department, School of Applied Mathematical and Physical Sciences, National Technical University of Athens, 157 80 Athens, Greece
2. Theoretical Particle Physics and Cosmology Group, Department of Physics, King's College London, Strand, London WC2R 2LS, UK
3. Instituto de Ciencias Físicas y Matemáticas, Universidad Austral de Chile, Casilla 567, Valdivia 5090000, Chile; pais@ipnp.troja.mff.cuni.cz
4. Faculty of Mathematics and Physics, Charles University, V Holešovičkách 2, 18000 Prague, Czech Republic; alfredo.iorio@mff.cuni.cz
* Correspondence: nikolaos.mavromatos@kcl.ac.uk

Abstract: The concept of torsion in geometry, although known for a long time, has not gained considerable attention from the physics community until relatively recently, due to its diverse and potentially important applications to a plethora of contexts of physical interest. These range from novel materials, such as graphene and graphene-like materials, to advanced theoretical ideas, such as string theory and supersymmetry/supergravity, and applications thereof in terms of understanding the dark sector of our Universe. This work reviews such applications of torsion at different physical scales.

Keywords: quantum gravity; torsion; supersymmetry and supergravity; analogs; Dirac materials

Citation: Mavromatos, N.E.; Pais, P.; Iorio, A. Torsion at Different Scales: From Materials to the Universe. *Universe* **2023**, *9*, 516. https://doi.org/10.3390/universe9120516

Academic Editors: Arundhati Dasgupta and José Velhinho

Received: 6 November 2023
Revised: 7 December 2023
Accepted: 12 December 2023
Published: 14 December 2023

Copyright: © 2023 by the authors. Licensee MDPI, Basel, Switzerland. This article is an open access article distributed under the terms and conditions of the Creative Commons Attribution (CC BY) license (https://creativecommons.org/licenses/by/4.0/).

1. Introduction

Torsion is as important a concept of differential geometry as curvature [1–3]. The latter plays a key role in General Relativity (GR), but the former plays no role at all there. Nonetheless, torsion enters into various contexts and formulations, directing to diverse physical predictions and realisations that span a huge range of length scales—from cosmology to condensed matter and particle physics. Therefore, the related literature is huge, and it is not possible to cover it all in the restricted space of this review.

Here, we focus on specific aspects of torsion, either in the emergent geometric description of the physics of various materials of great interest to condensed matter physics—mainly graphene—or in the spacetime geometry itself, in particular in the early Universe. These two situations correspond to scales that are separated by a huge amount, yet the mathematical properties of torsion appear to be universal. Torsion has important physical effects, in principle experimentally testable, in both scenarios.

Graphene and related materials provide a tabletop realisation of some high-energy scenarios where torsion is associated with (the continuum limit of) the appropriate dislocations in the material. A way to represent the effect of dislocations, in the long wave-length regime, through torsion tensor is to consider a continuum field-theoretic fermionic system in a $(2 + 1)$-dimensional space with a torsion-full spin-connection.

In the case of fundamental physics, torsion is associated with supergravity (SUGRA) theories or with the geometry of the early Universe (cosmology). We discuss physical aspects of torsion that may affect particle physics phenomenology. In such cases, the (totally antisymmetric component of) torsion corresponds to a dynamical pseudoscalar (axion-like) degree of freedom, which is responsible for giving the vacuum a form encountered in the so-called running vacuum model (RVM) cosmology, characterised by a dynamical inflation without external inflaton fields, but rather due to non-linearities of the underlying gravitational dynamics. Moreover, under some circumstances, the torsion-associated axions can lead to background configurations that spontaneously violate Lorentz (and CPT) symmetry, leading, in some models with right-handed neutrinos, to lepton asymmetry in the early radiation epoch, that succeeds the exit from inflation.

The structure of the review is as follows. First, in Section 2, we extensively discuss the concept of the torsion tensor in general geometric terms. This has the double scope of introducing our notations but also, and more importantly, of elucidating as many details as possible of the geometry and physics of torsion. The following Section 3 is dedicated to an important illustration of how torsion may affect well known theories, such as quantum electrodynamics (QED), while Section 4 focuses on some ambiguities of the Einstein–Cartan gravity theories and on the Barbero–Immirzi (BI) parameter. In Section 5, we discuss how torsion can be practically realised in a tabletop system, that is graphene. Then, after recalling in Section 6 how standard SUGRAs necessarily include torsion, we discuss in Section 7 a novel type of local supersymmetry (SUSY), without superpartners, whose natural realisation is in graphene. The rather extended Section 8 is dedicated to the important and hot topic of torsion in cosmology. Our concluding remarks and some brief description of other applications of torsion, which are not covered in this review, are given in the last Section 9.

2. Properties of Torsion

As already mentioned, torsion is an old subject [1–3] that goes beyond GR, as it constitutes a more general formalism in the sense that, to obtain Einstein's GR, one needs to impose a constraint to guarantee the absence (vanishing) of torsion tensor in a Riemannian spacetime. Specifically, let $\mathcal{M}$ be a (3 + 1)-dimensional Minkowski-signature curved world-manifold[1], parametrised by coordinates x^μ, where Greek indices $\mu, \nu = 0, \ldots 3$ are spacetime volume indices, raised and lowered by the curved metric $g_{\mu\nu} = \eta_{ab} e^a{}_\mu e^b{}_\nu$, with e^a_μ the vielbein (we also define the inverse vielbein as $E^\mu{}_a e^a{}_\nu = \delta^\mu_\nu$, and $E^\mu{}_a e^b{}_\mu = \delta^b_a$, such that $g^{\mu\nu} = \eta^{ab} E^\mu{}_a E^\nu{}_b$ gives the inverse metric tensor). In the above formulae, Latin indices $a, b, \cdots = 0, \ldots 3$ are (Lorentz) indices on the tangent hyperplane of $\mathcal{M}$ at a given point p (cf. Figure 1), and are raised and lowered by the Minkowski metric η_{ab} (and its inverse η^{ab}), which is the metric of the tangent space $T_p\mathcal{M}$.

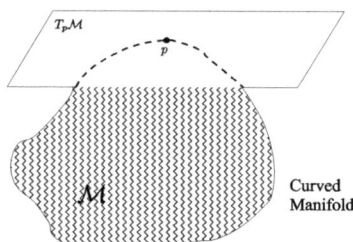

Figure 1. Tangent hyperplane $T_p\mathcal{M}$ at a point p of a curved $(d+1)$-dimensional manifold $\mathcal{M}$, used in the first order formalism of GR to define the vielbein $e^a{}_\mu$ map $\mathcal{M} \to T_p\mathcal{M}$.

In differential form language [4,5], which we use here often for notational convenience, the torsion two form is defined as [1–3,6]:

$$T^a = \frac{1}{2} T^a{}_{\mu\nu} dx^\mu \wedge dx^\nu \equiv de^a + \omega^a{}_b \wedge e^b, \qquad (1)$$

where in the first equality, we used the definition of a differential two form [4], and the $\wedge$ denotes the exterior product[2], and $\omega^a{}_{b\mu}$ is the generalised (contorted) spin connection one form, which can can be split into a part that is torsion-free, $\mathring{\omega}^a{}_{b\mu}$, and related to the standard Christoffel symbols of GR, and another part that involves the *contorsion one-form*[3] $\mathcal{K}^a{}_{b\mu}$ [2,3]:

$$\omega^a{}_{b\,\mu} = \mathring{\omega}^a{}_{b\,\mu} + \mathcal{K}^a{}_{b\,\mu}. \tag{2}$$

We can use the one-form $\omega^a{}_b$ to define the covariant derivative $D(\omega)$ acting on q-forms $Q^{a\ldots}_{b\ldots}$ in this contorted spacetime [6]:

$$D(\omega)\, Q^{a\ldots}_{b\ldots} = d\, Q^{a\ldots}_{b\ldots} + \omega^a{}_c \wedge Q^{c\ldots}_{b\ldots} + \cdots - (-1)^q\, Q^{a\ldots}_{d\ldots} \wedge \omega^d{}_b - \ldots \tag{3}$$

It can be readily seen, using the covariant constancy of the Minkowski tangent space metric η^{ab}

$$D(\omega)\, \eta_{ab} = 0, \tag{4}$$

that the spin connection (2) is antisymmetric in its Lorentz indices

$$\omega_{ab} = -\omega_{ba}. \tag{5}$$

We also have covariant constancy for the totally antisymmetric Levi–Civita tensor ϵ_{abcd}:

$$D(\omega)\, \epsilon_{abcd} = 0. \tag{6}$$

In this Section, we discuss the generalisation of Einstein–Hilbert action for spacetime geometries with torsion. To this end, we first note that the *generalised* Riemann curvature, or Lorentz curvature, two-form is defined as:

$$R^a{}_b = d\,\omega^a{}_b + \omega^a{}_c \wedge \omega^c{}_b. \tag{7}$$

We can write the components of the Lorentz curvature in terms of the Riemann curvature two-form $\mathring{R}^a{}_b$, defined only by the torsion-less spin-connection, i.e., $\mathring{R}^a{}_b = d\,\mathring{\omega}^a{}_b + \mathring{\omega}^a{}_c \wedge \mathring{\omega}^c{}_b$, and the contorsion $\mathcal{K}^a{}_b$,

$$R^a{}_b = \mathring{R}^a{}_b + D(\mathring{\omega})\,\mathcal{K}^a{}_b + \mathcal{K}^a{}_c\,\mathcal{K}^c{}_b, \tag{8}$$

where the quantity $D(\mathring{\omega})$ denotes the diffeomorphic covariant derivative of GR. From the definition of the covariant derivative (3), we therefore have that the torsion two form is just the covariant derivative of the vielbein,

$$T^a = d\,e^a + \omega^a{}_b \wedge e^b, \tag{9}$$

and [6]

$$D(\omega)\, T^a = R^a{}_b \wedge e^b,$$
$$D(\omega)\, R^a{}_b = 0, \tag{10}$$

where the Equation (10) are the generalisation of the usual Bianchi identity. The two Equations (7) and (9), are known as *Cartan structure equations* [5].

Taking into account that the full diffeomorphic covariant derivative on the vielbein is zero, $D(\omega, \Gamma)\, e^a = 0$, we obtain a relation between the affine connection, $\Gamma^\lambda{}_{\mu\nu}$, and the spin connection (2), $\omega_\mu{}^a{}_b$. In components [6]:

$$D_\mu(\Gamma)\, e^a_\nu = \partial_\mu e^a_\nu - \Gamma^\lambda{}_{\nu\mu}\, e^a_\lambda = -\omega_\mu{}^a{}_b\, e^b_\nu. \tag{11}$$

From (1), (7) and (11), we easily obtain

$$T^a{}_{\mu\nu} = e^a_\lambda \left(\Gamma^\lambda{}_{\mu\nu} - \Gamma^\lambda{}_{\nu\mu} \right) \equiv 2\, e^a_\lambda\, \Gamma^\lambda{}_{[\mu\nu]}, \tag{12}$$

where $[\mu\nu]$ indicates antisymmetrisation of the indices.

The relation (12) expresses the essence of torsion, namely that in its presence the affine connection loses its symmetry in its lower indices. The torsion *tensor* is associated with the antisymmetric part (in the lower indices) of the affine connection, which is its only part that transforms as a tensor under general coordinate transformations.

The spin connection then, in general, is torsion-full. If we want a torsion-free connection (that is the case of GR) we need to impose

$$d e^a + \omega^a_{\ b} \wedge e^b = 0, \qquad (13)$$

and we have that the antisymmetric (because of (4)) connection is $\omega_{ab} = \mathring{\omega}_{ab}$. In other words, covariant constancy of the metric is a separate request from zero torsion. In fact, in Riemann–Cartan spaces the metric is compatible, hence ω_{ab} is antisymmetric, but torsion is nonzero.

We next remark that the contorsion one-form coefficients $\mathcal{K}^a_{\ bc} = \mathcal{K}^a_{\ b\mu} E^\mu_c$ satisfy $\mathcal{K}^c_{\ ab} = -\mathcal{K}^c_{\ ba}$ and are related to the torsion tensor coefficients $T^a_{\ bc} = T^a_{\ \mu\nu} E^\mu_b E^\nu_c$ via [5]

$$\mathcal{K}_{abc} = -\frac{1}{2}(T_{cab} - T_{abc} - T_{bca}) \Rightarrow T_{[abc]} = -2\mathcal{K}_{[abc]}, \qquad (14)$$

where $[abc]$ denotes total antisymmetrisation. From (14) we can write the torsion tensor in term of contorsion as

$$T^a_{\ bc} = -2\mathcal{K}^a_{\ [bc]}. \qquad (15)$$

Equations (14) and (15) tell us that both torsion and contorsion tensors carry the same information.

2.1. Geometric Interpretation

Let us now discuss the geometric interpretation of torsion, by parallel transporting the vector v^μ along the direction dx^ν, using the connection $\Gamma^\lambda_{\ \mu\nu}$ that appears in (12)

$$\delta_\| v^\mu = v^\mu_\|(x + dx) - v^\mu(x) = \Gamma^\mu_{\ \rho\nu} v^\rho \, dx^\nu.$$

Then, the covariant derivative $D(\Gamma) v$ can be written, in its components, as the difference

$$v^\mu(x + dx) - v^\mu_\|(x + dx) = v^\mu(x + dx) - v^\mu(x) - \delta_\| v^\mu = \left(\partial_\nu v^\mu - \Gamma^\mu_{\ \rho\nu} v^\rho\right) dx^\nu$$
$$\equiv D_\nu(\Gamma) v^\mu \, dx^\nu. \qquad (16)$$

Both curvature and torsion measure the noncommutativity of covariant derivatives of a vector taken along two directions [7],

$$D_\nu(\Gamma) D_\rho(\Gamma) v^\mu - D_\rho(\Gamma) D_\nu(\Gamma) v^\mu = R^\mu_{\ \sigma\rho\nu} v^\sigma + T^\sigma_{\ \rho\nu} D_\sigma(\Gamma) v^\mu,$$

where the torsion components have been defined already and the curvature components can be written as

$$R^\mu_{\ \sigma\rho\nu} = \partial_\rho \Gamma^\mu_{\ \sigma\nu} - \partial_\nu \Gamma^\mu_{\ \sigma\rho} + \Gamma^\mu_{\ \lambda\nu} \Gamma^\lambda_{\ \sigma\rho} - \Gamma^\mu_{\ \lambda\rho} \Gamma^\lambda_{\ \sigma\nu}. \qquad (17)$$

It is remarkable that for a scalar field, φ, such noncommutativity is entirely due to torsion,

$$D_\nu(\Gamma) D_\rho(\Gamma) \varphi - D_\rho(\Gamma) D_\nu(\Gamma) \varphi = T^\sigma_{\ \rho\nu} \partial_\sigma \varphi.$$

We introduce the *metric tensor* $g_{\mu\nu}$ when it is necessary to measure angles and distances between events in a spacetime manifold. The line element is

$$ds^2 = g_{\mu\nu} \, dx^\mu \, dx^\nu. \qquad (18)$$

We can define the longitude of any curve on $\mathcal{M}$ by integrating (18).

A very reasonable assumption usually taken is that local distances do not change under parallel transport, i.e.,

$$D_\rho(\Gamma) g_{\mu\nu} = 0. \qquad (19)$$

The condition (19) for a linear connection Γ is called *metric compatibility*, which leads to the antisymmetry of the spin-connection (5) [5]. Figure 2 illustrates a geometric interpretation of torsion, with details in the caption.

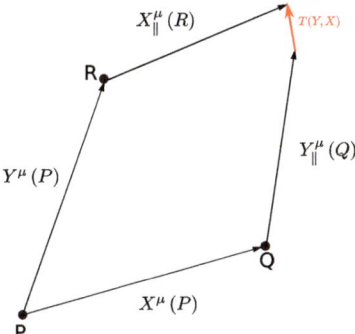

Figure 2. A geometric interpretation of torsion in Riemann–Cartan spaces. Consider two vector fields, X and Y, at a point P. First, parallel-transport X along Y to the infinitesimally close point R. Then, again from P, parallel-transport Y along X to reach a point Q. The failure of the closure of the parallelogram is the geometrical signal of torsion, and its value is the difference $T(X, Y)$ (here in red) between the two resulting vectors. An n-dimensional manifold M with a linear connection preserving local distances, i.e., fulfilling condition (19), is called a *Riemann–Cartan space*, denoted by U_n. In Riemannian spaces, V_n, this tensor is assumed to be zero. The picture was inspired by [8] but with the notation of [5], and was taken from [9].

The unique linear metric-compatible torsionless connection, called the *Levi–Civita connection*, can then be obtained from the metric $g_{\mu\nu}$ (see [5] for details)

$$\left\{ \begin{array}{c} \mu \\ \nu\rho \end{array} \right\} = \frac{1}{2} g^{\mu\sigma} \left(\partial_\nu g_{\rho\sigma} + \partial_\rho g_{\nu\sigma} - \partial_\sigma g_{\nu\rho} \right). \tag{20}$$

The quantities (20) are called *Christoffel symbols*, and the curvature associated with the Levi–Civita connection is the *Riemannian curvature tensor*, denoted by $\mathring{R}^\mu{}_{\nu\rho\sigma}$. In this way, the linear connection in a Riemann–Cartan space can be written as

$$\Gamma^\mu{}_{\nu\rho} = \left\{ \begin{array}{c} \mu \\ \nu\rho \end{array} \right\} + K^\mu{}_{\nu\rho}.$$

2.2. Gravitational Dynamics in Presence of Torsion

The Einstein–Hilbert scalar curvature term corresponding to the generalised contorted Riemann tensor is given by

$$\begin{aligned} \mathcal{S}_{\text{grav}} &= \frac{1}{2\kappa^2} \int d^4x \sqrt{-g}\, R = \frac{1}{2\kappa^2} \int R_{ab} \wedge \star(e^a \wedge e^b) \\ &= \frac{1}{2\kappa^2} \int \left(\mathring{R}_{ab} + D(\mathring{\omega}) \mathcal{K}_{ab} + \mathcal{K}_{ac} \wedge \mathcal{K}^c{}_b \right) \wedge \star(e^a \wedge e^b) \\ &= \frac{1}{2\kappa^2} \int \left(\mathring{R}_{ab} + \mathcal{K}_{ac} \wedge \mathcal{K}^c{}_b \right) \wedge \star(e^a \wedge e^b), \end{aligned} \tag{21}$$

where, in the last two equalities, we used form language and took into account the definition of the generalised curvature two form (7) in terms of the contorted spin connection (2). In (21), $\kappa^2 = 8\pi G = M_{\text{Pl}}^{-2}$ is the gravitational constant in four dimensions, which is the inverse of the square of the reduced Planck mass M_{Pl} in units $\hbar = c = 1$ we work throughout. In passing from the second to the third equality we used the fact that the term $D(\mathring{\omega}) \mathcal{K}_{ab} \wedge \star(e^a \wedge e^b)$ is a total derivative and thus yields, by means of Stoke theorem, a boundary term that we assume to be zero (we used also the metric compatibility of the spin-connection (5)).

For completeness, we give below the component form of the gravitational action (in the notation of [6]):

$$S_{\text{grav}} = \frac{1}{2\kappa^2} \int d^4x \sqrt{-g} \left(\mathring{R} + \Delta \right),$$

$$\Delta \equiv K^\lambda{}_{\mu\nu} K^{\nu\mu}{}_\lambda - K^{\mu\nu}{}_\nu K_{\mu\lambda}{}^\lambda = T^\nu{}_{\nu\mu} T^\lambda{}_\lambda{}^\mu - \frac{1}{2} T^\mu{}_{\nu\lambda} T^{\nu\lambda}{}_\mu + \frac{1}{4} T_{\mu\nu\lambda} T^{\mu\nu\lambda}. \quad (22)$$

Next, we decompose the torsion tensor in its irreducible parts under the Lorentz group [3,6,10],

$$T_{\mu\nu\rho} = \frac{1}{3}\left(T_\nu g_{\mu\rho} - T_\rho g_{\mu\nu}\right) - \frac{1}{6}\epsilon_{\mu\nu\rho\sigma} S^\sigma + q_{\mu\nu\rho}, \quad (23)$$

where

$$T_\mu \equiv T^\nu{}_{\mu\nu}, \quad (24)$$

is the *torsion trace vector*, transforming like a vector,

$$S_\mu \equiv \epsilon_{\mu\nu\rho\sigma} T^{\nu\rho\sigma}, \quad (25)$$

is the *pseudotrace axial vector* and the antisymmetric tensor $q_{\mu\nu\rho}$ satisfies

$$q^\nu{}_{\rho\nu} = 0 = \epsilon^{\sigma\mu\nu\rho} q_{\mu\nu\rho}. \quad (26)$$

Thus, we may write the contorsion tensor components as:

$$\mathcal{K}_{abc} = \frac{1}{2}\epsilon_{abcd} S^d + \widehat{\mathcal{K}}_{abc}, \quad (27)$$

being $\widehat{\mathcal{K}}_{abc}$, by definition, the difference of $\mathcal{K}_{abc}$ and the first term of (27). This yields for the quantity Δ in (22):

$$\Delta = \frac{3}{2} S_d S^d + \widehat{\Delta}, \quad (28)$$

with $\widehat{\Delta}$ being given by the combination appearing in the expression for Δ in (22) in terms of the contorsion tensor, but with K replaced by $\widehat{K}$ [6].

Using the decomposition (23) and the relations (24)–(26) and discarding total derivative terms, the gravitational part of the action can be written as:

$$S_{\text{grav}} = \frac{1}{2\kappa^2} \int d^4x \sqrt{-g} \left(E_a^\mu E_b^\nu R^{ab}{}_{\mu\nu}(\omega) + \frac{1}{24} S_\mu S^\mu - \frac{2}{3} T_\mu T^\mu + \frac{1}{2} q_{\mu\nu\rho} q^{\mu\nu\rho} \right)$$

$$\equiv \frac{1}{2\kappa^2} \int d^4x \sqrt{-g} \left(R + \widehat{\Delta} \right) + \frac{3}{4\kappa^2} \int S \wedge \star S, \quad (29)$$

where in the last line we used mixed notation components/form, following [6], as this will be more convenient for the discussion that follows. For future use, the reader should notice that $\widehat{\Delta}$ is independent of the pseudovector S^d.

An important part of our review will deal with fermionic torsion, that is torsion induced by fermion fields in the theory. Such a feature arises either in certain materials, such as graphene, to be discussed in Section 5, or in fundamental theories, which may play a role in particle physics, such as SUGRA (local SUSY, Section 6), unconventional supersymmetry (USUSY) (Section 7), and string theory (with applications to cosmology, Section 8.1). In the next Section we review such a (quantum) torsion in a fermionic theory corresponding to QED, as an instructive example, which can be generalised to non-Abelian gauge fields as well.

3. (Quantum) Torsion, Axions and Anomalies in Einstein–Cartan Quantum Electrodynamics

Our starting point is a $(3+1)$-dimensional QED with torsion (termed, from now on, "contorted QED"), describing the dynamics of a massless Dirac fermion field $\psi(x)$, coupled to a gauged (electromagnetic) U(1) field A_μ, in a curved spacetime with torsion[4]. The action of the model reads [6]:

$$\mathcal{S}_{\text{TorsQED}} = \frac{i}{2}\int d^4x \sqrt{-g}\left[\overline{\psi}(x)\gamma^\mu \mathcal{D}_\mu(\omega, A)\psi(x) - \overline{\mathcal{D}_\mu(\omega, A)\psi(x)}\gamma^\mu \psi(x)\right], \tag{30}$$

where $\mathcal{D}_\mu(\omega, A) \equiv D_\mu(\omega) - i e A_\mu$ is the diffeomorphic *and* gauge covariant derivative and [2,3]:

$$D_\mu(\omega) = \partial_\mu + i\omega^a{}_{b\mu}\sigma^b{}_a, \quad \sigma^{ab} \equiv \frac{i}{4}[\gamma^a, \gamma^b]. \tag{31}$$

The quantities γ^a and γ^μ denote the 4×4 Dirac matrices in the tangent space and in the manifold, respectively. On account of (31) and (2) (discussed in Section 2), the action (30) becomes:

$$\mathcal{S}_{\text{TorQED}} = \mathcal{S}_{\text{QED}}(\mathring{\omega}, A) + \frac{1}{8}\int d^4x \sqrt{-g}\,\overline{\psi}(x)\{\gamma^c, \sigma^{ab}\}\mathcal{K}_{abc}\psi(x), \tag{32}$$

where $\mathcal{S}_{\text{QED}}(\mathring{\omega}, A)$ is the standard QED action in a torsion-free curved spacetime and $\{\,,\,\}$ denotes the standard anticommutator. Since the Dirac γ-matrices obey

$$\{\gamma^c, \sigma^{ab}\} = 2\epsilon^{abc}{}_d\gamma^d\gamma^5,$$

where ϵ^{abcd} is the Levi–Civita tensor in $(3+1)$-dimensions, one can prove that it is only the totally antisymmetric part of the torsion that couples to fermionic matter [3]. Indeed, on using (14), we may write (32) in the form

$$\mathcal{S}_{\text{TorQED}} = \mathcal{S}_{\text{QED}}(\omega, A) - \frac{3}{4}\int d^4x \sqrt{-g}\, S_\mu \overline{\psi}\gamma^\mu\gamma^5\psi, \tag{33}$$

where $S_d = \frac{1}{3}\epsilon^{abc}{}_d T_{abc}$ (or in form language $S = \star T$) is the dual pseudovector constructed out of the totally antisymmetric part of the torsion. From (33) we thus observe that only the totally antisymmetric part of the torsion couples to the fermion axial current

$$j^{5\mu} = \overline{\psi}\gamma^\mu\gamma^5\psi. \tag{34}$$

The $(2+1)$-dimensional version of (33) will be our starting point to describe the conductivity electrons in graphene-like materials in a fixed spacetime with torsion (see Section 5). In contorted QED, the Maxwell tensor is defined with respect to the ordinary torsion-free geometry, $F_{\mu\nu} = \partial_\mu A_\nu - \partial_\nu A_\mu = D_\mu(\mathring{\omega})A_\nu - D_\nu(\mathring{\omega})A_\mu$. This way, the Maxwell tensor continues to satisfy the Bianchi identity (in form language $dF = 0$) even in the presence of torsion. Thus the standard Maxwell term, independent from torsion, is added to the action (30) to describe the dynamics of the photon field:

$$\mathcal{S}_{\text{Max}} = -\frac{1}{4}\int d^4x \sqrt{-g}\, F_{\mu\nu}F^{\mu\nu} = -\frac{1}{2}\int F \wedge \star F, \tag{35}$$

where $\star$ denotes the Hodge star [4,5].

The dynamics of the gravitational field is described by adding Einstein–Hilbert scalar curvature action (21) (or, equivalently, (22), in component form) of Section 2 to the above actions. By adding (33) to (22), so as to obtain the full gravitational action in a contorted geometry, with QED as its matter content, we obtain from the graviton equations of motion the stress-energy tensor of the theory, which can be decomposed into various components gauge, fermion and torsion-S (the reader should recall that only the totally antisymmetric part of the torsion S couples to matter in the theory):

$$T^A_{\mu\nu} = F_{\mu\lambda} F^{\lambda}{}_{\nu} - \frac{1}{2} g_{\mu\nu} F_{\alpha\beta} F^{\alpha\beta},$$

$$T^\psi_{\mu\nu} = -\left(\frac{i}{2} \overline{\psi} \gamma_{(\mu} \mathcal{D}_{\nu)} \psi - (\mathcal{D}_{(\mu} \overline{\psi}) \gamma_{\nu)} \psi \right) + \frac{3}{4} S_{(\mu} \overline{\psi} \gamma_{\nu)} \gamma^5 \psi,$$

$$T^S_{\mu\nu} = -\frac{3}{2\kappa^2} \left(S_\mu S_\nu - \frac{1}{2} g_{\mu\nu} S_\alpha S^\alpha \right), \tag{36}$$

where $(\ldots)$ denotes indices symmetrisation.

Variation of the above gravitational action with respect to the torsion components T^μ, $q^{\mu\nu\rho}$ and S^μ (cf. (23)), treated as *independent* field variables, leads to the equations of motion (in form language):

$$T^\mu = 0, \qquad q_{\mu\nu\rho} = 0, \qquad S = \frac{\kappa^2}{2} j^5, \tag{37}$$

respectively, where j^5 is the axial fermion current one form, which in components is given by Equation (34). Thus, classically, only the totally antisymmetric component of the torsion is non vanishing in this Einstein–Cartan theory with fermions. From (2), (14) and (23), we then obtain for the *on-shell* torsion-full spin connection:

$$\omega^{ab}_\mu = \mathring{\omega}^{ab}_\mu + \frac{\kappa^2}{4} \epsilon^{ab}{}_{cd} e^c_\mu j^{5d}, \tag{38}$$

thereby associating the torsion part of the connection, induced by the fermions, with the spinor axial current.

We next remark that the equations of motion for the fermion, stemming from (33), imply the gauged Dirac equation with the vector pseudovector S_μ, corresponding to the totally antisymmetric torsion component, playing the role of an axial source:

$$i\gamma^\mu \mathcal{D}_\mu(\omega, A) \psi = \frac{3}{4} S_\mu \gamma^\mu \gamma^5 \psi. \tag{39}$$

Classically, (37) implies a direct substitution of the torsion by the axial fermion current in (36) and (39). Moreover, as a result of the Dirac equation (39), a classical conservation of the axial current follows, $d \star j^5 = 0$. In view of (37), this, in turn, implies a classical conservation of the torsion pseudovector S, that is:

$$d \star S = 0. \tag{40}$$

Because the action is quadratic in S_μ, one could integrate it out exactly in a path integral, thus producing repulsive four fermion interactions

$$-\frac{3\kappa^2}{16} \int j^5 \wedge \star j^5, \tag{41}$$

which are a characteristic feature of Einstein–Cartan theories.

However, this would *not* be a self consistent procedure in view of the fact that, due to chiral anomalies, the axial fermion current conservation is violated at a quantum level [12–17]. Specifically at one loop one obtains for the divergence of the axial fermion current in a curved spacetime with torsion:

$$d \star j^5 = \frac{e^2}{8\pi^2} F \wedge F - \frac{1}{96\pi^2} R^a{}_b \wedge R^b{}_a \equiv \mathcal{G}(\omega, A). \tag{42}$$

It can be shown [14,18,19] that, by the addition of appropriate counterterms, the torsion contributions to $\mathcal{G}(\omega, A)$ can be removed, and hence one obtains

$$d \star j^5 = \frac{e^2}{8\pi^2} F \wedge F - \frac{1}{96\pi^2} \mathring{R}^a{}_b \wedge \mathring{R}^b{}_a \equiv \mathcal{G}(\mathring{\omega}, A), \tag{43}$$

where only torsion-free quantities appear in the anomaly equation.

Therefore, to consistently integrate over the torsion S_μ in the path integral of the contorted QED, we need to add appropriate counterterms order by order in perturbation theory. This will ensure the conservation law (40) in the quantum theory, despite the presence of the anomaly (43). This can be achieved [6] by implementing (40) as a δ-functional constraint in the path integral, represented by means of a Lagrange multiplier pseudoscalar field Φ:

$$\delta(d \star S) = \int D\Phi \exp\left(i \int \Phi d \star S\right), \qquad (44)$$

thus writing for the S-path integral

$$\begin{aligned}\mathcal{Z} &\propto \int DS\, \delta(d \star S) \exp\left(i \int \left[\frac{3}{4\kappa^2} S \wedge \star S - \frac{3}{4} S \wedge \star j^5\right]\right) \\ &= \int DS\, D\Phi \exp\left(i \int \left[\frac{3}{4\kappa^2} S \wedge \star S - \frac{3}{4} S \wedge \star j^5 + \Phi d \star S\right]\right).\end{aligned} \qquad (45)$$

The path integral over S can then be performed, making this way the field Φ dynamical. Normalising the kinetic term of Φ, requires the rescaling $\Phi = (3/(2\kappa^2))^{1/2} b$. We may write then for the result of the S path-integration [6]:

$$\begin{aligned}\mathcal{Z} &\propto \int Db \exp\left[i \int \left(-\frac{1}{2} db \wedge \star db - \frac{1}{f_b} db \wedge \star j^5 - \frac{1}{2f_b^2} j^5 \wedge \star j^5\right)\right], \\ f_b &\equiv (3\kappa^2/8)^{-1/2},\end{aligned} \qquad (46)$$

which demonstrates the emergence of a massless axion-like degree of freedom $b(x)$ from torsion. The characteristic shift-symmetric coupling of the axion to the axial fermionic current with f_b the corresponding coupling parameter [20]. Using the anomaly Equation (43) we may partially integrate this term to obtain:

$$\mathcal{Z} \propto \int Db \exp\left[i \int \left(-\frac{1}{2} db \wedge \star db + \frac{1}{f_b} b\, \mathcal{G}(\omega, A) - \frac{1}{2f_b^2} j^5 \wedge \star j^5\right)\right]. \qquad (47)$$

The repulsive four fermion interactions in (46) and (47) are characteristic of Einstein–Cartan theories, as already mentioned, but as we see from (47) this is not the only effect of torsion. One has also the coupling of torsion to anomalies, which induces a coupling of the axion to gauge and gravitational anomaly parts of the theory. The emergence of axionic degrees of freedom from torsion is an important result which will play a crucial role in our cosmological considerations. We have observed that, in the massless chiral QED case, torsion became dynamical, due to anomalies. We stress that the effective field theory (47) guarantees the conservation law (40), and hence the conservation of the axion charge

$$Q_S = \int \star S, \qquad (48)$$

order by order in perturbation theory.

Viewed as a gravitational theory, (47) corresponds to a Chern–Simons gravity [21–23], due to the presence of the gravitational anomaly. From a physical point of view, placing the theory on an expanding Universe Friedman–Lemaitre–Robertson–Walker (FLRW) background spacetime, we observe that the gravitational anomaly term vanishes [21,23]. However, the gauge chiral anomaly survives. This could have important consequences for the cosmology of the model.

In fact, although above we discussed QED, we could easily consider more general models, with several fermion species, some of which could couple to non-Abelian gauge fields, e.g., the SU(3) colour group of Quantum Chromodynamics (QCD). In such a case, torsion, being gravitational in origin, couples to all fermion species, in a similar way as in the aforementioned QED case, (33), but now the axial current (34) is generalised to include all the fermion species:

$$J_{\text{tot}}^{5\mu} = \sum_{i=\text{fermion species}} \overline{\psi}_i \gamma^\mu \gamma^5 \psi_i. \tag{49}$$

Chiral anomalies of the axial fermion current as a result of (non-perturbative) instanton effects of the non-Abelian gauge group, e.g., SU(3), during the QCD cosmological era of the Universe, will be responsible for inducing a breaking of the axion shift-symmetry, by generating a potential for the axion b of the generic form [20]

$$V(b) = \int d^4x \sqrt{-g} \Lambda_{\text{QCD}}^4 \left[1 - \cos\left(\frac{b}{f_b}\right)\right], \tag{50}$$

where Λ_{QCD} is the energy scale at which the instantons are a dominant configurations. As we observe from (50) one obtains this way a *mass* for the torsion-induced axion $m_b = \frac{\Lambda_{\text{QCD}}^2}{f_b}$, which can thus play a role of a dark matter component. In this way, we can have a geometric origin of the dark matter component in the Universe [24], which we discuss in Section 8.1, where we describe a more detailed scenario in which such cosmological aspects of torsion are realised in the context of string-inspired cosmologies.

4. Ambiguities in the Einstein–Cartan Theory —The Barbero–Immirzi Parameter

The contorted gravitational actions discussed in the previous Section can be modified by the addition of total derivative topological terms, which do not affect the equations of motion, and hence the associated dynamics. One particular form of such total derivative terms plays an important role in Loop quantum gravity [25,26], a non-perturbative approach to the canonical quantization of gravity. Below, we shall briefly mention such modifications, which, as we shall see, introduce an extra (complex) parameter, β, in the connection, termed "BI parameter", due to its discoverer [27,28]. This is a free parameter of the theory and it may be thought of as the analogue of the instanton angle θ of non-Abelian gauge theories, such as QCD, associated with strong CP violation.

Let us commence our discussion by presenting the case of pure gravity in the first-order formalism. In pure gravity, if torsion is absent, a term in the action linear in the *dual* of the Riemann curvature tensor, $\tilde{\mathring{R}}_{\mu\nu}^{ab} \equiv \epsilon^{ab}{}_{cd} \mathring{R}_{\mu\nu}^{cd}$, called the Holst term [29]

$$S_{\text{Holst}} = -\frac{\beta}{4\kappa^2} \int d^4x\, e\, E_a^\mu E_b^\nu \tilde{\mathring{R}}_{\mu\nu}^{ab}, \tag{51}$$

where $e = \sqrt{-g}$ is the vielbein determinant, and vanishes identically, as a result of the corresponding Bianchi identity of the Riemann curvature tensor:

$$\mathring{R}_{\alpha\mu\nu\rho} + \mathring{R}_{\alpha\nu\rho\mu} + \mathring{R}_{\alpha\rho\mu\nu} = 0. \tag{52}$$

However, if torsion is present, such a term yields non-trivial contributions, since in that case the Bianchi identity (52) is not valid. In the general case β is a complex parameter, and the reader might worry that in order to guarantee the reality of the effective action one should add the appropriate complex conjugate (i.e., impose reality conditions). As we shall discuss below, however, the effective action contributions in the second-order formalism, obtained from (51) upon decomposing the connection into torsion and torsion-free parts, and using the solutions for the torsion obtained by varying the Holst modification of the general relativity action with respect the independent torsion components, as in the Einstein–Cartan theory discussed previously, are independent of the BI parameter β, which can thus take on any value.

We mention for completeness that the term (51) has been added by Holst [29] to the standard first-order GR Einstein–Hilbert term in the action in order to derive a Hamiltonian formulation of canonical general relativity suggested by Barbero [30,31] from an action. This formulation made use of a real SU(2) connection in general relativity, as opposed to the complex connection introduced by Ashtekar in his canonical formulation of gravity [32]. The link between the two approaches was provided by Immirzi [27,28] who, by means of a canonical transformation, introduced a finite complex number $\beta \neq 0$ (the *BI parameter*, previously mentioned) in the definition of the connection. When the (otherwise free) parameter takes on the purely imaginary values $\beta = \pm i$, the theory reduces to the self (or anti-self) dual formulation of canonical quantum gravity proposed by Ashtekar [32,33] and Ashtekar–Romano–Tate [34]. The values $\beta = \pm 1$ lead to Barbero's real Hamiltonian formulation of canonical gravity. The Holst modification (51), can then be used to derive these formulations from an effective action, with the coefficient β in (51) playing the role of the complex BI parameter[5].

In the presence of fermions, the Holst modification (51) is *not* a total derivative. Therefore, if added it will lead to the false prediction of "observable effects" of the BI parameter. In particular, following exactly the same procedure as for the Einstein–Cartan theory of the previous Section, and using the decomposition (23) of the torsion in the Holst modification of the Einstein action, obtained by adding (51) to the combined actions (29), (33) and (35), one can derive the following extra contributions in the action (up to total derivatives) [35–38]

$$S_{\text{Holst}} = -\frac{1}{2\kappa^2} \int d^4 x e \left(\frac{\beta}{3} T_\mu S^\mu + \frac{\beta}{2} \epsilon_{\mu\nu\rho\sigma} q_\lambda{}^{\mu\rho} q^{\lambda\nu\sigma} \right). \tag{53}$$

By varying independently the combined actions (29), (33) and (53) with respect to the torsion components, as in the Einstein–Cartan theory, one arrives at the equations:

$$\frac{1}{24\kappa^2} S^\mu + \frac{\beta}{6\kappa^2} T^\mu - \frac{1}{8} j^{5\mu} = 0,$$
$$-4T^\mu + \beta S^\mu = 0,$$
$$q_{\mu\nu\rho} + \beta \epsilon_{\nu\sigma\rho\lambda} q_\mu^{\sigma\lambda} = 0. \tag{54}$$

The solutions of (54) are [35,37]

$$T^\mu = \frac{3\kappa^2}{4} \frac{\beta}{\beta^2 + 1} j^{5\mu}, \quad S^\mu = \frac{3\kappa^2}{\beta^2 + 1} j^{5\mu}, \quad q_{\mu\nu\rho} = 0. \tag{55}$$

Substituting these back into the action, and following the steps carried out for the Einstein–Cartan theory, would lead to a four-fermion induced interaction term of the form [35]

$$S_{j^5-j^5} = -\frac{3}{16(\beta^2 + 1)} \kappa^2 \int d^4 x e \, j^{5\mu} j_\mu^5. \tag{56}$$

The coupling of this term depends on the BI parameter β, which is in contradiction to its role in the canonical formulation of gravity [27,28], as a free parameter, being implemented by a canonical transformation in the connection field. Moreover, for purely imaginary values of β, such that $|\beta|^2 > 1$, the four fermion interaction is *attractive*. For values of $\beta \to \pm i$ (which corresponds to the well-defined Ashtekar–Romano–Tate theory [34]) the interaction diverges, which presents a puzzle. Furthermore, for values of $|\beta| \to 1^+$ the coupling of the four-fermion interaction is strong. Such strong couplings can lead to the formation of fermion condensates in flat spacetimes, given that the attractive four-fermion effective coupling of (56) in this case is much stronger than the weak gravitational coupling $\kappa^2 \propto G_N$. These features are all in contradiction with the allegedly topological nature of the BI parameter.

The above are indeed pathologies related to the mere addition of a Holst term in a theory with fermions. Such an addition is inconsistent with the first-order formalism, for the simple reason that the Holst term (51) alone is not a total derivative in the presence of fermions, and thus there is no surprise that its addition leads to "observable" effects (56) in the effective action. In addition, as observed in [37], the solution (55) of (54) is mathematically inconsistent, given that the first line of (55) equates a proper vector (T^μ) with an axial one (the axial spinor current $j^{5\mu}$).

The only consistent cases are those for which either $\beta \to 0$ or $\beta \to \infty$. The first is the Einstein–Cartan theory. The second means no torsion, in the sense that in a path integral formalism, where one integrates over all spin connection configurations, only the zero torsion contributions survive in the partition function, so as to compensate the divergent coefficient. In either case, $T^\mu \to 0$, and the solution (55) reduces to that of the Einstein–Cartan theory (33) and (41). However, this is in sharp contradiction with the arbitrariness of the BI parameter β of the canonical formulation of gravity, which is consistent for every (complex in general) β.

The resolution of the problem was provided by Mercuri [37], who noticed that an appropriate Holst-like modification of a gravity theory in the presence of fermions is possible, if the Holst modification contains additional fermionic-field dependent terms so as to become a total derivative and thus retains its topological nature that characterises such modifications in the torsion-free pure gravity case. The proposed Holst-like term for the torsion-full case of gravity in the presence of fermions contains the Holst term (51) and an *additional* fermion-piece of the form [37] (we ignore the electromagnetic interactions from now on, for brevity, as they do not play an essential role in our arguments):

$$S_{\text{Holst-fermi}} = \frac{\alpha}{2} \int d^4x e \left(\overline{\psi}\gamma^\mu \gamma_5 D_\mu(\omega)\psi + \overline{D_\mu(\omega)\psi}\gamma^\mu \gamma_5 \psi \right), \quad \alpha = \text{const.}, \quad (57)$$

so that the total Holst-like modification is given by the sum $S_{\text{Holst-total}} \equiv S_{\text{Holst}} + S_{\text{Holst-fermi}}$.

We next note that the fermionic Holst contributions (57) when combined with the Dirac kinetic terms of the QED action, yield terms of the form (in our relative normalisation with respect to the Einstein terms in the total action):

$$S_{\text{Dirac-Holst-fermi}} = \frac{i}{2} \int d^4x e \left[\overline{\psi}\gamma^\mu(1 - i\alpha\gamma_5)D_\mu(\omega)\psi + \overline{D_\mu(\omega)\psi}\gamma^\mu(1 - i\alpha\gamma_5)\psi \right]. \quad (58)$$

We thus observe that in the Ashtekar limit [32,33] $\beta = \pm i$, the terms in the parentheses in (58) containing the constant α become the chirality matrices $(1 \pm \gamma_5)/2$ and this is why the specific theory is chiral.

In general, the (complex) parameter α is to be fixed by the requirement that the integrand in $S_{\text{Holst-total}}$ is a *total derivative*, so that it does not contribute to the equations of motion. It can be readily seen that this is achieved when

$$\alpha = \beta. \quad (59)$$

In that case, one recovers the results of the Einstein–Cartan theory, as far as the torsion decomposition and the second-order final form of the effective action are concerned[6].

4.1. Holst Actions for Fermions and Topological Invariants

A final comment concerns the precise expression of the total derivative term that amounts to the total Holst-like modification $S_{\text{Holst-total}}$. As discussed in [37], this action can be cast in a form involving (in the integrand) a *topologically invariant density*, the so-called Nieh–Yan topological density [39], which is the only exact form invariant under local Lorentz transformations associated with torsion:

$$S_{\text{Holst-total}} = -i\frac{\beta}{2} \int d^4x \left[I_{\text{NY}} + \partial_\mu j^{5\mu} \right], \quad (62)$$

with I_{NY} the Nieh–Yan invariant density [39]:

$$I_{\text{NY}} \equiv \epsilon^{\mu\nu\rho\sigma} \left(T^a_{\mu\nu} T_{\rho\sigma a} - \frac{1}{2} e^a_\mu e^b_\nu R_{\rho\sigma ab}(\omega) \right). \quad (63)$$

Taking into account that, in our case, the torsion-full connection has the form (38), we observe that the first term in I_{NY}, quadratic in the torsion T, vanishes identically, as a result of appropriate Fierz identities. Thus, upon taking into account (38), the Holst-like modification of the gravitational action in this case becomes a total derivative of the form [40]:

$$S_{\text{Holst-total}} = \frac{i\beta}{4} \int d^4x \partial_\mu j^{5\mu} = -\frac{i\beta}{6} \int d^4x \epsilon^{\mu\nu\rho\sigma} \partial_\mu T_{\nu\rho\sigma}(\psi), \tag{64}$$

where the last equality stems from the specific form of torsion in terms of the axial fermion current, implying $2\epsilon^{\mu\nu\rho\sigma} T_{\nu\rho\sigma}(\psi) + 3j^{5\mu} = 0$. In general, the Nieh–Yan density is just the divergence of the pseudotrace axial vector associated with torsion, $I_{\text{NY}} = \epsilon^{\mu\nu\rho\sigma} \partial_\mu T_{\nu\rho\sigma}$.

The alert reader can notice that if the axial fermion current is conserved in a theory, then the Holst action (64) vanishes trivially. However, in the case of chiral anomalies, examined above, the axial current is not conserved but its divergence yields the mixed anomaly (42). In that case, by promoting the BI parameter to a canonical pseudoscalar field $\beta \to \beta(x)$ [38], the Holst term (64) becomes equivalent to the torsion–axion–$j^{5\mu}$ interaction term in (46), upon identifying $\beta(x) = \frac{b(x)}{f_b}$. In this case, the field-prompted BI parameter plays a role analogous to the QCD CP violating parameter [38]. As we have discussed in Section 3. Therefore, this is consistent with the association of torsion with an axion-like dynamical degree of freedom, and thus the works of [38] and [6] lead to equivalent physical results from this point of view [41].

Before closing this subsection, we remark that Holst modifications, along the lines discussed for the spin 1/2 fermions above, are known to exist for higher spin 3/2 fermions, ψ_μ, like gravitinos of SUGRA theories [40,42]. In fact, Holst-like modifications, including fermionic contributions, have been constructed in [40,43] for various supergravities (e.g., N = 1, 2, 4), non-trivially extending the spin 1/2 case discussed above. The total derivative nature of these Holst-like actions implies no modifications to the equations of motion. On-shell (local and global) supersymmetries are then preserved. We discuss such issues in Section 6.

4.2. Barbero–Immirzi Parameter as an Axion Field

The classical models described until now in this Section 4 lack the presence of a dynamical pseudoscalar (axion-like) degree of freedom, which, as we have seen in Section 3, is associated with quantum torsion.

Such a pseudoscalar degree of freedom arises in [42,44], which were the first works to promote the BI parameter to a dynamical field, the starting point is the so-called Holst action (51), which by itself is *not* a topological invariant, in contrast to the Nieh–Yan term (63). The work of [42,44] deals with matter free cases. If $\gamma(x)$ represents the BI field, the Holst term now reads (in form language)

$$S_{\text{Holst}} = \frac{1}{2\kappa^2} \int \overline{\gamma}(x) e^a \wedge e^b \wedge R_{ab}, \tag{65}$$

where R_{ab} is the curvature two-form, in the presence of torsion, and we used the notation of [44] for the inverse of the BI field $\overline{\gamma}(x) = 1/\gamma(x)$, to distinguish this case from the Kalab–Ramond (KR) axion $b(x)$ in our string-inspired one. The analysis of [42,44] showed that the gravitational sector results in the action

$$S^{\text{eff}}_{\text{grav+Holst+BI-field}} = \int d^4x \sqrt{-g} \left[-\frac{1}{2\kappa^2} R + \frac{3}{4\kappa^2 (\overline{\gamma}^2 + 1)} \partial_\mu \overline{\gamma} \partial^\mu \overline{\gamma} \right]. \tag{66}$$

Coupling the theory to fermionic matter [35,36,45] can be achieved by introducing a rather generic non-minimal coupling parameter α, for massless Dirac fermions in the form

$$\mathcal{S}_F = \frac{i}{12} \int \epsilon_{abcd} e^a \wedge e^b \wedge e^c \wedge \left[(1 - i\alpha) \overline{\psi} \gamma^d D(\omega) \psi - (1 + i\alpha) \overline{D(\omega) \psi} \gamma^d \psi \right], \tag{67}$$

where $\alpha \in \mathbb{R}$ is a constant parameter. The case of constant $\overline{\gamma}$ has been discussed in [35,36] (in fact, Ref. [35] deals with minimally-coupled fermions, i.e., the limit $\alpha = 0$), whilst the work of [37] extended the analysis to coordinate-dependent BI, $\overline{\gamma}(x)$.

The extension of the BI to a coordinate dependent quantity, which is assumed to be a *pseudoscalar field*, implies:

(i) The consistency of (55), given that now the BI parameter being a pseudoscalar field, reinstates the validity of the first of the Equation (55), since the product of its right-hand side is now parity even, and thus transforms as a vector, in agreement with the nature of the left-hand side of the equation.

(ii) Additional terms of interaction of the fermions with the derivative of the BI field $\partial_\mu \overline{\gamma}$:

$$\mathcal{S}_{F\partial\overline{\gamma}} = \frac{1}{2} \int \sqrt{-g} \left(\frac{3}{2(\overline{\gamma}^2 + 1)} \partial^\mu \overline{\gamma} \left[-j_\mu^5 + \alpha \, \overline{\gamma}(x) j_\mu \right] \right), \qquad (68)$$

with j_μ^5 being the axial current (34) and

$$j_\mu = \overline{\psi} \gamma_\mu \psi, \qquad (69)$$

the vector current.

(iii) Interaction terms of fermions with non-derivative $\overline{\gamma}(x)$ terms:

$$\mathcal{S}_{F-\text{non-deriv}\overline{\gamma}} = \frac{i}{2} \int \sqrt{-g} \left[\left[(1 - i\alpha) \, \overline{\psi} \gamma^d D(\overset{\circ}{\Gamma}) \, \psi - (1 + i\alpha) \overline{(D(\overset{\circ}{\Gamma})\psi)} \, \gamma^d \, \psi \right] \right.$$
$$\left. - \int \sqrt{-g} \, \frac{3}{16(\overline{\gamma}^2 + 1)} \left[j_\mu^5 j^{5\mu} - 2\alpha \, \overline{\gamma} j_\mu^5 j^\mu - \alpha^2 \, j_\mu j^\mu \right], \qquad (70)$$

with $D(\overset{\circ}{\Gamma})$ being the diffeomorphic covariant derivative, expressed in terms of the torsion-free Christoffel connection, which is the result of [36], as expected, because this term contains non derivative terms of the BI.

The action (70) involves four-fermion interactions with *attractive* channels among the fermions. Such features may play a role in the physics of the early Universe, as we shall discuss in Section 8.2.

We also observe from (70) that the case $\alpha = 0$ (minimal coupling), corresponds to a four-fermion axial-current (56), which however depends on the BI field. Thus, this limiting theory is not equivalent to our string-inspired model, in which the corresponding quantum-torsion-induced four-fermion axial-current-current interaction (56) is independent from the KR axion field $b(x)$, although both cases agree with the sign of that interaction.

A different fermionic action than (70), using non-minimal coupling of fermions with γ^5, has been proposed in [37] as a way to resolve an inconsistency of the Holst action, when coupled to fermions, in the case of constant γ. In that proposal, the $1 + i\alpha$ factor in (70) below, is replaced by the Dirac-self-conjugate quantity $1 - i\alpha \gamma^5$. The decomposition of the torsion into its irreducible components in the presence of the Holst action with arbitrary (constant) BI parameter, leads to an inconsistency, implying that the vector component of the torsion is proportional to the axial fermion current, and hence this does not transform properly under improper Lorentz transformations. With the aforementioned modification of the fermion action the problem is solved, as demonstrated in [37], upon choosing $\alpha = \overline{\gamma}$, which eliminates the vector component of the torsion. But this inconsistency is valid only if $\overline{\gamma}$ is considered as a constant. Promotion of the BI parameter $\overline{\gamma}$ to a *pseudoscalar* field, $\overline{\gamma}(x)$, resolves this issue, as discussed in [45], given that one obtains in that case consistent results, in the sense that the vector component of the torsion transforms correctly under parity, as a vector, since it contains now, apart from terms proportional to the vector fermionic current (69), also terms proportional to the product of the BI pseudoscalar with the axial fermionic current (34), as well as terms of the form $\overline{\gamma} \partial_\mu \overline{\gamma}$, all transforming properly as vectors under improper Lorentz transformations.

5. Torsion on Graphene

The use of graphene as a tabletop realisation of some high-energy scenarios is now considerably well developed, see, e.g., [46], the review [47] and the contribution [48] to this Special Issue. Let us here recall the main ideas and those features that make graphene a place where torsion is present.

Graphene is an allotrope of carbon and, being a one-atom-thick material, it is the closest to a two-dimensional object in nature. It is fair to say that was theoretically speculated [49,50] and, decades later, it was experimentally found [51]. Its honeycomb lattice is made of two intertwined triangular sub-lattices L_A and L_B, see Figure 3. As is by now well known, this structure is behind a natural description of the electronic properties of graphene in terms of massless, $(2+1)$-dimensional, Dirac quasi-particles.

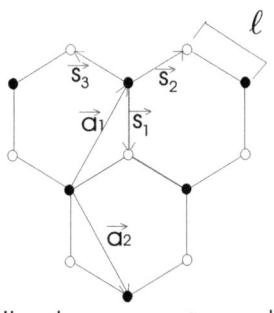

● = sublattice L_A ○ = sublattice L_B

Figure 3. The honeycomb lattice of graphene, and its two triangular sublattices L_A and L_B. The choice of the basis vectors, $(\vec{a}_1, \vec{a}_2)$ and $(\vec{s}_1, \vec{s}_2, \vec{s}_3)$, is, of course, not unique. Figure taken from [52].

Indeed, starting from the tight-binding Hamiltonian for the conductivity electrons, and considering only near-neighbours contribution[7]

$$H = -t \sum_{\vec{r} \in L_A} \sum_{i=1}^{i=3} \left(a^\dagger(\vec{r}) b(\vec{r} + \vec{s}_i) + b^\dagger(\vec{r} + \vec{s}_i) a(\vec{r}) \right), \tag{71}$$

where t is the nearest-neighbour hopping energy which is approximately 2.8 eV, and $a, a^\dagger$ ($b, b^\dagger$) are the anticommuting annihilation and creation operators for the planar electrons in the sub-lattice L_A (L_B).

If we Fourier-transform to momentum space, $\vec{k} = (k_x, k_y)$ annihilation and creation operators,

$$a(\vec{r}) = \sum_{\vec{k}} a_{\vec{k}} e^{i\vec{k} \cdot \vec{r}}, \quad b(\vec{r}) = \sum_{\vec{k}} b_{\vec{k}} e^{i\vec{k} \cdot \vec{r}}, \text{etc}, \tag{72}$$

then

$$H = -t \sum_{\vec{k}} \sum_{i=1}^{i=3} \left(a^\dagger_{\vec{k}} b_{\vec{k}} e^{i\vec{k} \cdot \vec{s}_i} + b^\dagger_{\vec{k}} a_{\vec{k}} e^{-i\vec{k} \cdot \vec{s}_i} \right).$$

Using the conventions for $\vec{s}_i$ of Figure 3, we find that

$$\mathcal{F}(\vec{k}) = -t \sum_{i=1}^{3} e^{i\vec{k} \cdot \vec{s}_i} = -t e^{-i\ell k_y} \left[1 + 2 e^{i\frac{3}{2} \ell k_y} \cos\left(\frac{\sqrt{3}}{2} \ell k_x \right) \right], \tag{73}$$

leading to

$$H = \sum_{\vec{k}} \mathcal{F}(\vec{k}) a^\dagger_{\vec{k}} b_{\vec{k}} + \mathcal{F}^*(\vec{k}) b^\dagger_{\vec{k}} a_{\vec{k}}.$$

For graphene, the conduction and valence bands touch at two points[8] $K_{D\pm} = (\pm \frac{4\pi}{3\sqrt{3}\ell}, 0)$, as one can check by finding the zeroes of (73). These points are called *Dirac points*. The dispersion relation $E(\vec{k}) = |f(\vec{k})|$, for $t\ell = 1$, is shown in Figure 4a.

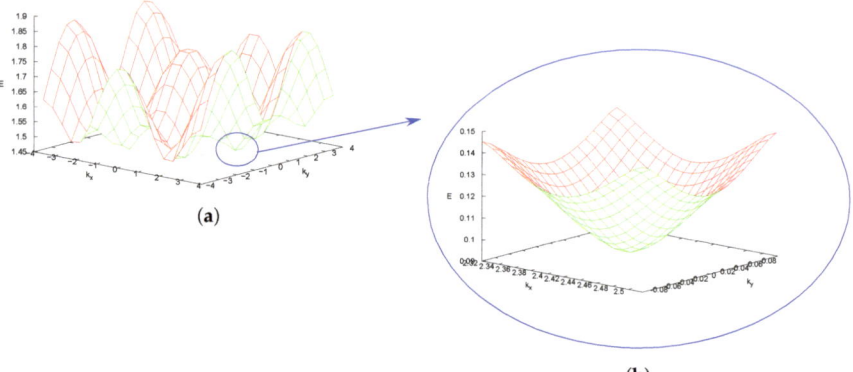

Figure 4. (**a**) The dispersion relation $E(\vec{k})$ for graphene, setting $t\ell = 1$. We only take into account the near neighbours contribution in (71). (**b**) A zoom near the Dirac point K_{D+} showing the linear approximation works well in the low energies regime.

If we expand $\mathcal{F}(\vec{k})$ around the Dirac points, $\vec{k}_\pm = \vec{K}_{D\pm} + \vec{p}$, assuming $|p| \ll |K_D|$, we have

$$\mathcal{F}_+(\vec{p}) \equiv f(\vec{k}_+) = v_F(p_x + i\, p_y),$$
$$\mathcal{F}_-(\vec{p}) \equiv f(\vec{k}_-) = -v_F(p_x - i\, p_y),$$

where $v_F \equiv \frac{3}{2}t\ell \sim c/300$ is the *Fermi velocity*. We can see from this that the dispersion relations around the Fermi point is

$$|E_\pm(\vec{p})| = v_F |\vec{p}|, \tag{74}$$

which is the dispersion relation for a v_F-relativistic massless particle (see Figure 4b).

Defining $a_\pm \equiv a(\vec{k}_\pm)$ and $b_\pm \equiv b(\vec{k}_\pm)$, and arranging the annihilation (creation) operators as a column (row) vector $\psi_\pm = \begin{pmatrix} b_\pm \\ a_\pm \end{pmatrix}$; $\psi_\pm^\dagger = \begin{pmatrix} b_\pm^\dagger & a_\pm^\dagger \end{pmatrix}$, then

$$H = v_F \sum_{\vec{p}} \left[\psi_+^\dagger \vec{\sigma} \cdot \vec{p}\, \psi_+ - \psi_-^\dagger \vec{\sigma}^* \cdot \vec{p}\, \psi_- \right], \tag{75}$$

where $\vec{\sigma} = (\sigma_1, \sigma_2)$ and $\vec{\sigma}^* = (\sigma_1, -\sigma_2)$, being σ_i the Pauli matrices.

Going back to the configuration space, which is equivalent to make the usual substitution $p^\mu \to -i\partial^\mu$,

$$H = -i v_F \int d^2x \left[\psi_+^\dagger \sigma^\mu \partial_\mu \psi_+ - \psi_-^\dagger \sigma^{*\mu} \partial_\mu \psi_- \right], \tag{76}$$

where sums turned into integrals because the continuum limit was assumed.

By including time to make the formalism fully relativistic, although with the speed of light c traded for the Fermi velocity v_F, and making the Legendre transform of (76), we obtain the action

$$S = i v_F \int d^3x\, \bar{\Psi} \gamma^a \partial_a \Psi, \tag{77}$$

here $x^a = (t, x, y)$, are the flat spacetime coordinates, $\Psi = (\psi_+, \psi_-)$ is a reducible representation for the Fermi field and the γ^a are Dirac matrices in the same reducible representation in three dimensions.

5.1. Torsion as Continuous Limit of Dislocations

Even if we will deal mainly with graphene, the considerations here apply to many other two-dimensional crystals [56]. For the purposes of this work, we can define a *topological defect* as a lattice configuration that cannot be undone by continuous transformations. These are obtained by cutting and sewing the pristine material through what is customarily called a *Volterra process* [57]. Probably, the easiest defects to visualise are the *disclinations*. For this hexagonal lattice, a disclination defect is an n-sided polygon with $n \neq 6$, characterised by a *disclination angle* s. When $n = 3, 4, 5$, the defect has a *positive* disclination angle $s = 180°, 120°, 60°$, respectively, whilst for $n = 7, 8, \ldots$, it has a *negative* disclination angle $s = -60°, -120°, \ldots$, respectively. These defects carry intrinsic positive or negative curvature, according to the sign of the corresponding angle s, localised at the tip of a conical singularity. In a continuum description, obtained for large samples in the large wave-length regime, one can associate[9] [64,65] to the disclination defect the spin-connection $\omega^{ab}{}_\mu$. Associated to ω is the curvature two-form tensor R^{ab},

$$R^{ab}{}_{\mu\nu} = \partial_\mu \omega^{ab}{}_\nu - \partial_\nu \omega^{ab}{}_\mu + \omega^a{}_{c\mu} \omega^{cb}{}_\nu - \omega^a{}_{c\nu} \omega^{cb}{}_\mu,$$

that we have already met in (7) and in (17).

A *dislocation* can be produced as a dipole of disclinations with zero total curvature. Figure 5 shows a heptagon–pentagon dipole, which in the Volterra process is equivalent to introducing a strip in the lower-half plane, whose width is the *Burgers vector* $\vec{b}$, that characterises this defect.

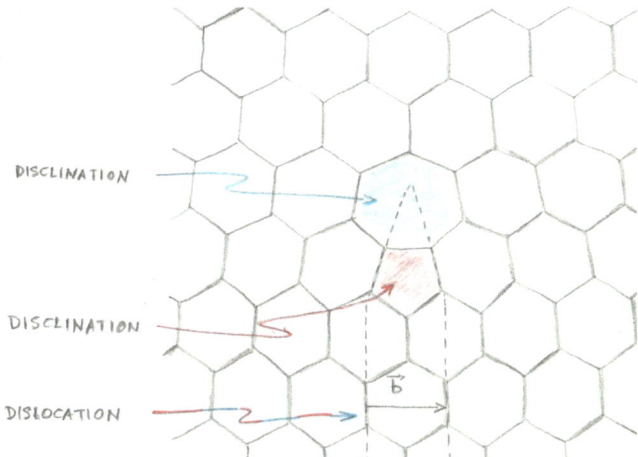

Figure 5. Edge dislocation from two disclinations. Two disclinations, a heptagon, and a pentagon add-up to zero total intrinsic curvature, and make a dislocation with Burgers vector $\vec{b}$, as indicated. In the continuous long wave-length limit, this configuration carries nonzero torsion. Figure taken from [66].

In the continuum limit one can associate the Burgers vector to the torsion tensor [64,65],

$$T^a{}_{\mu\nu} = \partial_\mu e^a{}_\nu - \partial_\nu e^a{}_\mu + \omega^a{}_{b\mu} e^b{}_\nu - \omega^a{}_{b\nu} e^b{}_\mu, \tag{78}$$

where $T^\rho{}_{\mu\nu} = E_a{}^\rho T^a{}_{\mu\nu}$. On this see our earlier discussion around (1) and (9).

The explicit relation between Burgers vectors and torsion can be written as [67]

$$b^i = \iint_\Sigma T^i_{\mu\nu} dx^\mu \wedge dx^\nu, \tag{79}$$

where the surface Σ has a boundary enclosing the defect. Roughly speaking, the torsion tensor is the surface density of the Burgers vector. Nonetheless, although the relation (79) looks simple, there are subtleties: given a distribution of Burgers vector, there is no simple procedure to assign a torsion tensor to it, even for the simple case of edge dislocations [68].

The smooth way to introduce the effect of dislocations in the long wave-length regime, through torsion tensor, is to consider an action in a $(2+1)$-dimensional space with a spin-connection that carries torsion, i.e., a Riemann–Cartan space U_3 [2]. Demanding only Hermiticity and local Lorentz invariance, starting by a simple action

$$S = \frac{i}{2} v_F \int d^3x \sqrt{|g|} \left(\overline{\Psi} \gamma^\mu \overrightarrow{D}_\mu(\omega) \Psi - \overline{\Psi} \overleftarrow{D}_\mu(\omega) \gamma^\mu \Psi \right), \tag{80}$$

where

$$\overrightarrow{D}_\mu(\omega) \Psi = \partial_\mu \Psi + \frac{1}{8} \omega_\mu^{ab} [\gamma_a, \gamma_b] \Psi, \tag{81}$$

$$\overline{\Psi} \overleftarrow{D}_\mu(\omega) = \partial_\mu \overline{\Psi} - \frac{1}{8} \overline{\Psi} [\gamma_a, \gamma_b] \omega_\mu^{ab}, \tag{82}$$

we obtain, besides possible boundary terms (see details in Appendix A of [69]),

$$S = i v_F \int d^3x \, |e| \, \overline{\Psi} \left(E_a^\mu \gamma^a \overrightarrow{D}_\mu(\mathring{\omega}) - \frac{i}{4} \gamma^5 \frac{\epsilon^{\mu\nu\rho}}{|e|} T_{\mu\nu\rho} \right) \Psi, \tag{83}$$

where $|e| = \sqrt{|g|}$, the covariant derivative is based only on the torsion-free connection, $\gamma^5 \equiv i \gamma^0 \gamma^1 \gamma^2 = \begin{pmatrix} I_{2\times 2} & 0 \\ 0 & -I_{2\times 2} \end{pmatrix}$ (we used the conventions for $\gamma^0, \gamma^1, \gamma^2$ that give a γ^5 that *commutes* with the other three gamma matrices[10]), and the contribution due to the torsion is all in the last term through its totally antisymmetric component [69].

We see that the last term couples torsion with the fermionic excitations describing the quasi-particles and is the three-dimensional version of (33), for $A_\mu = 0$. It can be also seen that, to have a nonzero effect, we need $\epsilon^{\mu\nu\rho} T_{\mu\nu\rho} \neq 0$, that requires at least three dimensions. This mathematical fact is behind the obstruction pointed out some time ago leading to the conclusion that, in two-dimensional Dirac materials, torsion can play no physical role [70–72].

To overcome this obstruction, in [69] the time dimension is included in the picture. With this in mind, we have two possibilities that a nonzero Burgers vector gives rise to $\epsilon^{\mu\nu\rho} T_{\mu\nu\rho} \neq 0$:

(i) a *time-directed* screw dislocation (only possible if the crystal has a time direction)

$$b_t \propto \int \int T_{012} dx \wedge dy, \tag{84}$$

or

(ii) an edge dislocation "felt" by an integration along a *spacetime circuit* (only possible if we can actually go around a loop in time), e.g,

$$b_x \propto \int \int T_{102} dt \wedge dy. \tag{85}$$

This last scenario is depicted in Figure 6.

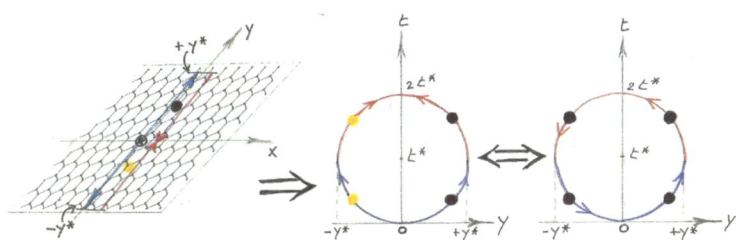

Figure 6. Idealised *time-loop*. At $t = 0$, the hole (yellow) and the particle (black) start their movements from $y = 0$, in opposite directions. At $t = t^* > 0$, the hole is at position $-y^*$, while the particle is at position $+y^*$, (the blue portion of the circuit). Then they come back to their original position, $y = 0$, at $t = 2t^*$ (red portion of the circuit). On the far right, is depicted an equivalent *time-loop*, where the hole moving forward in time is replaced by a particle moving backward. Figure taken from [69].

5.2. Time-Loops in Graphene

Scenario (i) could be explored in the context of the very intriguing time crystals introduced some time ago [73,74], and nowadays under intense experimental studies [75,76]. Lattices that are discrete in all dimensions, including time, would be an interesting playground to probe quantum gravity ideas [77]. In particular, it would have an impact to explore defect-based models of classical gravity/geometry, see for instance [64,65]. However, here we shall focus only on scenario (ii).

By assuming the Riemann curvature to be zero, $\mathring{R}^\mu{}_{\nu\rho\sigma} = 0$, but nonzero torsion (or contorsion $K^\mu{}_{\nu\rho} \neq 0$), and choosing a frame where $\mathring{\omega}^{ab}_\mu = 0$ (see Appendix B of [69]), the action (83) is

$$S = i\, v_F \int d^3x |e| \left(\overline{\Psi} \gamma^\mu \partial_\mu \Psi - \frac{i}{4} \overline{\psi}_+ \phi \psi_+ + \frac{i}{4} \overline{\psi}_- \phi \psi_- \right), \tag{86}$$

where $\Psi = (\psi_+, \psi_-)$ and $\phi \equiv \epsilon^{\mu\nu\rho} T_{\mu\nu\rho}/|e|$ is what we call *torsion field*; it is a pseudo-scalar and the three-dimensional version of the S_μ we discussed earlier. Even in the presence of torsion, the two irreducible spinors, ψ_+ and ψ_-, are decoupled (however, with opposite signs).

The pictures in Figure 6 refer to a defect-free honeycomb graphene-like sheet. The presence of a dislocation, with Burgers vector $\vec{b}$ directed along x, would result in a failure to close the loop proportional to $\vec{b}$ [69].

The idea of time-looping is fascinating. The challenge is to bring this idealised picture close to experiments. We present below the first steps in that direction, as taken in [69].

5.3. Towards Spotting Torsion in a Lab

The simplest way to realise the scenario just discussed is to have:

(i) the particle-hole pair required for the time-loop to be excited by an *external electromagnetic field*, and

(ii) that what we shall call *holonomy*—a proper disclination or torsion—provides the non-closure of the loop in the proper direction.

Stated differently, we are searching for *the quantifiable consequences of an holonomy, caused by disclination or torsion in a time-loop*. Only an appropriate combination of (i) and (ii) can yield the desired outcome.

With this in mind, the action governing such microscopic dynamics is

$$S = i \int d^3x |e| \left(\overline{\Psi} \gamma^\mu (\partial_\mu - i g_{em} A_\mu) \Psi - i g_{tor} \overline{\psi}_+ \phi \psi_+ + i g_{tor} \overline{\psi}_- \phi \psi_- \right) \tag{87}$$

$$\to i \int d^3x \left(\overline{\psi} \gamma^\mu \partial_\mu \psi - i g_{em} \hat{j}^\mu_{em} A_\mu - i g_{tor} \hat{j}_{tor} \phi \right) \equiv S_0[\overline{\psi}, \psi] + S_I[A, \phi], \tag{88}$$

where v_F is taken to be one, while g_{em} and g_{tor} are the electromagnetic and torsion coupling constants, respectively, the latter including the factor $1/4$. In (88) we only have one Dirac point, say $\psi \equiv \psi_+$, as this simplifies calculations, and we focus on flat space, $|e| = 1$. Finally, $\hat{j}^\mu_{em} \equiv \bar{\psi}\gamma^\mu\psi$ and $\hat{j}_{tor} \equiv \bar{\psi}\psi$.

The electromagnetic field is *external*, hence it is a four-vector $A_\mu \equiv (V, A_x, A_y, A_z)$. Nonetheless, the dynamics it induces on the electrons living on the membrane is two-dimensional. Therefore, the effective vector potential may be taken to be $A_\mu \equiv (V, A_x, A_y)$, see, e.g., [78,79]. There are two alternatives to this approach. One is the *reduced QED* of [80,81], where the gauge field propagates on a three-dimensional space and one direction is integrated out to obtain an effective interaction with the electrons, constrained to move on a two-dimensional plane. In this approach a Chern–Simons photon naturally appears (see, e.g., [82,83]). Another approach is to engineer a $(2+1)$-dimensional A_μ by suitably straining the material, see, e.g., [71,72], and [84]. In that case, one usually takes the temporal gauge and $A_x \sim u_{xx} - u_{yy}$, $A_x \sim 2u_{xy}$, where u_{ij} is the strain tensor.

As said, defects here are not dynamical, therefore the torsion field ϕ enters the action as an external field, just like the electromagnetic field. One could, as well, include the effects of the constant ϕ into the unperturbed action, as a mass term $S_0 \to S_m$, see, e.g., [85], where $S_m = i \int d^3x\, \bar{\psi}(\slashed{\partial} - m(\phi))\psi$.

We are in the situation described by the microscopic perturbation,

$$S_I[F_i] = \int d^3x\, \hat{X}_i(\vec{x},t) F_i(\vec{x},t), \tag{89}$$

with the system responding through $\hat{X}_i(\vec{x},t)$ to the external probes $F_i(\vec{x},t)$. The general goal is then to find

$$\hat{X}_i[F_i], \tag{90}$$

to the extent of predicting a measurable effect of the combined action of the two perturbations $F_i(\vec{x},t)$: $F_1^{em}(\vec{x},t) \propto A_\mu(\vec{x},t)$ that induces the response $\hat{j}^\mu_{em}$, and $F_2^{tor}(\vec{x},t) \propto \phi(\vec{x},t)$ inducing the response $\hat{j}_{tor}$:

$$S_I[A,\phi] = \int d^3x \left(\hat{j}^\mu_{em} A_\mu + \hat{j}_{tor} \phi \right), \tag{91}$$

where the couplings, g_{em} and g_{tor}, are absorbed in the respective currents.

With no explicit calculations, simply based on the charge conjugation invariance of the action (88), we can already predict that

$$\chi^{torem}_\mu(x,x') \sim \langle \hat{j}^{em}_\mu(x) \hat{j}^{tor}(x') \rangle \equiv 0, \tag{92}$$

which is just an instance of the Furry's theorem of quantum field theory [86], that in QED reads

$$\chi^{em}_{\mu_1\ldots\mu_{2n+1}}(x_1,\ldots,x_{2n+1}) \sim \langle \hat{j}^{em}_{\mu_1}(x_1) \cdots \hat{j}^{em}_{\mu_{2n+1}}(x_{2n+1}) \rangle = 0, \tag{93}$$

and for us implies

$$\chi^{torem}_{\mu_1\ldots\mu_{2n+1}}(x_1,\ldots,x_{2n+1},y_1,\ldots,y_m) \sim \langle \hat{j}^{em}_{\mu_1}(x_1) \cdots \hat{j}^{em}_{\mu_{2n+1}}(x_{2n+1}) \hat{j}^{tor}(y_1) \cdots \hat{j}^{tor}(y_m) \rangle = 0. \tag{94}$$

This finding indicates that entering the nonlinear response domain is necessary to observe the desired consequences. High-order harmonic generation (HHG) is a well-established technique that has been used to analyse structural changes in atoms, molecules, and more recently, bulk materials (see, e.g., ref. [87] for a recent overview). Thus, the presence or absence of dislocations will significantly alter the intra-band harmonics in our system, which are controlled by the intra-band (electron-hole) current.

5.4. On the Continuum Description of the Two Inequivalent Dirac Points

We have shown earlier that two Dirac points, associated to the reducible $\Psi = (\psi_+, \psi_-)$ [47], are important to treat torsion. Two such points are actually relevant in a broader set of cases. From the material point of view [88], this generally has to do with an extra "valley" degree of freedom in a pristine material, also called colour index [89]. Things change more drastically when topological defects are present. For instance, to make a fullerene C_{60} form pristine graphene we need twelve pentagons sitting at the vertices of an icosahedron, and this generates colour mismatches, see a discussion of these effects in [90]. There, different magnetic flux are added for each vertex which contain a colour line frustration, pointing out to a "magnetic monopole" at the centre of the molecule structure [90]. Such a "monopole" is associated with the SU(2) symmetry group stemming from the doublet structure of the valley degree of freedom (not to be confused with the doublet structure associated with each valley, which generates the irreducibles $\psi_\pm$).

Another instance where both Dirac points are needed for an effective description are *grain boundaries* (GBs). A GB is a line of disclinations of opposite curvature, pentagonal and heptagonal here, arranged in such a way that the two regions (grains) of the membrane match. The two grains have lattice directions that make an angle $\theta/2$ with respect to the direction the lattice would have in the absence of the GB. Different arrangements of the disclinations, always carrying zero total curvature, correspond to different θs, the allowed number of which is of course finite, and related to the discrete symmetries of the lattice (hexagonal here). The most common (stable) being $\theta = 21.8°$, and $\theta = 32.3°$, see, e.g., [91,92]. Other arrangements can be found in [93]. In general, one might expect that the angle of the left grain differs in magnitude from the angle of the right grain, $|\theta_L| \neq |\theta_R|$, nonetheless, high asymmetries are not common, and the symmetric situation depicted in Figure 7 is the one the system tends to on annealing [94].

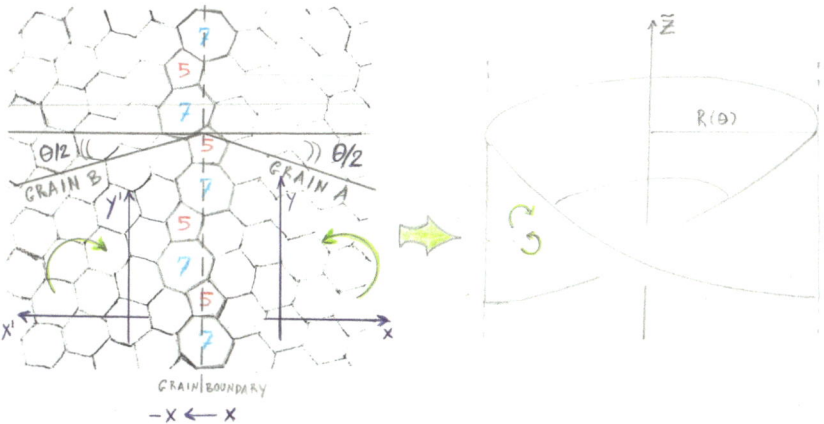

Figure 7. A grain boundary (**left**), and a possible modelling of its effects in a continuum (**right**). This is the prototypical GB, where grain A and grain B are related via a parity ($x \to -x$) transformation. With this, the right-handed frame in grain A is mapped to the left-handed frame in grain B, so that the net effect of a GB is that two orientations coexist on the membrane, and a discontinuous change happens at the boundary. If one wants to trade this discontinuous change for a continuous one, an equivalent coexistence is at work in the non-orientable Möbius strip. One way to quantify the effects of different θs is to relate a varying θ to a varying radius $R(\theta)$ of the Möbius strip. Notice that the third spatial axis is an abstract coordinate, $\tilde{z}$, whose relation with the real z of the embedding space is not specified. Figure taken from [66].

There exists [92,93] a relation (the Frank formula) between θ and the resultant Burgers vector, obtained by adding all Burgers vectors $\vec{b}$s cut by rotating a vector, laying on the GB, of an angle θ with respect to the reference crystal. A possible modelling for this kind of defects was put forward in [66]. That is a four-spinor living on a Möbius strip, see Figure 7, and [66].

6. Torsion in Standard Local Supersymmetry

As a prelude to the Section dedicated to cosmology, we should discuss fermionic (gravitino) torsion in SUGRA models, which can also lead to dynamical breaking of SUGRA. Such models can serve in inducing inflationary scenarios by providing sources for primordial gravitational waves which play a crucial role in inflation, to be discussed in detail in Section 8.

SUGRA theories are Einstein–Cartan theories with fermionic torsion, provided by the gravitino field, $\psi_\mu(x)$, the spin-3/2 (local) supersymmetric fermionic partner of the graviton.

We commence our discussion with the first local SUSY constructed historically, the $(3+1)$-dimensional $N=1$ SUGRA [95–97], which in fact finds a plethora of (conjectural) applications to the phenomenology of particle physics [98]. In the remainder of this Section we shall work in units of the gravitational constant $\kappa = 1$ for brevity.

The spectrum of the unbroken $(3+1)$-dimensional $N=1$ SUGRA is a massless spin 2 graviton field, described by the symmetric tensor field $g_{\mu\nu}(x) = g_{\nu\mu}(x)$, $\mu,\nu = 0,\ldots 3$ and a massless gravitino spin 3/2 Rarita–Schwinger Majorana fermion $\psi_\mu(x)$.

The standard action is given by [97]

$$S_{\text{SG1}} = \frac{1}{2}\int d^4x\,\sqrt{-g}\left(\Sigma^{\mu\nu}_{ab}\,R_{\mu\nu}{}^{ab}(\omega) - \epsilon^{\mu\nu\rho\sigma}\,\overline{\psi}_\mu\,\gamma^5\,\gamma_\nu\,D_\rho(\omega)\,\psi_\sigma\right), \tag{95}$$

where $\Sigma^{\mu\nu}_{ab} = \frac{1}{2}E^\mu_{[a}E^\nu_{b]}$ and $D_\mu(\omega) = \partial_\mu + \frac{1}{8}\omega_{ab\,\mu}[\gamma^a,\gamma^b]$ is the diffeomorphic covariant derivative, with respect to a spin connection $\omega^a{}_{b\mu}$ which, as we shall discuss below, necessarily contains fermionic (gravitino-induced) torsion.

As shown in [40,43], the action (95) can be augmented by adding to it a total derivative Holst type action, which preserves the on-shell $N=1$ SUSY for an arbitrary coefficient t:

$$S_{\text{Holst1}} = i\frac{\eta}{2}\int d^4x\,\sqrt{-g}\left(\Sigma^{\mu\nu}_{ab}\,\widetilde{R}_{\mu\nu}{}^{ab}(\omega) - \epsilon^{\mu\nu\rho\sigma}\,\overline{\psi}_\mu\,\gamma_\nu\,D_\rho(\omega)\,\psi_\sigma\right), \tag{96}$$

with $\widetilde{R}_{\mu\nu}{}^{ab}(\omega)$ the dual Lorentz curvature tensor.

Indeed, as demonstrated in [40,43], the combined action

$$S_{\text{total SG}} = S_{\text{SG1}} + S_{\text{Holst1}} =$$
$$\frac{1}{2}\int d^4x\left(\sqrt{-g}\left[E^\mu_a E^\nu_b R^{ab}{}_{\mu\nu} - \frac{t}{2}\epsilon^{ab}{}_{cd}R^{cd}{}_{\mu\nu}\right] + \epsilon^{\mu\nu\rho\sigma}\,\overline{\psi}_\mu\,\gamma^5\,\gamma_\rho\,\frac{1-i\eta\gamma^5}{2}\,D_\sigma(\omega)\,\psi_\nu\right), \tag{97}$$

is invariant under the *local* SUSY transformation with infinitesimal Grassmann parameter $\alpha(x)$:

$$\delta\psi_\mu = D_\mu(\omega)\,\alpha,\quad \delta e^a_\mu = \frac{i}{2}\overline{\alpha}\,\gamma^a\,\psi_\mu,\quad \delta B_{ab\mu} = \frac{1}{2}\left(C_{\mu ab} - e_{\mu[a}\,C^c{}_{cb]}\right), \tag{98}$$

where, by definition,

$$C^{\lambda\mu\nu} \equiv \frac{1}{\sqrt{-g}}\,\epsilon^{\mu\nu\rho\sigma}\,\overline{\alpha}\,\gamma^5\,\gamma^\lambda\,\frac{1-i\eta\gamma^5}{2}\,D_\rho(\omega)\,\psi_\sigma. \tag{99}$$

We remark for completion that in the special case where $\eta = \pm i$ we obtain Ashtekar's chiral SUGRA extension, while for $\eta = 0$ one recovers the standard $N=1$ SUGRA transformations.

We next remark that variation of the action (97) with respect to the spin connection, leads to the well-known gravitational equation of motion in first order formalism [95–97], which leads to an expression of the gravitino-induced torsion $T_{\rho\sigma}{}^\mu(\psi)$ in terms of the gravitino fields:

$$D_{[\mu}(\omega)\,e^a_{\nu]} \equiv 2T_{\mu\nu}{}^a(\psi) = \frac{1}{2}\overline{\psi}_\mu\,\gamma^a\,\psi_\nu, \tag{100}$$

with the contorted spin connection being given by:

$$\omega_\mu{}^{ab}(e,\psi) = \mathring{\omega}_\mu{}^{ab}(e) + K_\mu{}^{ab}(\psi), \qquad (101)$$

where $\mathring{\omega}_\mu{}^{ab}(e)$ is the torsion-free spin connection (expressible, as in standard GR, in terms of the vielbeins e_μ^a), and $K_\mu{}^{ab}(\psi)$ is the contorsion, given in terms of the gravitino field as:

$$K_{\mu\rho\sigma}(\psi) = \frac{1}{4}\left(\overline{\psi}_\rho \gamma_\mu \psi_\sigma + \overline{\psi}_\mu \gamma_\rho \psi_\sigma - \overline{\psi}_\mu \gamma_\sigma \psi_\rho\right). \qquad (102)$$

The parameter η does not enter the expression for the contorsion, which thus assumes the standard form of $N = 1$ SUGRA without the Holst terms.

Substitution of the solution of the torsion equations of motion into the first-order Lagrangian density, corresponding to the action (97), leads to a second-order Lagrangian density that can be written as the sum of the standard $N = 1$ SUGRA Lagrangian density [97] and a total derivative, depending on the gravitino fields only:

$$\mathcal{L}(\text{second order}) = \mathcal{L}_{\text{usual } N=1 \text{ SUGRA}}(\text{second order}) + \frac{i}{4}\eta\, \partial_\mu(\epsilon^{\mu\mu\rho\sigma}\overline{\psi}_\nu \gamma_\rho \psi_\sigma), \qquad (103)$$

where the standard $N = 1$ SUGRA in the second-order formalism includes four-gravitino terms,

$$\mathcal{L}_{\text{usual } N=1 \text{ SUGRA}} = \sqrt{-g}\,\frac{1}{2}R(e) + \frac{1}{4}\partial_\mu[E_a^\mu E_b^\nu \sqrt{-g}]\left(\overline{\psi}_\nu \gamma^a \psi^b - \overline{\psi}_\nu \gamma^b \psi^a + \overline{\psi}^a \gamma_n u\, \psi^b\right)$$

$$- \frac{1}{2}\epsilon^{\mu\nu\lambda\rho}\overline{\psi}_\mu \gamma^5 \gamma_\nu \left[\partial_\lambda + \frac{1}{2}\omega_\lambda^{ab}(e)\sigma_{ab}\right]\psi_\rho$$

$$- \frac{11}{16}\sqrt{-g}\left[(\overline{\psi}_a \psi^a)^2 - (\overline{\psi}_b \gamma^5 \psi^b)^2\right] + \frac{33}{64}\sqrt{-g}\,(\overline{\psi}_b \gamma^5 \gamma_c \gamma^b)^2$$

$$+ \text{ appropriate auxilliary} - \text{field terms,} \qquad \sigma_{ab} = \frac{i}{4}[\gamma_a, \gamma_b], \qquad (104)$$

and as standard [97] the Lagrangian density is computed by requiring the irreducibility condition:

$$\gamma^\mu \psi_\mu = 0, \qquad (105)$$

which ensures that the spin is exactly 3/2 and not a mixture of this and lower spins. We note that the four-gravitino terms of (104) have been used in [99,100] in order to discuss, upon appropriate inclusion of Goldstino terms [101][11], the possibility of dynamical breaking of SUGRA, via the formation of condensates of gravitino fields $\sigma_c = \langle \overline{\psi}_\mu \psi^\mu \rangle \neq 0$. The gravitino field becomes massive, with mass which can be close to Planck mass, which implies its eventual decoupling from the low-energy (non supersymmetric) theory.

Such scenarios have been used to discuss hill-top inflation, as a consequence of the double-well shape of the effective gravitino potential. Indeed, for small condensates $\kappa^6 \sigma_c(x) \ll 1$, one may obtain an inflationary epoch, not necessarily slow roll, as the gravitino rolls down towards one of the local minima of its double well potential [103] (cf. Figure 8). Such scenarios will be exploited further in Section 8.1, from the point of view of the generation of gravitational waves in the very early Universe, which can lead to a second inflationary era in such models, that could provide interesting—and compatible with the data—phenomenology/cosmology.

We complete the discussion on $N = 1$ SUGRA as in Einstein–Cartan theory, by noticing that, on using (62), (63) and (103), we may write for the super Holst term in this case [40,43]:

$$S_{\text{Super Holst } N=1 \text{ SUGRA}}(e, \psi) = -\frac{i\eta}{2}\int d^4x \left[T_{\text{NY}} + \partial_\mu J^\mu(\psi)\right], \qquad (108)$$

with

$$J^\mu(\psi) = \epsilon^{\mu\nu\rho\sigma}\overline{\psi}_\nu \gamma_\rho \psi_\sigma, \qquad (109)$$

the axial gravitino current, and the Nieh–Yan invariant is given by (63).

Finally, combining the Fierz identity $\epsilon^{\mu\nu\rho\sigma}\,(\overline{\psi}_\mu\,\gamma_a\,\psi_\nu)\,\gamma^a\,\psi_\rho = 0$, with the expression for the N=1 SUGRA torsion $T_{\mu\nu}{}^a(\psi)$ (100), we arrive at $\epsilon^{\mu\nu\rho\sigma}\,T_{\mu\nu a}(\psi)\,T_{\rho\sigma}{}^a(\psi) = 0$, we may write for the on-shell-local-SUSY-preserving Holst term (108):

$$S_{\text{Super Holst N=1 SUGRA}}(e,\psi) = -\frac{i\eta}{4}\int d^4x \partial_\mu J^\mu(\psi)] = \frac{i\eta}{2}\int d^4x\,\epsilon^{\mu\nu\rho\sigma}\,\partial_\mu T_{\nu\rho\sigma}(\psi). \qquad (110)$$

This section is concluded by mentioning that super Holst modifications have been constructed [40] for extended SUGRAs, such as $N = 2, 4$, following and extending appropriately the $N = 1$ case. The spectrum of the $N = 2$ SUGRA consists of a massless spin-2 graviton, two massless chiral spin-3/2 gravitinos, $\gamma^5\,\psi_\mu^I = +\psi_\mu^I$, $\gamma^5\,\psi_{I\mu} = -\psi_{I\mu}$, $I = 1, 2$, and an Abelian gauge field A_μ. This is also an Einstein–Cartan theory, with torsion

$$2T_{\mu\nu}{}^a = \frac{1}{2}\Big(\overline{\psi}_\mu^I\,\gamma^a\,\psi_{I\nu} + \overline{\psi}_{I\mu}\,\gamma^a\,\psi_\nu^I\Big), \qquad (111)$$

and contorsion

$$K_{\mu\rho\sigma} = \frac{1}{4}\Big[\overline{\psi}_\rho^I\,\gamma_\mu\,\psi_{I\sigma} + \overline{\psi}_\mu^I\,\gamma_\rho\,\psi_{I\sigma} - \overline{\psi}_\mu^I\,\gamma_\sigma\,\psi_{I\rho} + \text{c.c.}\Big], \qquad (112)$$

where c.c. denotes complex conjugate, whilst the super Holst term has the form [40]:

$$S_{\text{Super Holst N=2 SUGRA}}(e,\psi) = -\frac{i\eta}{4}\int d^4x \partial_\mu J^\mu(\psi)] = \frac{i\eta}{2}\int d^4x\,\epsilon^{\mu\nu\rho\sigma}\,\partial_\mu T_{\nu\rho\sigma}(\psi), \qquad (113)$$

with $J^\mu(\psi) = \epsilon^{\mu\nu\rho\sigma}\,\overline{\psi}_\nu^I\,\gamma_\rho\,\psi_{I\sigma}$ the axial gravitino current in this case. We observe from (112) that then contorsion is again independent, as in the $N = 1$ case, from the super Holst action parameter η.

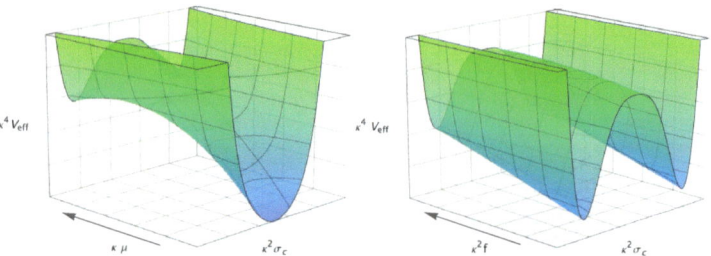

Figure 8. The effective potential of the torsion-induced gravitino condensate $\sigma_c = \langle\overline{\psi}_\mu\,\psi^\mu\rangle$ in the dynamical breaking of $N = 1$ SUGRA scenario of [99], in which, for simplicity, the one-loop-corrected cosmological constant $\Lambda \to 0^+$ (for an analysis with $\Lambda > 0$ see [100] and references therein). The figures show schematically the effect of tuning the inverse-proper-time (renormalisation-group like) scale μ and the scale of SUSY breaking f, whilst holding, respectively, f and μ fixed. The arrows in the respective axes correspond to the direction of increasing μ and f. The reader should note (see left panel) that the double-wall shape of the potential, characteristic of the super-Higgs effect (dynamical SUGRA breaking), appears for values of μ larger than a critical value, in the direction of increasing μ, that is as we flow from Ultraviolet (UV) to infrared (IR) regions. Moreover, as one observes from the right panel of the figure, tuning f allows us to shift the value of the effective potential V_{eff} appropriately so as to attain the correct vacuum structure, that is, non-trivial minima σ_c such that $V_{\text{eff}}(\sigma_c) = \Lambda \to 0^+$. Picture taken from [99].

Finally, we complete the discussion with the $N = 2$ gauged SU(4) SUGRA. For our discussion, we restrict our attention only to the relevant part of its spectrum, consisting of massless spin-2 gravitons, four chiral Majorana spin-3/2 gravitinos ψ_μ^I, $I = 1,\ldots 4$, in the 4 and 4* representations of SU(4), and four Majorana chiral gauginos Λ^I, $I = 1,\ldots 4$. The torsion of this theory depends on both the gravitino and gaugino fields [40],

$$2T_{\mu\nu}{}^a = 2T_{\mu\nu}{}^a(\psi) + 2T_{\mu\nu}{}^a(\psi) = \frac{1}{2}\overline{\psi}^I_{[\mu}\gamma^a\psi_{\nu]I} + \frac{1}{2\sqrt{-g}}e^{a\rho}\epsilon_{\mu\nu\rho\sigma}\overline{\Lambda}_I\gamma^\sigma\lambda^I, \quad (114)$$

and the contorsion reads

$$K_{\mu\nu\rho} = \frac{1}{4}\left(\overline{\psi}^I_\nu\gamma_\mu\psi_{\rho I} + \overline{\psi}^I_\mu\gamma_\nu\psi_{\rho I} - \overline{\psi}^I_\mu\gamma_\rho\psi_{\nu I} + \text{c.c.}\right) - \frac{1}{4\sqrt{-g}}\epsilon_{\mu\nu\rho\sigma}\overline{\Lambda}_I\gamma^\sigma\Lambda^I, \quad (115)$$

which again is independent of the parameter η of the super Holst term, which has the form [40]:

$$S_{\text{Super Holst N=2 SUGRA}}(e,\psi) = -\frac{i\eta}{4}\int d^4x\,\partial_\mu[J^\mu(\psi) - J^\mu(\Lambda)]$$

$$= \frac{i\eta}{2}\int d^4x\,\epsilon^{\mu\nu\rho\sigma}\partial_\mu\left(T_{\nu\rho\sigma}(\psi) - \frac{1}{3}T_{\nu\rho\sigma}(\Lambda)\right), \quad (116)$$

where $J^\mu(\Lambda) = \sqrt{-g}\,\overline{\Lambda}_I\gamma^\mu\Lambda^I$, and the torsion quantities have been defined in (114).

7. Torsion in Unconventional Supersymmetry

USUSY is an appealing theory where all the fields belong to a one-form connection $\mathbb{A}$, in $(2+1)$ dimensions, and the vielbein is realised in a different way than in standard SUGRA models [104]. It has nontrivial dynamics, and leads to a scenario where local SUSY is absent (although there is still diffeomorphism invariance), but rigid SUSY can survive for certain background geometries. Because there is no local SUSY, there are no SUSY pairings. Likewise, no gauginos are present. The only propagating degrees of freedom are fermionic [105], and the parameters that appear in the model are either dictated by gauge invariance, or arise as integration constants. We take the one-form connection spanned by the Lorentz generators $\mathbb{J}_a$, the SU(2) generators corresponding to the internal gauge symmetry $\mathbb{T}_I$ (or a other internal group generator, including the Abelian U(1)), the supercharges $\overline{\mathbb{Q}}^i$ and $\mathbb{Q}_i$ (note that these last generators contain the index corresponding to the fundamental group of SU(2) as well as the spinors)[12] [106]

$$\mathbb{A} = A^I\mathbb{T}_I + \overline{\psi}^i\slashed{e}\mathbb{Q}_i + \overline{\mathbb{Q}}^i\slashed{e}\psi_i + \omega^a\mathbb{J}_a, \quad (117)$$

where $A^I = A^I_\mu dx^\mu$ is the one-form SU(2) connection, $\omega^a = \omega^a{}_\mu dx^\mu$ is the one-form Lorentz connection in $(2+1)$ dimensions, and we defined the one-form $\slashed{e} \equiv e^a{}_\mu\gamma_a dx^\mu$.

We can construct a three-form Chern–Simons Lagrangian from (117), namely[13]

$$L = \frac{\kappa}{2}\langle\mathbb{A}d\mathbb{A} + \frac{2}{3}\mathbb{A}^3\rangle, \quad (118)$$

where $\langle\ldots\rangle$ is the invariant supertrace of $\mathfrak{usp}(2,1|2)$ graded Lie algebra (for the case of internal SU(2) group) and κ is a dimensionless constant. This way, the Lagrangian can be written simply as

$$L = \frac{\kappa}{4}\left(A^IdA_I + \frac{1}{3}\epsilon_{IJK}A^IA^JA^K\right) + \frac{\kappa}{4}\left(\omega^ad\omega_a + \frac{1}{3}\epsilon_{abc}\omega^a\omega^b\omega^c\right) + L_\psi, \quad (119)$$

where the fermionic part is

$$L_\psi = \kappa\overline{\psi}\left(\gamma^\mu\overrightarrow{D}_\mu - \overleftarrow{D}_\mu\gamma^\mu - \frac{i}{2}\epsilon_a{}^{bc}T^a{}_{bc}\right)\psi|e|d^3x.$$

We can see the action (119) possesses also a local scale (Weyl) symmetry. Indeed, by scaling the dreibein and the fermions as

$$e^a{}_\mu \to e^a{}_\mu{}' = \lambda e^a{}_\mu, \quad \psi \to \psi' = \lambda^{-1}\psi,$$

where $\lambda = \lambda(x)$ is a non-singular function on the spacetime manifold, the action (119) is invariant. This is a consequence of the particular construction of the connection (117), where the fermions always appear along with the dreibein field, forming a composite field.

For the case of the internal group SU(2), the internal index can be interpreted as valley index, making USUSY another good scenario with which to describe the continuous limit of both Dirac points (see details in [66]).

The action of USUSY in $(2+1)$ dimensions, for fixed background bosonic fields, apart from possible boundary terms, is obtained from the Chern–Simons three-form for $\mathbb{A}$ with an SU(2) internal gauge group [106]

$$S_{USUSY} = \kappa \int \overline{\psi}^i \left(\gamma^\mu \overset{\leftrightarrow}{D}_\mu - \frac{i}{8} \epsilon_a{}^{bc} T^a{}_{bc} \right) \psi_i |e| d^3x, \tag{120}$$

where lower case Latin letters, $a, b, \ldots$, represent tangent space Lorentz indices, and $T^a{}_{bc} = T^a{}_{\mu\nu} E^\mu_b E^\nu_c$.

This action immediately points to (83), that is, the action with torsion we have seen emerging in graphene, where we only need to fix the dimensionfull κ to include v_F rather than c. Notice that, as discussed at length in [66] the two Dirac points are both necessary, so that the emergent action is made of two parts, one per Dirac point. This makes possible to have both the internal SU(2) symmetry and torsion, that is necessary for the USUSY description. So far as for similarities between (83) and (120). There are differences, though. The first is the coefficient of the torsion term, which appears in USUSY as an integration constant [104]. The second difference is the index i (here taken as an internal colour index, considering both Dirac points in the model). Both differences are due to the starting point to obtain (83), which is an Hermitian action with local Lorentz invariance in a Riemann–Cartan space. In contrast, the starting point of USUSY is an action with a supergroup $USp(2,1|2)$ invariance, which is allowed by using another representation for ψ and the Dirac matrices (see details in Appendix B of [66]). In addition, it is also possible to take into account the two Dirac points by using other internal supergroups, such as $OSp(p|2) \times OSp(q|2)$ in this USUSY context [85]. In any case, (83) and (120) and the model proposed in [85] are top-down approaches to describe the ψ electrons in graphene-like systems. Therefore, we should keep in mind these (and others) models to compare them with the results of a real experiment in the lab.

Finally, let us comment that the Bañados–Zanelli–Teitelboim (BTZ) black hole [107], in a pure bosonic vacuum state ($\psi = 0$), is a solution of USUSY [104]. This follows from the fact that the BTZ black hole, whose metric in cylindrical coordinates ($-\infty < t < +\infty$, $0 < r < +\infty$, and $0 \leq \phi \leq 2\pi$) is

$$ds^2 = -N^2(r) dt^2 + N^2(r) dr^2 + r^2 \left(N^\phi(r) dt + d\phi \right)^2, \tag{121}$$

$$N^2(r) = -M + \frac{r^2}{\ell^2} + \frac{J^2}{4r^2}, \quad N^\phi(r) = \frac{J^2}{2r^2}, \tag{122}$$

can be obtained from a Lorentz-flat connection with torsion [108]. The spectrum of these black holes is given in terms of their mass, M, and angular momentum, J, including the extremal, $M\ell = |J|$ and $M = 0$ cases[14]. We also mention here that the $M = 0$ case could play a very important role in the generalised Uncertainty Principle induced by gravity [110,111], and in Hawking–Unruh phenomenon on graphene and graphene-like materials [112].

8. Torsion in Cosmology

A plethora of precision cosmological data [113], over the past twenty-five years, have indicated that the energy budget of the current cosmological epoch of our (observable) Universe is dominated (by ∼95%) by a dark sector of unknown, at present, microscopic origin. If one fits the available data at large scales, corresponding to the modern era of the Universe, within the so-called ΛCDM framework, which consists of a de Sitter Universe (dominated by a positive cosmological constant Λ) and a Cold Dark Matter (CDM) component, then one obtains excellent agreement. On the other hand, there appear to be tensions to such data at smaller scales [114–116], arising either from discrepancies between the value of the Hubble parameter in the modern era obtained from direct observations of nearby galaxies and that inferred by ΛCDM fits ("H_0 tension"), or from discrepancies in the value of the parameter σ_8 characterising galactic growth data between direct observations and ΛCDM fits ("σ_8 tension").

To these tensions, provided of course the latter do not admit more mundane astrophysical explanations or are mere artefacts of relatively low statistics [117], and thus will be absent from future data, one should add theoretical obstacles to the self consistency of the ΛCDM framework, when viewed as a viable gravity model embeddable in microscopic models of quantum gravity, such as string theory [118,119] and its brane extensions [120]. Indeed, the existence of eternal de Sitter horizons, in spacetimes with a constant $\Lambda > 0$, prohibits the definition of asymptotic states, and thus a perturbative scattering S-matrix, which is the cornerstone of perturbative string theories, appears not to be well defined, thus posing problems with the compatibility of a de Sitter spacetime as a consistent background of perturbative strings [121,122]. Such problems extend to fully quantum gravity considerations, when one attempts to embed de Sitter spacetimes in microscopic ultraviolet complete models such as strings or branes, due to the so-called swampland conjectures [123–128], which are violated by the ΛCDM framework.

Barring the (important) possibility of misinterpretation of the Planck data as far as dark energy is concerned, by, e.g., *relaxing* the assumption of homogeneity and isotropy of the Universe at cosmological scales [129,130], one is therefore tempted to seek for theoretical alternatives to ΛCDM, which will not be characterised by a positive constant Λ, but rather having the de Sitter vacuum as a *metastable one*, in such a way that there are no asymptotic in future time de Sitter horizons. The current literature has a plethora of potential theoretical resolutions to the de Sitter Λ problem [131], which simultaneously alleviate the aforementioned tensions in small-scale cosmological data. What we would like to discuss below, in the context of our review, is the potential role of a purely geometric origin of such a metastable dark sector, including both Dark Energy (DE) and Dark Matter (DM), which is associated with the existence of torsion in the geometry of the early Universe [24,41,132].

To this end, we consider as a first example, in the next Subsection, string-inspired cosmologies with chiral anomalies. Our generic discussion in Section 2 on the role of (quantum) torsion in Einstein–Cartan QED [6], will find interesting application in this case. There we argued that, as a generic feature, the torsion degrees of freedom implied the existence of pseudoscalar (axion-like) massless dynamical fields in the spectrum, coupled to chiral anomalies.

8.1. Quantum Torsion in String-Inspired Cosmologies and the Universe Dark Sector

We have seen that in Einstein–Cartan theories, which have been exemplified here by massless contorted QED, torsion conservation (40) introduces an axionic degree of freedom to the system, associated with the totally antisymmetric part of the torsion which is the only part that couples to matter (fermions). The axion-like field becomes a dynamical part of the theory as a result of (chiral) anomalies, otherwise it would decouple from the quantum path integral. A similar situation characterises string-inspired theories in which anomalies are not supposed to be cancelled in the (3 + 1)-dimensional spacetime after string compactification, which, as we shall review below, provide interesting cosmological models [133–136] in which the dark sector of the Universe, including the origin of its inflationary epoch, admits a geometric interpretation.

The starting point of such an approach to cosmology is that the early Universe is described by the (bosonic) gravitational theory of the degrees of freedom that constitute the massless gravitational multiplet of the string (which in the case of superstring is also their ground state). The latter consists of spin-0 dilatons, Φ, spin-2 gravitons $g_{\mu\nu}$, and the spin-1 antisymmetric KR tensor field [118,119] $B_{\mu\nu} = -B_{\nu\mu}$.

Due to an Abelian gauge symmetry that characterises the closed string sector $B_{\mu\nu} \to B_{\mu\nu} + \partial_{[\mu}\theta_{\nu]}$, the $(3+1)$-dimensional effective target spacetime action arising in the low-energy limit of strings (compared to the string mass scale M_s) depends only on the totally antisymmetric field strength of the KR field $B_{\mu\nu}$,

$$H_{\mu\nu\rho} = \partial_{[\mu}B_{\nu\rho]}. \tag{123}$$

As explained in [134], one can assume self consistently a constant dilaton, so that the low-energy particle phenomenology is not affected. In this case, to lowest non-trivial order in a derivative expansion, or equivalent to $\mathcal{O}((\alpha')^0)$, with $\alpha' = M_s^2$ the Regge slope, the effective gravitational action reads [137,138]:

$$S_B = \int d^4x \sqrt{-g} \left(\frac{1}{2\kappa^2} R - \frac{1}{6} \mathcal{H}_{\lambda\mu\nu} \mathcal{H}^{\lambda\mu\nu} + \dots \right), \tag{124}$$

where $\mathcal{H}_{\mu\nu\rho} \equiv \kappa^{-1} H_{\mu\nu\rho}$ has dimension [mass]^2, and the ... represent higher derivative terms

Comparing (124) with (21), one observes that the quadratic in the H-field terms can be viewed as a contorsion, in such a way that the effective action (124) can be expressed in terms of a generalised scalar curvature in a contorted geometry, with a generalised Christoffel symbols:

$$\overline{\Gamma}^{\rho}_{\mu\nu} = \mathring{\Gamma}^{\rho}_{\mu\nu} + \frac{\kappa}{\sqrt{3}} \mathcal{H}^{\rho}_{\mu\nu} \neq \overline{\Gamma}^{\rho}_{\nu\mu}, \tag{125}$$

where $\mathring{\Gamma}^{\rho}_{\mu\nu} = \mathring{\Gamma}^{\rho}_{\nu\mu}$ is the torsion-free Christoffel symbols[15].

The requirement of cancellation of gauge versus gravitational anomalies lead Green and Schwarz [140] to add appropriate counterterms in the effective target space action of strings, expressed by the modification of the field strength of the KR field (123) by the Lorentz (L) and Yang–Mills (Y) gauge Chern–Simons (CS) terms [119]:

$$\mathcal{H} = dB + \frac{\alpha'}{8\kappa} \left(\Omega_{3L} - \Omega_{3Y} \right),$$

$$\Omega_{3L} = \omega^a{}_c \wedge d\omega^c{}_a + \frac{2}{3}\omega^a{}_c \wedge \omega^c{}_d \wedge \omega^d{}_a, \quad \Omega_{3Y} = A \wedge dA + A \wedge A \wedge A, \tag{126}$$

where ω is the standard torsion-free spin connection, and A the non-Abelian gauge fields that characterise strings.

The modification (126) of the KR field strength (123) leads to the following Bianchi identity [119]

$$d\mathcal{H} = \frac{\alpha'}{8\kappa} \text{Tr} \left(R \wedge R - F \wedge F \right), \tag{127}$$

with $F = dA + A \wedge A$ the Yang–Mills field strength two form and $R^a{}_b = d\omega^a{}_b + \omega^a{}_c \wedge \omega^c{}_b$ the curvature two form and the trace (Tr) is over gauge and Lorentz group indices. The non zero quantity on the right hand side of (127) is the "mixed (gauge and gravitational) quantum anomaly" we have seen previously in the non-conservation of the axial fermion current (43)[16].

In [133], the crucial assumption made was the $(3+1)$-dimensional gravitational anomalies are not cancelled in the very early Universe. This was the consequence of the assumption that only fields from the massless gravitational string multiplets characterised the early universe gravitational theory, appearing as external fields. Chiral fermionic matter, radiation and in general gauge fields, which constitute the physical content of the low-energy particle physics models derived from strings, appear as the result of the decay of the false vacuum at the end of inflation in the scenario of [133–136].

In this sense, the gauge fields A in (126) can be set to zero, $A = 0$. In such a case, the Bianchi identity (127) becomes (in component form):

$$\varepsilon_{abc}{}^{\mu} \mathcal{H}^{abc}{}_{;\mu} = \frac{\alpha'}{32\kappa} \sqrt{-g} R_{\mu\nu\rho\sigma} \widetilde{R}^{\mu\nu\rho\sigma} \equiv -\sqrt{-g}\, \mathcal{G}(\omega), \tag{128}$$

where the semicolon denotes covariant derivative with respect to the standard Christoffel connection, and

$$\varepsilon_{\mu\nu\rho\sigma} = \sqrt{-g}\,\epsilon_{\mu\nu\rho\sigma}, \qquad \varepsilon^{\mu\nu\rho\sigma} = \frac{\text{sgn}(g)}{\sqrt{-g}}\,\epsilon^{\mu\nu\rho\sigma}, \qquad (129)$$

with $\epsilon^{0123} = +1$, etc., are the gravitationally covariant Levi–Civita tensor densities, totally antisymmetric in their indices, and the dual is defined as

$$\widetilde{R}_{\mu\nu\rho\sigma} = \frac{1}{2}\varepsilon_{\mu\nu\lambda\pi} R^{\lambda\pi}{}_{\rho\sigma}. \qquad (130)$$

The alert reader should have observed similarities between the contorted QED model, examined in the previous Section 3, and the string inspired gravitational theory, insofar as the constraints imposed by the torsion conservation (40) in the QED case, and the Bianchi constraint (128). They are both exact results that are valid in the quantum theory (the Bianchi (128) is an exact one-loop result due to the nature of the chiral anomalies). In fact the dual of $H^{\mu\nu\rho}$, $\varepsilon_{\mu\nu\rho\sigma} H^{\nu\rho\sigma}$ plays a role analogous with the pseudovector S_μ of the contorted QED case, associated with the totally antisymmetric component of the torsion. In the string theory example, this is all there is from torsion, as we infer from (125).

Following the contorted QED case, one may implement the Bianchi constraint (128) via a δ-functional in the corresponding path integral, represented by means of an appropriate Lagrange multiplier pseudoscalar field $b(x)$, canonically normalised:

$$\Pi_x \delta\left(\varepsilon^{\mu\nu\rho\sigma}\mathcal{H}_{\nu\rho\sigma}(x)_{;\mu} + \mathcal{G}(\omega)\right) \Rightarrow$$

$$\int \mathcal{D}b \exp\left[i \int d^4x \sqrt{-g}\,\frac{1}{\sqrt{3}} b(x)\left(\varepsilon^{\mu\nu\rho\sigma}\mathcal{H}_{\nu\rho\sigma}(x)_{;\mu} - \mathcal{G}(\omega)\right)\right]$$

$$= \int \mathcal{D}b \exp\left[-i \int d^4x \sqrt{-g}\,\left(\partial^\mu b(x)\frac{1}{\sqrt{3}}\epsilon_{\mu\nu\rho\sigma}\mathcal{H}^{\nu\rho\sigma} + \frac{b(x)}{\sqrt{3}}\mathcal{G}(\omega)\right)\right], \qquad (131)$$

where to arrive at the second equality we performed partial integration, upon assuming that fields die out properly at spatial infinity, so that no boundary terms arise. We remark at this point that the similarity [41] of the exponent in the right-hand side of the last equality in (131), upon performing a partial integration of the first term, and identifying the anomaly with $\partial_\mu j^{5\mu}$, with the total Holst action (including the Nieh–Yan invariant) (64), in the case where the BI parameter is promoted to a pseudoscalar field [38].

Inserting the identity (131) in the path integral over H of the theory (124), we observe that the equations of motion of the (non-derivative) field H yield $\epsilon_{\mu\nu\rho\sigma} H^{\nu\rho\sigma} \propto \partial_\mu b$, implying an analogy of the pseudovector field S_μ with $\partial_\mu b$. After path-integrating out the H-torsion, one obtains an effective target space action with a dynamical torsion-induced axion b:

$$S_B^{\text{eff}} = \int d^4x \sqrt{-g}\left[\frac{1}{2\kappa^2} R + \frac{1}{2}\partial_\mu b\,\partial^\mu b + \sqrt{\frac{2}{3}}\frac{\alpha'}{96\kappa} b(x)\,R_{\mu\nu\rho\sigma}\widetilde{R}^{\mu\nu\rho\sigma} + \ldots\right], \qquad (132)$$

where the dots ... denote higher derivative terms appearing in the target-space string effective action [6,137,138].

With the exception of the four-fermion interactions, which are absent here, as the theory is bosonic, the action (132) has the same form as the effective action (47), with the pseudoscalar field b having similar origin related to torsion as its contorted QED counterpart. But the action (132) is purely bosonic, and the anomalies here arise from the Green–Schwarz counterterms (126). In the model of [133] these are primordial anomalies, unrelated to chiral matter fermions as in the QED case, but because of the presence of such anomalies, the torsion (through its dual axion field $b(x)$) maintains its non trivial role via its coupling to the gravitational anomaly CS term. The gravitational model (132) is a Chern–Simons modified gravity model [21,23].

The massless axion field $b(x)$ is the so-called string-model independent axion [141], and is one of the many axion fields that string models have. The other axions are due to compactification. The string axions lead to a rich phenomenology and cosmology [142,143].

From our point of view we restrict ourselves to the role of the KR axion in implying a geometric origin of the dark sector of the Universe, including non conventional inflation. Indeed, in [133–136] it was argued that condensation of primordial gravitational waves (GW) leads to a non-vanishing contribution of the gravitational Chern–Simons term $\langle R_{\mu\nu\rho\sigma}\widetilde{R}^{\mu\nu\rho\sigma}\rangle$, where $\langle \ldots \rangle$ denote weak graviton condensates associated with primordial chiral GW [144,145]. If one assumes a density of sources for primordial GW, which have been formed in he very early Universe, before the inflationary stage in the model of [135,136], then, the weak quantum graviton calculation of [145], adopted to include densities of GW sources, leads to [146]:

$$\langle R_{\mu\nu\rho\sigma}\widetilde{R}^{\mu\nu\rho\sigma}\rangle_{\text{condensate }\mathcal{N}} = \frac{\mathcal{N}(t)}{\sqrt{-g}}\frac{1.1}{\pi^2}\left(\frac{H}{M_{\text{Pl}}}\right)^3 \mu^4 \frac{\dot{b}(t)}{M_s^2} \equiv n_\star \frac{1.1}{\pi^2}\left(\frac{H}{M_{\text{Pl}}}\right)^3 \mu^4 \frac{\dot{b}(t)}{M_s^2}. \quad (133)$$

In the above expression, μ is an ltraviolet (UV) cutoff for the graviton modes entering the chiral GW, and $n_\star \equiv \frac{\mathcal{N}(t)}{\sqrt{-g}}$ denotes the number density (over the proper de Sitter volume) of the sources of GW. Without loss of generality, we may take this density to be (approximately) time independent during the very early universe. The parameter $H(t)$ is the Hubble parameter of a FLRW Universe, which is assumed slowly varying with the cosmic time[17]. The analysis of [133,135] then, shows that there is a metastable de Sitter spacetime emerging, given that the condensate (133) is only mildly depending on cosmic time through $H(t)$ mainly, and thus can be considered approximately constant. It can be shown [133] that, as a consequence of the axion b equations of motion, the existence of a condensate leads to approximately constant $\dot{b}$ during the inflationary period (for which $H \simeq$ constant)

$$\dot{b} \simeq \epsilon H M_{\text{Pl}}, \quad (134)$$

where the overdot denotes a derivative with respect to the cosmic time t. The parameter ϵ is phenomenological and, to satisfy the Planck data [113] on slow-roll inflation, one should set it to $\epsilon = \mathcal{O}(10^{-2})$ [135]. Then, conditions for an approximately constant

$$\langle b(t) R_{\mu\nu\rho\sigma}\widetilde{R}^{\mu\nu\rho\sigma}\rangle_{\text{condensate }\mathcal{N}} \simeq \text{constant}, \quad (135)$$

for some period Δt, can be ensured, which then leads to a *metastable* de Sitter spacetime (inflation), with Δt the duration of inflation. Taking into account that the scale of inflation, set by the current Planck data [113] is

$$H_I \lesssim 10^{-5} M_{\text{Pl}}, \quad (136)$$

and that the the number of e-foldings is estimated to be (in single-field models of inflation) $\mathcal{N} = \mathcal{O}(60 - 70)$, these conditions can be stated as:

$$|\overline{b}(t_0)| \gtrsim N_e \sqrt{2\epsilon}\, M_{\text{Pl}} = \mathcal{O}(10^2)\sqrt{\epsilon}\, M_{\text{Pl}}, \quad (137)$$

with $\overline{b}(t_0)$ being the initial value of the axion field at the onset ($t = t_0$) of inflation.

In view of the H-dependence of the condensate the inflation is of the so-called Running-Vacuum-Model (RVM) type [148–153], which involves a time-dependent, rather than a constant de Sitter parameter $\Lambda(t) \propto H^2(t)$, but with a de Sitter equation of state for the vacuum,

$$p_{\text{rvm}} = -\rho_{\text{rvm}}, \quad (138)$$

where p (ρ) denotes pressure (energy) density. In the model of [135], detailed calculations have shown that, in the phase of the GW-induced condensate (133) and (135), the de Sitter–RVM equation of state (138) is satisfied. The corresponding energy density, comprising of contributions from b field (superscript b), the gravitational CS terms (superscript gCS) and the condensate term (superscript) Λ), acquires [133,135,136,146] the familiar RVM form [151–153]

$$\rho^{\text{total}} = \rho^b + \rho^{\text{gCS}} + \rho^{\Lambda}_{\text{condensate}} = -\frac{1}{2}\epsilon M_{\text{Pl}}^2 H^2 + 4.3 \times 10^{10} \sqrt{\epsilon} \frac{|\overline{b}(0)|}{M_{\text{Pl}}} H^4. \tag{139}$$

The important point to notice is that the RVM inflation does not require a fundamental inflaton scalar field, but is due to the non-linear H^4 terms in the respective vacuum energy density (139) [151–153], arising in our case by the form of the condensate (133). Such terms are dominant in the early Universe and drive inflation. During the RVM inflation in our string-inspired CS gravity the H^2 term is negative in contrast to standard RVM formalisms with a smooth evolution from inflation to the current era [151,152]. In our case, it is the CS quadratic curvature corrections to GR that leads to such negative contributions tom the stress-energy tensor, in full analogy to the dilaton–Gauss–Bonnet string-inspired theories [154]. Nevertheless, the dominance of the condensate (i.e., $\mathcal{O}(H^4)$) terms in (139) ensures the positivity of the vacuum energy density during the RVM inflationary era. We stress that the H^4 term in the vacuum energy density (139) arises exclusively from the gravitational anomaly condensate in our string-inspired cosmology. In standard quantum field theories in curved spacetime, RVM energy densities arise after appropriate renormalisation of the quantum matter fields in the FLRW spacetime background but, in such cases, an H^4 term is *not* generated in the vacuum energy density. Instead, one has the generation of order H^6 terms and higher [153,155–158]. Such non linear terms, which will be dominant in the early Universe, can still, of course, drive RVM inflation.

During the final stages of RVM inflation, the decay of the RVM metastable vacuum [151,152] results in the generation of chiral matter fermions in the cosmology model of [133–136] we are analysing here. The chiral fermions would generate their own mixed (gauge and gravitational) chiral anomaly terms through the non conservation of the chiral current (49) over the various chiral fermion species (43). The effective action during such an era will, therefore, contain fermions, which will couple universally to the torsion $H_{\mu\nu\rho}$ via the diffeomorphic covariant derivative. After integrating out the H-field, we arrive at the following effective action including fermions [133]:

$$S^{\text{eff}} = \int d^4x \sqrt{-g} \Big[\frac{1}{2\kappa^2} R + \frac{1}{2} \partial_\mu b \, \partial^\mu b + \sqrt{\frac{2}{3}} \frac{\alpha'}{96\kappa} b(x) R_{\mu\nu\rho\sigma} \widetilde{R}^{\mu\nu\rho\sigma} \Big]$$
$$+ S^{\text{Free}}_{\text{Dirac or Majorana}} + \int d^4x \sqrt{-g} \Big[\Big(\mathcal{F}_\mu - \frac{\alpha'}{2\kappa} \sqrt{\frac{3}{2}} b J^{5\mu}_{;\mu} \Big) - \frac{3\alpha'^2}{16\kappa^2} J^5_\mu J^{5\mu} \Big] + \dots, \tag{140}$$

where the $S^{\text{Free}}_{\text{Dirac or Majorana}}$ fermionic terms denote the standard Dirac or Majorana fermion kinetic terms in a curved spacetime without torsion, and $\mathcal{F}^a = \varepsilon^{abcd} e_{b\lambda} \partial_d e^\lambda_c$.

The gravitational part of the anomaly is assumed in [133] to *cancel* the primordial gravitational anomalies, but the chiral gauge anomalies remain in general. Thus, in [133], we assumed that, at the exit phase from RVM inflation, one has the condition,

$$\partial_\mu \Big[\sqrt{-g} \Big(\sqrt{\frac{3}{8}} \kappa J^{5\mu} - \sqrt{\frac{2}{3}} \frac{\kappa}{96} \mathcal{K}^\mu \Big) \Big] = \sqrt{\frac{3}{8}} \frac{\alpha'}{\kappa} \frac{e^2}{8\pi^2} \sqrt{-g} F^{\mu\nu} \widetilde{F}_{\mu\nu}$$
$$+ \sqrt{\frac{3}{8}} \frac{\alpha'}{\kappa} \frac{\alpha_s}{8\pi} \sqrt{-g} G^a_{\mu\nu} \widetilde{G}^{a\mu\nu}, \tag{141}$$

where we used the fact that the gravitational CS anomaly is a total derivative of an appropriate topological current $\mathcal{K}^\mu$ [15–17],

$$R_{\mu\nu\rho\sigma} \widetilde{R}^{\mu\nu\rho\sigma} = \mathcal{K}^\mu_{;\mu}. \tag{142}$$

$F_{\mu\nu}$ denotes the electromagnetic U(1) Maxwell tensor, which corresponds to radiation fields in the post inflationary epoch, and $G^a_{\mu\nu}$, $a = 1, \dots, 8$ is the gluon tensor associated with the SU(3) (of colour) strong interactions with (squared) coupling $\alpha_s = g_s^2/(4\pi)$, which dominate the Universe during the QCD epoch, and the $\widetilde{(\ldots)}$ denotes the corresponding duals, as usual (cf. (130)), with $\widetilde{F}^{\mu\nu} = \frac{1}{2}\epsilon^{\mu\nu\rho\sigma}F_{\rho\sigma}$.

At the exit from RVM inflation, it was assumed in [133–136] that no chiral gauge anomalies are dominant. Such dominance comes much later in the post inflationary Universe evolution. In such a case, it can be shown [133] that the b-field equation of motion implies a scaling of $\dot{b}$ with the temperature as

$$\dot{b} \propto T^3. \tag{143}$$

In this case, one may obtain an unconventional leptogenesis of the type discussed in [159,160] in theories involving massive sterile right handed neutrinos, as a result of the decay of the latter to standard-model particles in the presence of the Lorentz-violating background (143). Hence, in such scenarios the torsion is also linked to matter-antimatter asymmetry, given that the so-generated lepton asymmetry can be communicated to the baryon sector vial Baryon (B) and Lepton number (L) violating, but B-L conserving sphaleron processes in the standard-model sector [161].

Connection of torsion to DM might be obtained by noting that the QCD dominance era (which in the models of [133,134] comes much after the leptogenesis epoch) might be characterised by SU(3) instanton effects, which in turn break the axionic shift symmetry by inducing appropriate potential, and mass terms, (cf. (50)) for the torsion-induced axion field b, which could play a role as a DM component. The electromagnetic U(1) chiral anomalies may be dominant in the modern eras, and their effects have been discussed in detail in [133].

We also mention for completion that, as a result of the (anomalous) coupling $\dot{b}J^{50}$ (cf. (140)), one obtains a Standard-Model-Extension (SME) situation, with the Lorentz and CPT Violating SME background being provided by $\dot{b}$. It is the latter that is constrained by a plethora of precision experiments, which provide stringent bounds for Lorentz and CPT violation [162]. Using the chiral gauge anomalies at late eras of the Universe, as appearing in (141), the thermal evolution of the Lorentz- and CPT- symmetry-violating torsion-induced background $\dot{b}(T)$ at late eras of the Universe, including the current epoch, has also been estimated in [133], and found to be comfortably consistent with the aforementioned existing bounds of Lorentz and CPT Violation, as well as torsion today [162].

In the above cosmological scenarios, the entire dark sector of the Universe and its cosmological evolution are one way or another linked to some sort of torsion in the geometry. During the very early epochs after the Big bang, it is the gravitino torsion of a SUGRA theory, which the effective string cosmology model of [133,135] is embedded to, that leads to a first inflationary epoch [103], whilst it is the stringy torsion associated with the field strength of the antisymmetric spin-one KR field, which in turn gives rise to the KR axion $b(x)$, that is responsible for the second RVM type inflation, and the eventual cosmological evolution until the present era, during which the field $b(x)$ can also develop a mass, thus becoming a dark-matter candidate. Schematically, such a cosmological evolution is depicted in Figure 9 [136].

Before closing this section, we would like to mention the very recent related work of [163], which explores further the cosmology of Kalb–Ramond-like particles (KRLP), which one encounters in string models, and which contain also massive pseudovector excitations, in addition to the massless pseudoscalar ones, discussed in this review in connection with the totally antisymmetric part of torsion. Although the non-interacting KRLP are related to either pseudoscalar or pseudovector excitations, the interacting massive KRLP can be distinguished from its scalar and vector counterparts, and can have important phenomenological implications for the dark sector of the Universe, which are described in detail in [163].

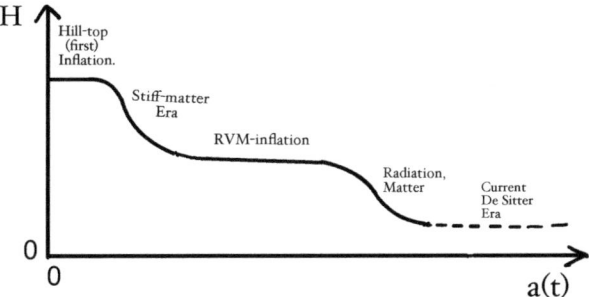

Figure 9. Schematic representation of the RVM cosmological evolution of the contorted cosmological model of [133–136]. The figure depicts the evolution of the Hubble parameter with the scale factor of an expanding stringy-RVM Universe, involving two torsion-induced inflationary eras, interpolated by a stiff KR-axion "matter" epoch: a first hill-top first inflation, which exists immediately after the Big-Bang, and is due to dynamical breaking of SUGRA, as a result of gravitino-torsion-induced condensates of the gravitino field, and second an RVM inflation, due to gravitational anomaly condensates, that are coupled to the torsion-induced KR axion field $b(x)$. The latter can also play the role of a dark matter component during post-RVM inflationary eras. Picture taken from [136].

8.2. Comments on Other Contorted Cosmological Models with a Spin

In the previous Subsection we discussed cosmological models corresponding to the standard generic type of Einstein–Cartan theories with fermionic torsion, involving in their Lagrangian densities repulsive four fermion interactions, of axial-current-current terms $j^5 \cdot j^5$, with fixed coefficient depending on the theory, proportional to the gravitational coupling κ^2. Condensates of such repulsive terms, when formed, have been interpreted as providers of dark energy components in both the early [164] and the late [165] Universe, thus leading to a current-era acceleration of the Universe.

In this subsection, we shall briefly discuss generalisations involving more general four-fermi structures among *chiral* (Weyl) spinors [166], which include vector fermionic currents in addition to the axial ones, in similar spirit to the models (70), but with more general coefficients (on the other, hand, unlike the situation encountered in (70), the BI parameter in [166] is assumed constant, which, as we have discussed in Section 2, and mention below as well, is a problematic feature). Depending on the couplings considered, such fermion self-interactions may conserve or break parity invariance, while they may contribute positively or negatively to the energy density, thus having the feature that they could also be attractive. Thus, such "cosmologies with a spin" [166] exhibit a broad spectrum of possibilities, ranging from cases for which no significant cosmological novelties arise, to cases in which the fermion self-interaction can turn a mass potential into an upside-down Mexican hat potential, leading to cosmologies with a bounce [166,167], without a cosmic singularity.

However, as we shall discuss below, there are some subtleties in the treatment of [166], which, in view of what we discussed in Section 2, require some discussion. Let us first describe the approach of [166]. On defining Dirac spinors $\Psi(x)$ from the chiral ones ξ, χ as

$$\Psi(x) = \begin{pmatrix} \xi(x) \\ \chi(x) \end{pmatrix}, \qquad (144)$$

the authors of [166] constructed fermionic field theories in a contorted curved spacetime, with action given by:

$$\mathcal{S}_\Psi[e, \omega^{ab}, \Psi] = \frac{1}{3} \int d^4x\, \epsilon_{abcd}\, e^b\, e^c \left[\frac{1}{2} e^a\, \overline{\Psi}\, \gamma^d D(\omega)\, \Psi - \overline{D(\omega)\Psi}\, \gamma^d\, \Psi \right) + \frac{3}{2} T^a (\alpha\, V^d + \beta A^d) \right]$$
$$- \frac{1}{4} \int d^4x\, U\, \epsilon_{abcd}\, e^a\, e^b\, e^c\, e^d + S_{\text{int}}[\xi, \chi, A], \qquad (145)$$

where T^a is the torsion two-form, (1), U is a fermion-self-interaction potential which is assumed to be a function of scalars constructed from $\overline{\Psi}\Psi$ and $\overline{\Psi}\gamma^5\Psi$, while S_{int} denotes an interaction term of the chiral spinors ξ, χ with (in general, non-Abelian) gauge fields A. We also defined $V^d = \overline{\Psi}\gamma^d\Psi$ as the vector chiral current, and $A^d = \overline{\Psi}\gamma^5\gamma^d\Psi$ its axial counterpart. Finally, the quantities $\alpha, \beta \in \mathbb{R}$ are real couplings that characterise the model.

The gravitational dynamics, on the other hand, is described by the standard Einstein–Hilbert term plus the Holst action, this is the combination (29) and (51), which in the parametrization and normalisation of [166] is written as:

$$S_{\text{grav}+\text{Holst}} = \frac{1}{2\kappa^2} \int d^4x \left(\epsilon_{abcd} + \frac{1}{\gamma} \eta_{ac}\eta_{bd} \right) e^a e^b R^{cd}, \tag{146}$$

with R^{ab} being the Riemann curvature two-form, and $\gamma \in \mathbb{R}$ being related to the BI parameter $\beta = -1/\gamma$ (51).

This is not a minimal torsion model, as the generic Einstein–Cartan theories examined before, given that it includes several postulated interaction potentials. Because of this, this model leads to more general four-fermion interactions than the standard Einstein–Cartan theory. The effective four fermion interaction is found by using, as in the standard Einstein–Cartan theories, the Euler–Lagrange equations of motion for the fermions, torsion and gravity fields. By varying the action with respect to the contorted spin connection, we determine the torsion T^a and contorsion K_{abc} for this model [166]:

$$\frac{1}{\kappa^2}\left(\epsilon_{abcd} + \frac{2}{\gamma}\eta_{a[c}\eta_{d]b}\right)T^a e^b = \frac{1}{4}\epsilon_{amnp} e^a e^m e^n e^{dp}{}_{cd} A_d - \frac{1}{4}\epsilon_{[c|mnq} e^m e^n e_{|d]}\left(\alpha V^q + \beta A^q\right),$$

$$K_{abc} = \kappa^2 \frac{\gamma^2}{4(\gamma^2+1)}\left[\epsilon^d{}_{abc}\frac{1}{2}\left(A_d + \frac{1}{\gamma}(\alpha V_d + \beta A_d)\right) - \frac{1}{\gamma}A_{[b}\eta_{a]c} + \alpha V_{[b}\eta_{a]c} + \beta A_{[b}\eta_{a]c}\right]. \tag{147}$$

From the graviton (vielbein) and fermion equations of motion, on the other hand, we obtain, respectively:

$$\frac{2}{\kappa^2}\mathring{G}_{\mu\nu} = -\frac{i}{2}e_{d\mu}(\overline{\Psi}\gamma^d \mathring{D}_\nu\Psi - \overline{\mathring{D}_\nu\Psi}\gamma^d\Psi) + \frac{i}{2}e^\sigma_d(\overline{\Psi}\gamma^d \mathring{D}_\sigma\Psi - (\overline{\mathring{D}_\sigma\Psi})\gamma^d\Psi) - g_{\mu\nu}W,$$

$$i\gamma^d e^\mu_d \mathring{D}_\mu \Psi = \frac{\delta W}{\delta \overline{\Psi}}, \tag{148}$$

where $\mathring{G}_{\mu\nu}$ is the standard Einstein tensor of GR; to ease the notation we used $\mathring{D}_\mu \equiv D(\mathring{\omega})_\mu$ and W is the effective four-fermion interaction potential, which depends on the contorsion:

$$W = U + \frac{3\kappa^2}{16}\frac{\gamma^2}{\gamma^2+1}\left[(1-\beta^2 + \frac{2}{\gamma}\beta)A_a A^a - \alpha^2 V_a V^a - 2\alpha(\beta - \frac{1}{\gamma})A_a V^a\right]. \tag{149}$$

The mixed axial-vector current term in (149) breaks parity. One should compare these four-fermion interactions with the ones in the models (70), discussed in Section 4.2.

However, the analysis of [166], leading to (149), is not entirely formally correct, as we have explained in Section 2, following the careful analysis of [37]. The presence of the (constant) BI parameter in the effective potential (149) would imply that a parameter that appears in a total derivative term does affect physics at the end. As explained above, this paradox leads also to another inconsistency, that of Equation (55), in which, for non-zero $1/\gamma$, one obtains the inconsistent result that the vector component of torsion is proportional to the pseudovector of the axial current. As we discussed in Section 2, the resolution of this paradox is achieved by considering the addition of the Nieh–Yan topological invariant [39] (62).

We do mention at this stage that, naively, the independence of the potential W on the (constant) BI parameter γ can be achieved in the specific cases

$$\beta = \frac{1}{\gamma} \quad \text{and} \quad \alpha^2 = c_0^2 \frac{\gamma^2+1}{\gamma^2}, \tag{150}$$

where $c_0 \in \mathbb{R}$ is an arbitrary real constant. This case preserves parity, since the mixed term $A_a V^a$ in the potential W (149) is absent. In such a case the effective four-fermion interactions become

$$W = U + \frac{3\kappa^2}{16}\left(A_a A^a - c_0^2 V_a V^a\right). \tag{151}$$

This model contains, in addition to the potential term U, the standard repulsive axial-current-current four-fermion interactions of the Einstein–Cartan theory, augmented by vector-current-current four fermion interactions.

Superficially looking at (151), one may think that the contributions to the vacuum energy density due to such interactions could be positive or negative, depending on the relative magnitude of the parameter c_0^2, and in general the terms in (151). However, this is not the case. Indeed, as discussed in [166], for classical spinors, as appropriate for solutions of Euler–Lagrange equations of motion, one may argue that

$$\langle A_a A^a \rangle = -\langle V_a V^a \rangle, \tag{152}$$

given that the axial term is always space-like, while the vector time-like. From (151) and (152) we obtain that in this case $W = U + \frac{\kappa^2}{16}(1 + c_0^2) A_a A^a$ and, due to the space-like nature of the classical axial-current-current term $\langle A_a A^a \rangle$, the four-fermion interaction is always repulsive, as in the standard Einstein–Cartan theory, but with a coefficient whose magnitude is unconstrained, given the phenomenological nature of the parameter c_0. In that case, one can show that there are no bouncing cosmologies or other effects, such as for instance turning a positive mass potential into a Higgs one, which arose in the treatment of [166]. Nonetheless, doubt is cast on the mathematical consistency of such solutions in view of (55), which is still valid in such special cases, even if the potential (151) is independent from the BI parameter.

The above criticisms, however, may be bypassed in the case one promotes the BI parameter to a pseudoscalar (axion-like) field $1/\gamma \to a(x)$, as discussed previously in Section 4.2. Indeed in such a case, the corresponding effective four-fermion interactions (149) have to be reworked in accordance with the fact that the BI parameter is now a fully fledged pseudoscalar field, as in the case of the action (70). Thus, cosmologies based on such models, with four-fermion interactions that may include *attractive* fermion channels, may justify (some of) the expectations of [168] on the role of torsion-induced *fermion condensates* in the early universe cosmology, which cannot characterise the repulsive terms (56). In this latter respect, in the context of SUGRA theories (cf. Section 6), the torsion-connected four gravitino interactions can also lead, due to the existence of attractive channels, to the formation of appropriate condensates [99,100], which, as we have discussed in Section 8.1, may play an important role in the early eras of string-inspired cosmologies.

9. Concluding Remarks: Other Observational Effects of Torsion

We reviewed various aspects of torsion, both in emergent geometric descriptions of graphene or other Dirac materials, and in fundamental theories of spacetime, especially cosmology. These two scenarios have enormously different scales, yet the physical properties of torsion appear to be universal, and can in principle be appreciated in experiments in both frameworks.

On the cosmological side, we focused on specific string-inspired models in which the totally antisymmetric component of torsion is represented as an axion-like field. Condensates associated with torsion can lead, as we have discussed, to inflationary physics of RVM type, characterised under some conditions, by torsion-induced-axion background that spontaneously violates Lorentz symmetry. Such a situation may leave imprints in the early Universe cosmic microwave background.

In general, however, in generic Einstein–Cartan theories, torsion has more components. In [169], a plethora of tests involving coupling of the various torsion components to fermions in combination with Lorentz violation, in the context of the Standard Model

Extension [162], have been discussed which exhibit sensitivity for some of the pertinent Lorentz-violating parameters down to 10^{-1} GeV.

The presence of torsion may also have important consequences for cosmological observations independent of Lorentz violation. For instance, as discussed in [170], non-zero torsion affects the relation between the angular-diameter (D_A) and luminosity (D_L) distances used in astrophysical/cosmological measurements, such that the quantity $\eta = \frac{D_L}{D_A(1+z)^2} - 1$ is linked to various types of torsion. This may affect low-redshift measurements, and thus contribute to the observed Hubble-parameter (H_0) tensions [171]. Of course, contributions to such tensions, including the growth of structure ones (σ_8) [114–116], can also come, as we discussed in Section 8.1, from the late-Universe RVM cosmology, which the contorted string-inspired models lead to, but the combination of the plethora of late-time cosmological measurements, and details of structure formation [172] can provide information that can distinguish between the quantum string-inspired RVM cosmology and generic torsion models.

Other constraints on late Universe torsion of relevance to our discussion here, namely of associating axions to torsion, come from CP (rather than Lorentz) violation effects in axion-photon cosmic plasma through dynamo primordial-magnetic-field amplification [173] (see also [174] on the role of axion fields), which torsion is a specific species of for cosmic magnetic helicity generation).

An alternative way to probe experimentally the role of torsion is to realise in graphene, or other Dirac materials, the scenarios described in this review. At this time, there is still nothing going on in that direction. There are two steps that will make this enterprise possible. On the theory side, we should identify the best experimental setting to have a precise correspondence between the specific dislocation defects (the nonzero Burgers vectors) and the torsion term in the Dirac action. On the experimental side, we should be able to realise, with the help of suitable external electromagnetic fields, the time-loop that will spot the nonzero torsion in the time direction.

We mention for completeness that we have not covered here certain interesting aspects of torsion, such as those characterising teleparallel theories [175], in which torsion replaces the metric, or the so-called $f(Q)$ gravity theories [176], which involve the non-metricity tensor $Q_{\alpha\mu\nu} = D_\alpha(\omega) g_{\mu\nu} \neq 0$. The interested reader is referred to the rich relevant literature for more details on the formal, phenomenological and cosmological aspects of such models.

We would like also to mention here that, in the current literature, there are several works which deal with topics partially overlapping with those of our review, but from a different perspective than ours.

In a revisited Einstein–Cartan approach to graphene dislocations, in particular wedge disclination in a planar graphene sheet, the authors of [177], studied the properties of its electronic degrees of freedom in a novel approach which relates to elasticity theory, given that the aforementioned disclination is found there. An important novel result, as these authors claim, is the demonstrated explicit dependence of the energy on the elasticity (Poisson's) constant. The works [178,179] examine effects of the thermal Nieh–Yan anomaly terms of the axial fermion current, of the form $\partial_\mu J^{5\mu} \propto T^2 T^a \wedge T_a$, where T is the temperature, and T^a the effective/emergent torsion, where the proportionality constant is determined by the geometry and topology of the material, and the number of chiral quantum fields. In the case of Weyl superfluids, the authors show that such anomalous terms characterise the hydrodynamics of a chiral p-wave superfluid, such as ^{3}He-A, or a chiral superconductor.

The role of torsion, when induced by the BI field within Holst and Nieh–Yan formulations, in modified general relativity and bounce cosmology has been studied in a series of works [180–184], which complement our treatment of the H-torsion in this review and related references. In this context, the role of spacetime torsion, sourced from antisymmetric tensor (Kalb–Ramond) fields in various modified gravity theories is discussed in [185–189], including phenomenological aspects, providing potential explanation for the invisibility of torsion in late eras of the Universe, due to dynamical suppression in its couplings with standard model fields.

Claims on potential connection between Lorentz symmetry breaking and torsion are provided in [190], where the one-loop fermionic effective action in Einstein–Cartan theories, computed by the proper time method, results in a contact interaction term between the two topological terms of the Nieh–Yan topological current (axial vector torsion S_μ in our review) and the Chern–Simons topological current, which is thoroughly determined by the metric. Such terms may lead to spontaneous breaking of Lorentz symmetry, through appropriate vacuum expectation values of S_μ. We note that similar mechanisms for spontaneous breaking of Lorentz symmetry arise in our stringy RVM model [133,135], where the time derivative of the Kalb–Ramond axion acquires a constant vacuum expectation value.

The effects of torsion on gravitational waves in extended theories of gravity, in particular in Einstein–Cartan gravity using the post-Newtonian formalism devised by Blanchet–Damour, that goes beyond the linearised gravitational theory is discussed in the works [191–196].

In [197], the Lorentzian gravitational path integral has been evaluated in the presence of non-vanishing torsion (with the application of the Picard–Lefschetz theory for minisuperspaces corresponding to a number of phenomenological bouncing cosmological models as well as for the inflationary paradigm). In addition, in [198], it was demonstrated that, unlike any other non-trivial modifications of the Einstein gravity, the presence of spacetime torsion does not affect the entropy of a black hole. In [199], a shift-symmetric Galileon model in the presence of spacetime torsion has been constructed, with applications to the study of the evolution of the universe at a cosmological scale. For a wide class of torsional structures, the model leads to late time cosmic acceleration, while the standard results are obtained in the limit of vanishing torsion which is a smooth one.

The role of metric-scalar-torsion couplings and their impact on the growth of matter perturbations in the Universe has been discussed in [200] within the context of an interacting dark-energy scenario in which the matter density of a scalar field that sources a torsion mode ceases to be self-conserved, thereby affecting not only the background cosmological evolution but also the perturbative spectrum of the local inhomogeneities, thus leading to cosmic growth. As argued in [200], the model can become phenomenologically viable.

A rather surprising feature of spacetimes with torsion was pointed out in [201], where the authors, on considering the coupling of fermions in the presence of torsion, have demonstrated the emergence of a possibly new length scale (in analogy to the electroweak theory, as we shall explain below), which turns out to be transplanckian, and actually much larger than the Planck length. The new scale arises as a result of the non-renormalisable, gravitational four-fermion contact interaction, which characterises generic Einstein–Cartan theories, as we discussed repeatedly in this review. The authors of [201], argued that, by augmenting the Einstein–Cartan Lagrangian with suitable kinetic terms quadratic in the torsion and curvature, gives rise to new, massive propagating gravitational degrees of freedom. The whole situation is to be viewed in close analogy to the Fermi's effective four-fermion weal interaction, which is the effective low energy theory of the standard model and arises from virtual exchange of the (emergent) W and Z weak bosons of the electroweak theory.

In an interesting recent work [202], torsion was associated with potentially measurable properties of the electroweak vacuum, in the sense that the latter can be stabilised provided one assumes the metric-affine framework instead of the usual metric formulation of gravity. In this framework the Holst invariant is present since in general the torsion does not vanish and this leads to important physical consequences, according to the claims of the authors of that work. Specifically, by using measured quantities such as the Higgs and top quark masses, the authors claim that, in principle, the Einstein–Cartan theory can be differentiated from the standard General Relativity.

Last but not least, we mention the work of ref. [203], where the authors, with the help of appropriate conformal transformations, explored the use of non-symmetric contorted connections in "Fisher information geometry". As is well known, the latter corresponds to a probability distribution function ubiquitous in the study of the effective "geometry" entering information theory. They introduced the idea of both metric and torsion playing equal roles in such a context, and studied the corresponding scalar curvature for a few statistical systems, which served as concrete examples for pointing out the relevant properties. As the authors claimed, this study helps to solve some long-standing problems in the field of information geometry, concerning the uniqueness of the Fisher information metric.

Our report would not be complete if we did not mention the role of torsion in the hydrodynamics of a fluid system with spin currents, as discussed in [204]. This could be of interest in the case of, say, heavy-ion collisions in particle physics, where there is experimental evidence for correlations between the spin polarisation of Λ-hyperons and the angular momentum of the quark-gluon plasma in off-centre collisions [205,206] or in the case of liquid metals, where an experimental realisation of spin currents has been demonstrated [207].

A fully consistent theory of spin-current hydrodynamics is currently lacking. In constructing such a theory, the first open issue to be addressed is identifying a canonical spin current. At this stage, we remind the reader that, in a relativistic theory, on a flat background without torsion, Lorentz invariance dictates that energy and momentum are conserved, which, as a result, implies also the conservation of angular momentum. In the absence of torsion, it is always possible to add an improvement term to the energy momentum tensor, $T_{\mu\nu}$ such that the symmetry property $T_{\mu\nu} = T_{\nu\mu}$ follows from an additional equation of motion, implying angular momentum conservation from energy-momentum conservation. Recalling that the angular momentum tensor $\mathcal{J}^{\mu\nu\rho}$ is related to the spin current $S^{\mu\nu\rho}$ via $\mathcal{J}^{\mu\nu\rho} = x^\nu T^{\mu\rho} - x^\rho T^{\mu\nu} - S^{\mu\nu\rho}$, it follows that the spin current suffers from ambiguities due to the possibility of adding improvement terms to the stress tensor. Specifically, by a judicious choice of such terms, it can be set to zero. In the work of [204], it has been argued that one way of dealing with such ambiguity is to couple the theory to an external spin connection with torsion (which is thus independent of the vielbeins). As discussed in that work, the presence of such a background torsion leads to a uniqueness of the spin current by precluding the addition of improvement terms to the stress tensor. After the computation of the spin current, one can set the background torsion to zero, going back to Minkowski spacetime. The presence of torsion plays an important role in ensuring uniqueness, i.e., the absence of ambiguities, in the so-called entropy current that enters the local version of the second law of thermodynamics in the pertinent fluid. The formalism of turning on the background torsion, and eventually turning it off, ensures that the total entropy current is independent of the choice of improvement terms, which in turn resolves some issues regarding the effect of the improvement terms (called *pseudo-gauge transformations*) on the entropy production in the system.

Funding: N.E.M. is supported in part by the UK Science and Technology Facilities research Council (STFC) under the research grants ST/T000759/1 and ST/X000753/1, and UK Engineering and Physical Sciences Research Council (EPSRC) under the research grant EP/V002821/1; he also acknowledges participation in the COST Association Action CA18108 *Quantum Gravity Phenomenology in the Multimessenger Approach (QG-MM)*. P. P. thanks Fondo Nacional de Desarrollo Científico y Tecnológico–Chile (Fondecyt Grant No. 3200725). A.I. and P.P. gladly acknowledge support from Charles University Research centre (UNCE/SCI/013).

Data Availability Statement: Data are contained within the article.

Acknowledgments: A.I. and P.P. are indebted to Jorge Zanelli, for explaining to them the special role of torsion in general, and in USUSY and graphene in particular.

Conflicts of Interest: The authors declare no conflict of interest.

Notes

1. Although the contorted geometry formalism can be generic and valid in $(d+1)$-dimensional spacetime, nonetheless for the sake of concreteness, in this work we shall present the analysis for $d = 3$, and, in the case of graphene, for $d = 2$.
2. The action of $\wedge$ on forms is expressed as [4,5]: $f^{(k)} \wedge g^{(\ell)} = (-1)^{k\ell} g^{(\ell)} \wedge f^{(k)}$, where $f^{(k)}$ and $g^{(\ell)}$ are k-forms and ℓ-forms, respectively.
3. Some references refer to it as a *contortion tensor* [2]. However, as we are more closely following the terminology of [5], we keep the name *contorsion*. As far as we know, there is no consensus yet about the name of this quantity.
4. For a recent study of the massive case, where the focus is on neutrino mixing and oscillations, see [11].
5. In the original formulation of Barbero and Immirzi, the BI parameter is $\gamma = 1/\beta$, but this is not important for our purposes.
6. Indeed, by applying the decomposition (23) onto (57), prior to imposing (59), we obtain the following extra contribution in the effective action, as compared to the terms discussed previously in the case $\alpha = 0$ [37]:

$$\int d^4 x e \frac{\alpha}{2} T_\mu j^{5\mu}. \tag{60}$$

Including such contributions, and considering the vanishing variations of the total action with respect to the (independent) torsion components, T^μ, S^μ and $q^{\mu\nu\rho}$, we obtain the solution

$$T^\mu = \frac{3\kappa^2}{4}\left(\frac{\beta - \alpha}{\beta^2 + 1}\right) j^{5\mu}, \qquad S^\mu = 3\kappa^2 \frac{1 + \alpha\beta}{1 + \beta^2} j^{5\mu}, \quad q_{\mu\nu\rho} = 0. \tag{61}$$

Clearly, as we discussed above, the first equation is problematic from the point of view of leading to a proportionality relation between a vector and a pseudovector, except in the Einstein–Cartan case $\beta = 0$ and the limit $\alpha = \beta$, where the situation is reduced again to the Einstein–Cartan theory, given that in such a case the Holst-like modification is a total derivative.

7. It is possible to include in the description next-to-near neighbour contributions, while keeping a modified Dirac structure [53]. In fact, such modifications allow for the reproduction of scenarios related to generalised uncertainty principles both for commuting coordinates [54] and noncommuting coordinates [55].
8. Actually, there are six such points, but the only two shown above are inequivalent under lattice discrete symmetry.
9. A deep study of how curvature and torsion emerge in a geometrical approach to quantum gravity, along the lines of how classical elastic-theory emerges from QED, can be found in [58], see also [59]. In those papers, the authors elaborate on a model of quantum gravity inspired by graphene, but independent from it [60,61], see also [62,63]. A review can be found in [48]
10. This is due to the reducible, rather than irreducible, representation of the Lorentz group we use
11. The Goldstino λ is a Majorana spin 1/2 fermion which plays the role of the Goldstone-type fermionic mode arising from the spontaneous breaking of global SUSY. To incorporate the relevant dynamics into the dynamically-broken SUGRA scenario, one adds to the SUGRA Lagrangian (104) the terms

$$\mathcal{L}_{\text{golds}} = -f^2 \det\left(\delta^\mu_\nu + i \frac{1}{2f^2} \bar{\lambda} \gamma^\mu \partial_\nu \lambda \right) = -f^2 - \frac{1}{2} i \bar{\lambda} \gamma^\mu \partial_\mu \lambda + \ldots \tag{106}$$

where $f \in \mathbb{R}$ is the energy scale of SUSY breaking, and the ... denote higher order self-interaction terms of λ. Such a term realises SUSY non linearly in the sense of Volkov and Akulov [102]. After an appropriate gauge fixing (105) the derivative $\partial_\mu \lambda$ can then be absorbed, by a suitable redefinition of the gravitino field ψ_μ in the schematic combination $\psi'_\mu = \psi_\mu + \partial_\mu \lambda$, so that the gravitino field acquires a non zero mass, proportional to the gravitino condensate σ. Then, all that is left from the lagrangian density (106) is a negative cosmological constant term $-f^2 < 0$, and thus the final, gauge fixed, SUGRA lagrangian encoding dynamical breaking of local SUSY, is given by:

$$\mathcal{L}_{\text{total}} = -f^2 + \mathcal{L}_{N=1 \text{ SUGRA}}. \tag{107}$$

We shall not give further details here on this dynamical mechanism for SUGRA breaking, referring the interested reader to the literature (see refs. [99,100] and references therein).

12. It is possible to add a central extension generator $\mathbb{Z}$ and its corresponding one-form coefficient b [106]. However, we shall not consider this extension in the present work.
13. Here, we omitted the wedge notation for the exterior product. For instance, $\mathbb{A}^3$ stands for the three-form $\mathbb{A} \wedge \mathbb{A} \wedge \mathbb{A}$.
14. The case $M = -1$ is the globally anti-De Sitter space, while the other cases are conical singularities [109].
15. We note for completeness that, by exploiting local field redefinition ambiguities [6,137–139], which do not affect the perturbative scattering amplitudes, one may extend the above conclusion to the fourth order in derivatives, that is, to the $\mathcal{O}(\alpha'^2)$ effective low-energy action, which includes quadratic curvature terms.

16 We stress once again that the modifications (126) and the right-hand-side of the Bianchi (127) contain the *torsion-free* spin connection, given that, as explained previously, any H-torsion contribution can be removed by an appropriate addition of counterterms [18,19].

17 To ensure homogeneity and isotropy conditions, the authors of [135] assumed the existence of a stiff-axion-b-dominated era (i.e., with equation of state $w_b = +1$) that succeeds a first hill-top inflation [103] (cf. Figure 9), which is the result of dynamical breaking of local SUSY (SUGRA) right after the Big Bang, that is assumed to characterise the superstring inspired theories. This breaking is achieved by a condensation of the gravitino (supersymmetric partner of gravitons) as a result of the existence of attractive channels in the four-gravitino interactions that characterise the SUGRA Lagrangian due to fermionic torsion [99,100], as discussed in Section 6. As argued in [135,136], unstable domain walls (DW) are formed as a result of the gravitino condensate double well potential (Figure 8), whose degeneracy can be lifted by percolation effects [147]. The non-spherical collapse of such DW leads to primordial GW, which then condense leading to (133).

References

1. Cartan, E. *Riemannian Geometry in an Orthogonal Frame*; World Scientific: Singapore, 2001.
2. Hehl, F.W.; Von Der Heyde, P.; Kerlick, G.D.; Nester, J.M. General Relativity with Spin and Torsion: Foundations and Prospects. *Rev. Mod. Phys.* **1976**, *48*, 393–416. [CrossRef]
3. Shapiro, I.L. Physical aspects of the space-time torsion. *Phys. Rep.* **2002**, *357*, 113. [CrossRef]
4. Eguchi, T.; Gilkey, P.B.; Hanson, A.J. Gravitation, Gauge Theories and Differential Geometry. *Phys. Rep.* **1980**, *66*, 213. [CrossRef]
5. Nakahara, M. *Geometry, Topology and Physics*; CRC Press: Boca Raton, FL, USA, 2003.
6. Duncan, M.J.; Kaloper, N.; Olive, K.A. Axion hair and dynamical torsion from anomalies. *Nucl. Phys. B* **1992**, *387*, 215–235. [CrossRef]
7. Santaló, L.A. *Vectores y Tensores con sus Aplicaciones*; Editorial Universitaria de Buenos Aires: Buenos Aires, Argentina, 1973.
8. Hehl, F.W.; Obukhov, Y.N. Elie Cartan's torsion in geometry and in field theory, an essay. *Ann. Fond. Broglie* **2007**, *32*, 157.
9. Iorio, A.; Pais, P. Time-loops to spot torsion on bidimensional Dirac materials with dislocations. In Proceedings of the Spacetime, Matter, Quantum Mechanics, Castiglioncello, Italy, 19–23 September 2022.
10. Capozziello, S.; Lambiase, G.; Stornaiolo, C. Geometric classification of the torsion tensor in space-time. *Ann. Phys.* **2001**, *10*, 713–727. [CrossRef]
11. Capolupo, A.; Maria, G.D.; Monda, S.; Quaranta, A.; Serao, R. Quantum Field Theory of neutrino mixing in spacetimes with torsion. *arXiv* **2023**, arXiv:2310.09309.
12. Adler, S.L. Axial-Vector Vertex in Spinor Electrodynamics. *Phys. Rev.* **1969**, *177*, 2426–2438. [CrossRef]
13. Bell, J.S.; Jackiw, R. A PCAC puzzle: $\pi 0 \to \gamma\gamma$ in the σ-model. *Il Nuovo C. A (1965–1970)* **1969**, *60*, 47–61. [CrossRef]
14. Bardeen, W.A.; Zumino, B. Consistent and Covariant Anomalies in Gauge and Gravitational Theories. *Nucl. Phys. B* **1984**, *244*, 421–453. [CrossRef]
15. Zumino, B.; Wu, Y.S.; Zee, A. Chiral Anomalies, Higher Dimensions, and Differential Geometry. *Nucl. Phys. B* **1984**, *239*, 477–507. [CrossRef]
16. Fujikawa, K. Comment on Chiral and Conformal Anomalies. *Phys. Rev. Lett.* **1980**, *44*, 1733. [CrossRef]
17. Alvarez-Gaume, L.; Witten, E. Gravitational Anomalies. *Nucl. Phys. B* **1984**, *234*, 269. [CrossRef]
18. Hull, C.M. Anomalies, Ambiguities and Superstrings. *Phys. Lett. B* **1986**, *167*, 51–55. [CrossRef]
19. Mavromatos, N.E. A Note on the Atiyah-singer Index Theorem for Manifolds With Totally Antisymmetric H Torsion. *J. Phys. A* **1988**, *21*, 2279. [CrossRef]
20. Kim, J.E.; Carosi, G. Axions and the Strong CP Problem. *Rev. Mod. Phys.* **2010**, *82*, 557–602; Erratum in *Rev. Mod. Phys.* **2019**, *91*, 049902. [CrossRef]
21. Jackiw, R.; Pi, S.Y. Chern–Simons modification of general relativity. *Phys. Rev. D* **2003**, *68*, 104012. [CrossRef]
22. Guralnik, G.; Iorio, A.; Jackiw, R.; Pi, S.Y. Dimensionally reduced gravitational Chern–Simons term and its kink. *Ann. Phys.* **2003**, *308*, 222–236. [CrossRef]
23. Alexander, S.; Yunes, N. Chern–Simons Modified General Relativity. *Phys. Rep.* **2009**, *480*, 1–55. [CrossRef]
24. Mavromatos, N.E. Geometrical origins of the universe dark sector: String-inspired torsion and anomalies as seeds for inflation and dark matter. *Phil. Trans. A Math. Phys. Eng. Sci.* **2022**, *380*, 20210188. [CrossRef]
25. Ashtekar, A.; Lewandowski, J. Background independent quantum gravity: A Status report. *Class. Quantum Gravity* **2004**, *21*, R53. [CrossRef]
26. Rovelli, C. *Quantum Gravity*; Cambridge Monographs on Mathematical Physics; Cambridge University Press: Cambridge, UK, 2004. [CrossRef]
27. Immirzi, G. Real and complex connections for canonical gravity. *Class. Quantum Gravity* **1997**, *14*, L177–L181. [CrossRef]
28. Immirzi, G. Quantum gravity and Regge calculus. *Nucl. Phys. B Proc. Suppl.* **1997**, *57*, 65–72. [CrossRef]
29. Holst, S. Barbero's Hamiltonian derived from a generalised Hilbert-Palatini action. *Phys. Rev. D* **1996**, *53*, 5966–5969. [CrossRef]
30. Barbero, G.J.F. Real Ashtekar variables for Lorentzian signature space times. *Phys. Rev. D* **1995**, *51*, 5507–5510. [CrossRef]
31. Barbero, G.J.F. Reality conditions and Ashtekar variables: A Different perspective. *Phys. Rev. D* **1995**, *51*, 5498–5506. [CrossRef]
32. Ashtekar, A. New Variables for Classical and Quantum Gravity. *Phys. Rev. Lett.* **1986**, *57*, 2244–2247. [CrossRef] [PubMed]
33. Ashtekar, A. New Hamiltonian Formulation of General Relativity. *Phys. Rev. D* **1987**, *36*, 1587–1602. [CrossRef]

34. Ashtekar, A.; Romano, J.D.; Tate, R.S. New Variables for Gravity: Inclusion of Matter. *Phys. Rev. D* **1989**, *40*, 2572. [CrossRef]
35. Perez, A.; Rovelli, C. Physical effects of the Immirzi parameter. *Phys. Rev. D* **2006**, *73*, 044013. [CrossRef]
36. Freidel, L.; Minic, D.; Takeuchi, T. Quantum gravity, torsion, parity violation and all that. *Phys. Rev. D* **2005**, *72*, 104002. [CrossRef]
37. Mercuri, S. Fermions in Ashtekar-Barbero connections formalism for arbitrary values of the Immirzi parameter. *Phys. Rev. D* **2006**, *73*, 084016. [CrossRef]
38. Calcagni, G.; Mercuri, S. The Barbero–Immirzi field in canonical formalism of pure gravity. *Phys. Rev. D* **2009**, *79*, 084004. [CrossRef]
39. Nieh, H.T.; Yan, M.L. An Identity in Riemann–Cartan Geometry. *J. Math. Phys.* **1982**, *23*, 373. [CrossRef]
40. Kaul, R.K. Holst Actions for Supergravity Theories. *Phys. Rev. D* **2008**, *77*, 045030. [CrossRef]
41. Mavromatos, N.E. Torsion in String-Inspired Cosmologies and the Universe Dark Sector. *Universe* **2021**, *7*, 480. [CrossRef]
42. Castellani, L.; D'Auria, R.; Fre, P. *Supergravity and Superstrings: A Geometric Perspective. Vol. 1: Mathematical Foundations*; World Scientific: Singapore, 1991.
43. Tsuda, M. generalised Lagrangian of N = 1 supergravity and its canonical constraints with the real Ashtekar variable. *Phys. Rev. D* **2000**, *61*, 024025. [CrossRef]
44. Taveras, V.; Yunes, N. The Barbero–Immirzi Parameter as a Scalar Field: K-Inflation from Loop Quantum Gravity? *Phys. Rev. D* **2008**, *78*, 064070. [CrossRef]
45. Torres-Gomez, A.; Krasnov, K. Remarks on Barbero–Immirzi parameter as a field. *Phys. Rev. D* **2009**, *79*, 104014. [CrossRef]
46. Iorio, A. Weyl-gauge symmetry of graphene. *Ann. Phys.* **2011**, *326*, 1334–1353. [CrossRef]
47. Iorio, A. Curved Spacetimes and Curved Graphene: A status report of the Weyl-symmetry approach. *Int. J. Mod. Phys. D* **2015**, *24*, 1530013. [CrossRef]
48. Acquaviva, G.; Iorio, A.; Pais, P.; Smaldone, L. Hunting Quantum Gravity with Analogs: The case of graphene. *Universe* **2022**, *8*, 455. [CrossRef]
49. Wallace, P.R. The Band Theory of Graphite. *Phys. Rev.* **1947**, *71*, 622–634. [CrossRef]
50. Semenoff, G.W. Condensed-Matter Simulation of a Three-Dimensional Anomaly. *Phys. Rev. Lett.* **1984**, *53*, 2449–2452. [CrossRef]
51. Novoselov, K.S.; Geim, A.K.; Morozov, S.V.; Jiang, D.; Zhang, Y.; Dubonos, S.V.; Grigorieva, I.V.; Firsov, A.A. Electric Field Effect in Atomically Thin Carbon Films. *Science* **2004**, *306*, 666–669. [CrossRef] [PubMed]
52. Iorio, A.; Lambiase, G. Quantum field theory in curved graphene spacetimes, Lobachevsky geometry, Weyl symmetry, Hawking effect, and all that. *Phys. Rev. D* **2014**, *90*, 025006. [CrossRef]
53. Iorio, A.; Pais, P.; Elmashad, I.A.; Ali, A.F.; Faizal, M.; Abou-Salem, L.I. generalised Dirac structure beyond the linear regime in graphene. *Int. J. Mod. Phys.* **2018**, *D27*, 1850080. [CrossRef]
54. Iorio, A.; Ivetić, B.; Mignemi, S.; Pais, P. Three "layers" of graphene monolayer and their analog generalised uncertainty principles. *Phys. Rev. D* **2022**, *106*, 116011. [CrossRef]
55. Iorio, A.; Ivetić, B.; Pais, P. Turning graphene into a lab for noncommutativity. *arXiv* **2023**, arXiv:2306.17196.
56. Wehling, T.; Black-Schaffer, A.; Balatsky, A. Dirac materials. *Adv. Phys.* **2014**, *63*, 1–76. [CrossRef]
57. Ruggiero, M.L.; Tartaglia, A. Einstein–Cartan theory as a theory of defects in space-time. *Am. J. Phys.* **2003**, *71*, 1303–1313. [CrossRef]
58. Iorio, A.; Smaldone, L. Quantum black holes as classical space factories. *Int. J. Mod. Phys. D* **2023**, *32*, 2350063. [CrossRef]
59. Iorio, A.; Smaldone, L. Classical space from quantum condensates. *J. Phys. Conf. Ser.* **2023**, *2533*, 012030. [CrossRef]
60. Acquaviva, G.; Iorio, A.; Scholtz, M. On the implications of the Bekenstein bound for black hole evaporation. *Ann. Phys.* **2017**, *387*, 317–333. [CrossRef]
61. Acquaviva, G.; Iorio, A.; Smaldone, L. Bekenstein bound from the Pauli principle. *Phys. Rev. D* **2020**, *102*, 106002. [CrossRef]
62. Acquaviva, G.; Iorio, A.; Scholtz, M. Quasiparticle picture from the Bekenstein bound. *PoS* **2017**, *CORFU2017*, 206. [CrossRef]
63. Acquaviva, G.; Iorio, A.; Smaldone, L. Bekenstein bound from the Pauli principle: A brief introduction. *PoS* **2021**, *ICHEP2020*, 681. [CrossRef]
64. Kleinert, H. *Gauge Fields in Condensed Matter*; World Scientific: Singapore, 1989. [CrossRef]
65. Katanaev, M.; Volovich, I. Theory of defects in solids and three-dimensional gravity. *Ann. Phys.* **1992**, *216*, 1–28. [CrossRef]
66. Iorio, A.; Pais, P. (Anti-)de Sitter, Poincaré, Super symmetries, and the two Dirac points of graphene. *Ann. Phys.* **2018**, *398*, 265–286. [CrossRef]
67. Katanaev, M.O. Geometric theory of defects. *Phys. Usp.* **2005**, *48*, 675–701. [CrossRef]
68. Lazar, M. A Nonsingular solution of the edge dislocation in the gauge theory of dislocations. *J. Phys. A* **2003**, *36*, 1415. [CrossRef]
69. Ciappina, M.F.; Iorio, A.; Pais, P.; Zampeli, A. Torsion in quantum field theory through time-loops on Dirac materials. *Phys. Rev. D* **2020**, *101*, 036021. [CrossRef]
70. de Juan, F.; Cortijo, A.; Vozmediano, M.A.H. Dislocations and torsion in graphene and related systems. *Nucl. Phys. B* **2010**, *828*, 625. [CrossRef]
71. Vozmediano, M.A.H.; Katsnelson, M.I.; Guinea, F. Gauge fields in graphene. *Phys. Rep.* **2010**, *496*, 109. [CrossRef]
72. Amorim, B.; Cortijo, A.; de Juan, F.; Grushin, A.G.; Guinea, F.; Gutiérrez-Rubio, A.; Ochoa, H.; Parente, V.; Roldán, R.; San-Jose, P.; et al. Novel effects of strains in graphene and other two dimensional materials. *Phys. Rep.* **2016**, *617*, 1. [CrossRef]
73. Wilczek, F. Quantum Time Crystals. *Phys. Rev. Lett.* **2012**, *109*, 160401. [CrossRef]
74. Shapere, A.; Wilczek, F. Classical Time Crystals. *Phys. Rev. Lett.* **2012**, *109*, 160402. [CrossRef]

75. Li, T.; Gong, Z.X.; Yin, Z.Q.; Quan, H.T.; Yin, X.; Zhang, P.; Duan, L.M.; Zhang, X. Space-Time Crystals of Trapped Ions. *Phys. Rev. Lett.* **2012**, *109*, 163001. [CrossRef] [PubMed]
76. Smits, J.; Liao, L.; Stoof, H.T.C.; van der Straten, P. Observation of a Space-Time Crystal in a Superfluid Quantum Gas. *Phys. Rev. Lett.* **2018**, *121*, 185301. [CrossRef] [PubMed]
77. Loll, R. Discrete approaches to quantum gravity in four-dimensions. *Living Rev. Rel.* **1998**, *1*, 13. [CrossRef]
78. Heide, C.; Higuchi, T.; Weber, H.B.; Hommelhoff, P. Coherent Electron Trajectory Control in Graphene. *Phys. Rev. Lett.* **2018**, *121*, 207401. [CrossRef]
79. Higuchi, T.; Heide, C.; Ullmann, K.; Weber, H.B.; Hommelhoff, P. Light-field-driven currents in graphene. *Nature* **2017**, *550*, 224. [CrossRef]
80. Marino, E.C. Quantum electrodynamics of particles on a plane and the Chern–Simons theory. *Nucl. Phys. B* **1993**, *408*, 551. [CrossRef]
81. Gorbar, E.V.; Gusynin, V.P.; Miransky, V.A. Dynamical chiral symmetry breaking on a brane in reduced QED. *Phys. Rev. D* **2001**, *64*, 105028. [CrossRef]
82. Dudal, D.; Mizher, A.J.; Pais, P. Remarks on the Chern–Simons photon term in the QED description of graphene. *Phys. Rev. D* **2018**, *98*, 065008. [CrossRef]
83. Dudal, D.; Mizher, A.J.; Pais, P. Exact quantum scale invariance of three-dimensional reduced QED theories. *Phys. Rev. D* **2019**, *99*, 045017. [CrossRef]
84. Iorio, A.; Pais, P. Revisiting the gauge fields of strained graphene. *Phys. Rev. D* **2015**, *92*, 125005. [CrossRef]
85. Andrianopoli, L.; Cerchiai, B.L.; D'Auria, R.; Gallerati, A.; Noris, R.; Trigiante, M.; Zanelli, J. $\mathcal{N}$-extended $D=4$ supergravity, unconventional SUSY and graphene. *J. High Energy Phys.* **2020**, *1*, 084. [CrossRef]
86. Peskin, M.E.; Schroeder, D.V. *An Introduction to Quantum Field Theory*; Addison-Wesley: Reading, PA, USA, 1995.
87. Kruchinin, S.Y.; Krausz, F.; Yakovlev, V.S. Colloquium: Strong-field phenomena in periodic systems. *Rev. Mod. Phys.* **2018**, *90*, 021002. [CrossRef]
88. Castro Neto, A.H.; Guinea, F.; Peres, N.M.R.; Novoselov, K.S.; Geim, A.K. The electronic properties of graphene. *Rev. Mod. Phys.* **2009**, *81*, 109–162. [CrossRef]
89. Gusynin, V.P.; Sharapov, S.G.; Carbotte, J.P. AC conductivity of graphene: From light-binding model to 2 + 1-dimensional quantum electrodynamics. *Int. J. Mod. Phys. B* **2007**, *21*, 4611–4658. [CrossRef]
90. Gonzalez, J.; Guinea, F.; Vozmediano, M.A.H. The Electronic spectrum of fullerenes from the Dirac equation. *Nucl. Phys. B* **1993**, *406*, 771. [CrossRef]
91. Yazyev, O.V.; Chen, Y.P. Polycrystalline graphene and other two-dimensional materials. *Nat. Nanotechnol.* **2014**, *9*, 755–767. [CrossRef]
92. Yazyev, O.V.; Louie, S.G. Topological defects in graphene: Dislocations and grain boundaries. *Phys. Rev. B* **2010**, *81*, 195420. [CrossRef]
93. Hirth, J.; Lothe, J. *Theory of Dislocations*; McGraw-Hill Series in Electrical Engineering: Electronics and Electronic Circuits; McGraw-Hill: New York, NY, USA, 1967.
94. Zhang, X.; Xu, Z.; Yuan, Q.; Xin, J.; Ding, F. The favourable large misorientation angle grain boundaries in graphene. *Nanoscale* **2015**, *7*, 20082–20088. [CrossRef] [PubMed]
95. Freedman, D.Z.; van Nieuwenhuizen, P.; Ferrara, S. Progress Toward a Theory of Supergravity. *Phys. Rev. D* **1976**, *13*, 3214–3218. [CrossRef]
96. Ferrara, S.; van Nieuwenhuizen, P. Simplifications of Einstein Supergravity. *Phys. Rev. D* **1979**, *20*, 2079. [CrossRef]
97. Van Nieuwenhuizen, P. Supergravity. *Phys. Rep.* **1981**, *68*, 189–398. [CrossRef]
98. Nilles, H.P. Supersymmetry, Supergravity and Particle Physics. *Phys. Rep.* **1984**, *110*, 1–162. [CrossRef]
99. Alexandre, J.; Houston, N.; Mavromatos, N.E. Dynamical Supergravity Breaking via the Super-Higgs Effect Revisited. *Phys. Rev. D* **2013**, *88*, 125017. [CrossRef]
100. Alexandre, J.; Houston, N.; Mavromatos, N.E. Inflation via Gravitino Condensation in Dynamically Broken Supergravity. *Int. J. Mod. Phys. D* **2015**, *24*, 1541004. [CrossRef]
101. Deser, S.; Zumino, B. Broken Supersymmetry and Supergravity. *Phys. Rev. Lett.* **1977**, *38*, 1433–1436. [CrossRef]
102. Volkov, D.V.; Akulov, V.P. Possible universal neutrino interaction. *JETP Lett.* **1972**, *16*, 438–440.
103. Ellis, J.; Mavromatos, N.E. Inflation induced by gravitino condensation in supergravity. *Phys. Rev. D* **2013**, *88*, 085029. [CrossRef]
104. Alvarez, P.D.; Valenzuela, M.; Zanelli, J. Supersymmetry of a different kind. *J. High Energy Phys.* **2012**, *1204*, 058. [CrossRef]
105. Guevara, A.; Pais, P.; Zanelli, J. Dynamical Contents of Unconventional Supersymmetry. *J. High Energy Phys.* **2016**, *08*, 085. [CrossRef]
106. Alvarez, P.D.; Pais, P.; Rodríguez, E.; Salgado-Rebolledo, P.; Zanelli, J. Supersymmetric 3D model for gravity with $SU(2)$ gauge symmetry, mass generation and effective cosmological constant. *Class. Quantum Gravity* **2015**, *32*, 175014. [CrossRef]
107. Bañados, M.; Teitelboim, C.; Zanelli, J. Black hole in three-dimensional spacetime. *Phys. Rev. Lett.* **1992**, *69*, 1849–1851. [CrossRef]
108. Alvarez, P.D.; Pais, P.; Rodríguez, E.; Salgado-Rebolledo, P.; Zanelli, J. The BTZ black hole as a Lorentz-flat geometry. *Phys. Lett. B* **2014**, *738*, 134–135. [CrossRef]
109. Miskovic, O.; Zanelli, J. On the negative spectrum of the 2 + 1 black hole. *Phys. Rev. D* **2009**, *79*, 105011. [CrossRef]

110. Iorio, A.; Lambiase, G.; Pais, P.; Scardigli, F. generalised uncertainty principle in three-dimensional gravity and the BTZ black hole. *Phys. Rev. D* **2020**, *101*, 105002. [CrossRef]
111. Iorio, A.; Pais, P. generalised uncertainty principle in graphene. *J. Phys. Conf. Ser.* **2019**, *1275*, 012061. [CrossRef]
112. Iorio, A.; Lambiase, G. The Hawking–Unruh phenomenon on graphene. *Phys. Lett.* **2012**, *B716*, 334–337. [CrossRef]
113. Aghanim, N. et al. [Planck Collaboration] Planck 2018 results. VI. Cosmological parameters. *Astron. Astrophys.* **2020**, *641*, A6; Erratum in *Astron. Astrophys.* **2021**, *652*, C4. [CrossRef]
114. Verde, L.; Treu, T.; Riess, A.G. Tensions between the Early and the Late Universe. *Nat. Astron.* **2019**, *3*, 891. [CrossRef]
115. Perivolaropoulos, L.; Skara, F. Challenges for ΛCDM: An update. *New Astron. Rev.* **2022**, *95*, 101659. [CrossRef]
116. Abdalla, E.; Abellán, G.F.; Aboubrahim, A.; Agnello, A.; Akarsu, Ö.; Akrami, Y.; Alestas, G.; Aloni, D.; Amendola, L.; Anchordoqui, L.A.; et al. Cosmology intertwined: A review of the particle physics, astrophysics, and cosmology associated with the cosmological tensions and anomalies. *J. High Energy Astrophys.* **2022**, *34*, 49–211. [CrossRef]
117. Freedman, W.L. Cosmology at a Crossroads. *Nat. Astron.* **2017**, *1*, 0121. [CrossRef]
118. Green, M.B.; Schwarz, J.H.; Witten, E. *Superstring Theory Vol. 1: 25th Anniversary Edition*; Cambridge Monographs on Mathematical Physics; Cambridge University Press: Cambridge, UK, 2012. [CrossRef]
119. Green, M.B.; Schwarz, J.H.; Witten, E. *Superstring Theory Vol. 2: 25th Anniversary Edition*; Cambridge Monographs on Mathematical Physics; Cambridge University Press: Cambridge, UK, 2012. [CrossRef]
120. Polchinski, J. *String Theory. Vol. 2: Superstring Theory and Beyond*; Cambridge Monographs on Mathematical Physics; Cambridge University Press: Cambridge, UK, 2007. [CrossRef]
121. Hellerman, S.; Kaloper, N.; Susskind, L. String theory and quintessence. *J. High Energy Phys.* **2001**, *06*, 003. [CrossRef]
122. Fischler, W.; Kashani-Poor, A.; McNees, R.; Paban, S. The Acceleration of the universe, a challenge for string theory. *J. High Energy Phys.* **2001**, *07*, 003. [CrossRef]
123. Palti, E. The Swampland: Introduction and Review. *Fortsch. Phys.* **2019**, *67*, 1900037. [CrossRef]
124. Palti, E. The swampland and string theory. *Contemp. Phys.* **2022**, *62*, 165–179. [CrossRef]
125. Obied, G.; Ooguri, H.; Spodyneiko, L.; Vafa, C. De Sitter Space and the Swampland. *arXiv* **2018**, arXiv:1806.08362.
126. Agrawal, P.; Obied, G.; Steinhardt, P.J.; Vafa, C. On the Cosmological Implications of the String Swampland. *Phys. Lett. B* **2018**, *784*, 271–276. [CrossRef]
127. Garg, S.K.; Krishnan, C. Bounds on Slow Roll and the de Sitter Swampland. *J. High Energy Phys.* **2019**, *11*, 075. [CrossRef]
128. Ooguri, H.; Palti, E.; Shiu, G.; Vafa, C. Distance and de Sitter Conjectures on the Swampland. *Phys. Lett. B* **2019**, *788*, 180–184. [CrossRef]
129. Mohayaee, R.; Rameez, M.; Sarkar, S. Do supernovae indicate an accelerating universe? *Eur. Phys. J. Spec. Top.* **2021**, *230*, 2067–2076. [CrossRef]
130. Secrest, N.J.; von Hausegger, S.; Rameez, M.; Mohayaee, R.; Sarkar, S. A Challenge to the Standard Cosmological Model. *Astrophys. J. Lett.* **2022**, *937*, L31. [CrossRef]
131. Di Valentino, E.; Mena, O.; Pan, S.; Visinelli, L.; Yang, W.; Melchiorri, A.; Mota, D.F.; Riess, A.G.; Silk, J. In the realm of the Hubble tension—A review of solutions. *Class. Quantum Gravity* **2021**, *38*, 153001. [CrossRef]
132. Mavromatos, N.E. Anomalies, the Dark Universe and Matter-Antimatter asymmetry. In Proceedings of the DICE 2022: Spacetime, Matter, Quantum Mechanics, Castiglioncello, Italy, 19–23 September 2022.
133. Basilakos, S.; Mavromatos, N.E.; Solà Peracaula, J. Gravitational and Chiral Anomalies in the Running Vacuum Universe and Matter-Antimatter Asymmetry. *Phys. Rev. D* **2020**, *101*, 045001. [CrossRef]
134. Basilakos, S.; Mavromatos, N.E.; Solà Peracaula, J. Quantum Anomalies in String-Inspired Running Vacuum Universe: Inflation and Axion Dark Matter. *Phys. Lett. B* **2020**, *803*, 135342. [CrossRef]
135. Mavromatos, N.E.; Solà Peracaula, J. Stringy-running-vacuum-model inflation: From primordial gravitational waves and stiff axion matter to dynamical dark energy. *Eur. Phys. J. Spec. Top.* **2021**, *230*, 2077–2110. [CrossRef]
136. Mavromatos, N.E.; Solà Peracaula, J. Inflationary physics and trans-Planckian conjecture in the stringy running vacuum model: From the phantom vacuum to the true vacuum. *Eur. Phys. J. Plus* **2021**, *136*, 1152. [CrossRef]
137. Gross, D.J.; Sloan, J.H. The Quartic Effective Action for the Heterotic String. *Nucl. Phys. B* **1987**, *291*, 41–89. [CrossRef]
138. Metsaev, R.R.; Tseytlin, A.A. Order alpha-prime (Two Loop) Equivalence of the String Equations of Motion and the Sigma Model Weyl Invariance Conditions: Dependence on the Dilaton and the Antisymmetric Tensor. *Nucl. Phys. B* **1987**, *293*, 385–419. [CrossRef]
139. Bento, M.C.; Mavromatos, N.E. Ambiguities in the Low-energy Effective Actions of String Theories With the Inclusion of Antisymmetric Tensor and Dilaton Fields. *Phys. Lett. B* **1987**, *190*, 105–109. [CrossRef]
140. Green, M.B.; Schwarz, J.H. Anomaly Cancellation in Supersymmetric D = 10 Gauge Theory and Superstring Theory. *Phys. Lett. B* **1984**, *149*, 117–122. [CrossRef]
141. Svrcek, P.; Witten, E. Axions in String Theory. *J. High Energy Phys.* **2006**, *6*, 051. [CrossRef]
142. Arvanitaki, A.; Dimopoulos, S.; Dubovsky, S.; Kaloper, N.; March-Russell, J. String Axiverse. *Phys. Rev. D* **2010**, *81*, 123530. [CrossRef]
143. Marsh, D.J.E. Axion Cosmology. *Phys. Rep.* **2016**, *643*, 1–79. [CrossRef]
144. Alexander, S.H.S.; Peskin, M.E.; Sheikh-Jabbari, M.M. Leptogenesis from gravity waves in models of inflation. *Phys. Rev. Lett.* **2006**, *96*, 081301. [CrossRef]

145. Lyth, D.H.; Quimbay, C.; Rodriguez, Y. Leptogenesis and tensor polarisation from a gravitational Chern–Simons term. *J. High Energy Phys.* **2005**, *3*, 016. [CrossRef]
146. Mavromatos, N.E. Lorentz Symmetry Violation in String-Inspired Effective Modified Gravity Theories. In Proceedings of the 740. WE-Heraeus-Seminar: Experimental Tests and Signatures of Modified and Quantum Gravity Workshop, Bad Honnef, Germany, 1–5 February 2023.
147. Lalak, Z.; Lola, S.; Ovrut, B.A.; Ross, G.G. Large scale structure from biased nonequilibrium phase transitions: Percolation theory picture. *Nucl. Phys. B* **1995**, *434*, 675–696. [CrossRef]
148. Shapiro, I.L.; Sola, J. Scaling behavior of the cosmological constant: Interface between quantum field theory and cosmology. *J. High Energy Phys.* **2002**, *2*, 006. [CrossRef]
149. Shapiro, I.L.; Sola, J. On the possible running of the cosmological 'constant'. *Phys. Lett. B* **2009**, *682*, 105–113. [CrossRef]
150. Shapiro, I.L.; Sola, J. Cosmological constant, renormalisation group and Planck scale physics. *Nucl. Phys. B Proc. Suppl.* **2004**, *127*, 71–76. [CrossRef]
151. Perico, E.L.D.; Lima, J.A.S.; Basilakos, S.; Sola, J. Complete Cosmic History with a dynamical $\Lambda = \Lambda(H)$ term. *Phys. Rev. D* **2013**, *88*, 063531. [CrossRef]
152. Lima, J.A.S.; Basilakos, S.; Sola, J. Expansion History with Decaying Vacuum: A Complete Cosmological Scenario. *Mon. Not. R. Astron. Soc.* **2013**, *431*, 923–929. [CrossRef]
153. Sola Peracaula, J. The cosmological constant problem and running vacuum in the expanding universe. *Phil. Trans. R. Soc. Lond. A* **2022**, *380*, 20210182. [CrossRef]
154. Kanti, P.; Mavromatos, N.E.; Rizos, J.; Tamvakis, K.; Winstanley, E. Dilatonic black holes in higher curvature string gravity. *Phys. Rev. D* **1996**, *54*, 5049–5058. [CrossRef]
155. Moreno-Pulido, C.; Sola, J. Running vacuum in quantum field theory in curved spacetime: Renormalizing ρ_{vac} without $\sim m^4$ terms. *Eur. Phys. J. C* **2020**, *80*, 692. [CrossRef]
156. Moreno-Pulido, C.; Sola Peracaula, J. Renormalizing the vacuum energy in cosmological spacetime: Implications for the cosmological constant problem. *Eur. Phys. J. C* **2022**, *82*, 551. [CrossRef]
157. Moreno-Pulido, C.; Sola Peracaula, J. Equation of state of the running vacuum. *Eur. Phys. J. C* **2022**, *82*, 1137. [CrossRef]
158. Moreno-Pulido, C.; Sola Peracaula, J.; Cheraghchi, S. Running vacuum in QFT in FLRW spacetime: The dynamics of $\rho_{vac}(H)$ from the quantized matter fields. *Eur. Phys. J. C* **2023**, *83*, 637. [CrossRef]
159. Bossingham, T.; Mavromatos, N.E.; Sarkar, S. Leptogenesis from Heavy Right-Handed Neutrinos in CPT Violating Backgrounds. *Eur. Phys. J. C* **2018**, *78*, 113. [CrossRef] [PubMed]
160. Bossingham, T.; Mavromatos, N.E.; Sarkar, S. The role of temperature dependent string-inspired CPT violating backgrounds in leptogenesis and the chiral magnetic effect. *Eur. Phys. J. C* **2019**, *79*, 50. [CrossRef] [PubMed]
161. Mavromatos, N.E.; Sarkar, S. Curvature and thermal corrections in tree-level CPT-Violating Leptogenesis. *Eur. Phys. J. C* **2020**, *80*, 558. [CrossRef]
162. Kostelecky, V.A.; Russell, N. Data Tables for Lorentz and CPT Violation. *Rev. Mod. Phys.* **2011**, *83*, 11–31. [CrossRef]
163. Capanelli, C.; Jenks, L.; Kolb, E.W.; McDonough, E. Cosmological Implications of Kalb–Ramond-Like-Particles. *arXiv* **2023**, arXiv:2309.02485.
164. Popławski, N.J. Cosmology with torsion: An alternative to cosmic inflation. *Phys. Lett. B* **2010**, *694*, 181–185; Erratum in *Phys. Lett. B* **2011**, *701*, 672–672. [CrossRef]
165. Poplawski, N.J. Cosmological constant from quarks and torsion. *Ann. Phys.* **2011**, *523*, 291–295. [CrossRef]
166. Magueijo, J.; Zlosnik, T.G.; Kibble, T.W.B. Cosmology with a spin. *Phys. Rev. D* **2013**, *87*, 063504. [CrossRef]
167. Poplawski, N.J. Nonsingular, big-bounce cosmology from spinor-torsion coupling. *Phys. Rev. D* **2012**, *85*, 107502. [CrossRef]
168. Giacosa, F.; Hofmann, R.; Neubert, M. A model for the very early Universe. *J. High Energy Phys.* **2008**, *02*, 077. [CrossRef]
169. Kostelecky, V.A.; Russell, N.; Tasson, J. New Constraints on Torsion from Lorentz Violation. *Phys. Rev. Lett.* **2008**, *100*, 111102. [CrossRef] [PubMed]
170. Bolejko, K.; Cinus, M.; Roukema, B.F. Cosmological signatures of torsion and how to distinguish torsion from the dark sector. *Phys. Rev. D* **2020**, *101*, 104046. [CrossRef]
171. Aluri, P.K.; Cea, P.; Chingangbam, P.; Chu, M.; Clowes, R.G.; Hutsemékers, D.; Kochappan, J.P.; Lopez, A.M.; Liu, L.; Martens, N.C.M.; et al. Is the observable Universe consistent with the cosmological principle? *Class. Quantum Gravity* **2023**, *40*, 094001. [CrossRef]
172. Gómez-Valent, A.; Mavromatos, N.E.; Solà Peracaula, J. Stringy Running Vacuum Model and current Tensions in Cosmology. *arXiv* **2023**, arXiv:2305.15774.
173. Garcia de Andrade, L.C. Torsion bounds from CP violation alpha(2)-dynamo in axion-photon cosmic plasma. *Mod. Phys. Lett. A* **2011**, *26*, 2863–2868. [CrossRef]
174. Campanelli, L.; Giannotti, M. Magnetic helicity generation from the cosmic axion field. *Phys. Rev. D* **2005**, *72*, 123001. [CrossRef]
175. Cai, Y.F.; Capozziello, S.; De Laurentis, M.; Saridakis, E.N. f(T) teleparallel gravity and cosmology. *Rep. Prog. Phys.* **2016**, *79*, 106901. [CrossRef] [PubMed]
176. D'Ambrosio, F.; Fell, S.D.B.; Heisenberg, L.; Kuhn, S. Black holes in f(Q) gravity. *Phys. Rev. D* **2022**, *105*, 024042. [CrossRef]
177. Fernández, N.; Pujol, P.; Solís, M.; Vargas, T. Revisiting the electronic properties of disclinated graphene sheets. *Eur. Phys. J. B* **2023**, *96*, 68. [CrossRef]

178. Nissinen, J.; Volovik, G.E. On thermal Nieh–Yan anomaly in topological Weyl materials. *JETP Lett.* **2019**, *110*, 789–792. [CrossRef]
179. Nissinen, J.; Volovik, G.E. Thermal Nieh–Yan anomaly in Weyl superfluids. *Phys. Rev. Res.* **2020**, *2*, 033269. [CrossRef]
180. Bombacigno, F.; Cianfrani, F.; Montani, G. Big-Bounce cosmology in the presence of Immirzi field. *Phys. Rev. D* **2016**, *94*, 064021. [CrossRef]
181. Bombacigno, F.; Montani, G. Big bounce cosmology for Palatini R^2 gravity with a Nieh–Yan term. *Eur. Phys. J. C* **2019**, *79*, 405. [CrossRef]
182. Bombacigno, F.; Boudet, S.; Montani, G. generalised Ashtekar variables for Palatini $f(\mathcal{R})$ models. *Nucl. Phys. B* **2021**, *963*, 115281. [CrossRef]
183. Boudet, S.; Bombacigno, F.; Montani, G.; Rinaldi, M. Superentropic black hole with Immirzi hair. *Phys. Rev. D* **2021**, *103*, 084034. [CrossRef]
184. Bombacigno, F.; Boudet, S.; Olmo, G.J.; Montani, G. Big bounce and future time singularity resolution in Bianchi I cosmologies: The projective invariant Nieh–Yan case. *Phys. Rev. D* **2021**, *103*, 124031. [CrossRef]
185. Elizalde, E.; Odintsov, S.D.; Paul, T.; Sáez-Chillón Gómez, D. Inflationary universe in $F(R)$ gravity with antisymmetric tensor fields and their suppression during its evolution. *Phys. Rev. D* **2019**, *99*, 063506. [CrossRef]
186. Paul, T.; Banerjee, N. Cosmological quantum entanglement: A possible testbed for the existence of Kalb–Ramond field. *Class. Quantum Gravity* **2020**, *37*, 135013. [CrossRef]
187. Paul, T. Antisymmetric tensor fields in modified gravity: A summary. *Symmetry* **2020**, *12*, 1573. [CrossRef]
188. Paul, T.; SenGupta, S. Dynamical suppression of spacetime torsion. *Eur. Phys. J. C* **2019**, *79*, 591. [CrossRef]
189. Das, A.; Paul, T.; Sengupta, S. Invisibility of antisymmetric tensor fields in the light of $F(R)$ gravity. *Phys. Rev. D* **2018**, *98*, 104002. [CrossRef]
190. Nascimento, J.R.; Petrov, A.Y.; Porfírio, P.J. Induced gravitational topological term and the Einstein–Cartan modified theory. *Phys. Rev. D* **2022**, *105*, 044053. [CrossRef]
191. Battista, E.; De Falco, V. First post-Newtonian generation of gravitational waves in Einstein–Cartan theory. *Phys. Rev. D* **2021**, *104*, 084067. [CrossRef]
192. Battista, E.; De Falco, V. Gravitational waves at the first post-Newtonian order with the Weyssenhoff fluid in Einstein–Cartan theory. *Eur. Phys. J. C* **2022**, *82*, 628. [CrossRef] [PubMed]
193. Battista, E.; De Falco, V. First post-Newtonian N-body problem in Einstein–Cartan theory with the Weyssenhoff fluid: Equations of motion. *Eur. Phys. J. C* **2022**, *82*, 782. [CrossRef]
194. De Falco, V.; Battista, E.; Antoniadis, J. Analytical coordinate time at first post-Newtonian order. *Europhys. Lett.* **2023**, *141*, 29002. [CrossRef]
195. Battista, E.; De Falco, V.; Usseglio, D. First post-Newtonian N-body problem in Einstein–Cartan theory with the Weyssenhoff fluid: Lagrangian and first integrals. *Eur. Phys. J. C* **2023**, *83*, 112. [CrossRef]
196. De Falco, V.; Battista, E. Analytical results for binary dynamics at the first post-Newtonian order in Einstein–Cartan theory with the Weyssenhoff fluid. *Phys. Rev. D* **2023**, *108*, 064032. [CrossRef]
197. Mondal, V.; Chakraborty, S. Lorentzian quantum cosmology with torsion. *arXiv* **2023**, arXiv:2305.01690.
198. Chakraborty, S.; Dey, R. Noether Current, Black Hole Entropy and Spacetime Torsion. *Phys. Lett. B* **2018**, *786*, 432–441. [CrossRef]
199. Banerjee, R.; Chakraborty, S.; Mukherjee, P. Late-time acceleration driven by shift-symmetric Galileon in the presence of torsion. *Phys. Rev. D* **2018**, *98*, 083506. [CrossRef]
200. Sharma, M.K.; Sur, S. Growth of matter perturbations in an interacting dark energy scenario emerging from metric-scalar-torsion couplings. *Phys. Sci. Forum* **2021**, *2*, 51.
201. Boos, J.; Hehl, F.W. Gravity-induced four-fermion contact interaction implies gravitational intermediate W and Z type gauge bosons. *Int. J. Theor. Phys.* **2017**, *56*, 751–756. [CrossRef]
202. Gialamas, I.D.; Veermäe, H. Electroweak vacuum decay in metric-affine gravity. *Phys. Lett. B* **2023**, *844*, 138109. [CrossRef]
203. Pal, K.; Pal, K.; Sarkar, T. Conformal Fisher information metric with torsion. *J. Phys. A* **2023**, *56*, 335001. [CrossRef]
204. Gallegos, A.D.; Gürsoy, U.; Yarom, A. Hydrodynamics of spin currents. *SciPost Phys.* **2021**, *11*, 041. [CrossRef]
205. Adamczyk, L. et al. [The STAR Collaboration] Global Λ hyperon polarisation in nuclear collisions: evidence for the most vortical fluid. *Nature* **2017**, *548*, 62–65. [CrossRef]
206. Adam, J. et al. [STAR Collaboration] Global polarisation of Λ hyperons in Au+Au collisions at $\sqrt{s_{NN}}$ = 200 GeV. *Phys. Rev. C* **2018**, *98*, 014910. [CrossRef]
207. Takahashi, R. Spin hydrodynamic generation. *Nat. Phys.* **2015**, *12*, 52. [CrossRef]

Disclaimer/Publisher's Note: The statements, opinions and data contained in all publications are solely those of the individual author(s) and contributor(s) and not of MDPI and/or the editor(s). MDPI and/or the editor(s) disclaim responsibility for any injury to people or property resulting from any ideas, methods, instructions or products referred to in the content.

Communication

Analogue Quantum Gravity in Hyperbolic Metamaterials

Igor I. Smolyaninov [1,*] and Vera N. Smolyaninova [2]

[1] Department of Electrical and Computer Engineering, University of Maryland, College Park, MD 20742, USA
[2] Department of Physics, Astronomy, and Geosciences, Towson University, Towson, MD 21252, USA; vsmolyaninova@towson.edu
* Correspondence: smoly@umd.edu

Abstract: It is well known that extraordinary photons in hyperbolic metamaterials may be described as living in an effective Minkowski spacetime, which is defined by the peculiar form of the strongly anisotropic dielectric tensor in these metamaterials. Here, we demonstrate that within the scope of this approximation, the sound waves in hyperbolic metamaterials look similar to gravitational waves, and therefore the quantized sound waves (phonons) look similar to gravitons. Such an analogue model of quantum gravity looks especially interesting near the phase transitions in hyperbolic metamaterials where it becomes possible to switch quantum gravity effects on and off as a function of metamaterial temperature. We also predict strong enhancement of sonoluminescence in ferrofluid-based hyperbolic metamaterials, which looks analogous to particle creation in strong gravitational fields.

Keywords: analogue quantum gravity; hyperbolic metamaterials; sonoluminescence

1. Introduction

Hyperbolic metamaterials are a special class of electromagnetic metamaterials, which exhibit extremely strong anisotropy. These metamaterials exhibit metallic behavior in one direction and dielectric behavior in the orthogonal direction. The original purpose of these metamaterials was to overcome the diffraction limit of optical microscopy [1,2]. However, very soon it was realized that these materials exhibit a very large number of strikingly interesting physical properties, which result from the singular behavior of their photonic density of states [3,4]. In addition to super resolution imaging [1,5,6], it was shown that these materials exhibit enhanced quantum electrodynamic effects [7–9], which may be used, for example, in stealth technologies [10]. The transport properties of hyperbolic metamaterials may also be quite unusual, resulting in such effects as thermal hyperconductivity [11], and high Tc superconductivity [12]. It was also pointed out that hyperbolic metamaterials may be used to create very interesting laboratory analogues of gravitational effects [3,13–16]. While initially it was believed that hyperbolic properties may only be observed in artificial structures, soon it was discovered that many natural materials may also exhibit hyperbolic properties [12,17]. Strikingly enough, even the physical vacuum may potentially exhibit hyperbolic properties [18] when it is subjected to a very strong magnetic field [19].

Essential electromagnetic properties of hyperbolic metamaterials may be understood by considering a nonmagnetic uniaxial anisotropic material with dielectric permittivities $\varepsilon_x = \varepsilon_y = \varepsilon_1 > 0$ and $\varepsilon_z = \varepsilon_2 < 0$. Any electromagnetic field propagating in this material may be expressed as a sum of ordinary and extraordinary contributions, each of these being a sum of an arbitrary number of plane waves polarized in the ordinary ($E_z = 0$) and extraordinary ($E_z \neq 0$) directions. Let us assume that an extraordinary photon wave function is $\varphi = E_z$ so that the ordinary portion of the electromagnetic field does not contribute to φ. Maxwell equations in the frequency domain result in the following wave equation for φ_ω if ε_1 and ε_2 are kept constant inside the metamaterial [3]:

Citation: Smolyaninov, I.I.; Smolyaninova, V.N. Analogue Quantum Gravity in Hyperbolic Metamaterials. *Universe* **2022**, *8*, 242. https://doi.org/10.3390/universe8040242

Academic Editors: Arundhati Dasgupta and Alfredo Iorio

Received: 22 March 2022
Accepted: 12 April 2022
Published: 14 April 2022

Publisher's Note: MDPI stays neutral with regard to jurisdictional claims in published maps and institutional affiliations.

Copyright: © 2022 by the authors. Licensee MDPI, Basel, Switzerland. This article is an open access article distributed under the terms and conditions of the Creative Commons Attribution (CC BY) license (https://creativecommons.org/licenses/by/4.0/).

$$-\frac{\partial^2 \phi_\omega}{\varepsilon_1 \partial z^2} + \frac{1}{(-\varepsilon_2)}\left(\frac{\partial^2 \phi_\omega}{\partial x^2} + \frac{\partial^2 \phi_\omega}{\partial y^2}\right) = \frac{\omega_0^2}{c^2}\phi_\omega = \frac{m^{*2}c^2}{\hbar^2}\phi_\omega \qquad (1)$$

This wave equation coincides with the Klein–Gordon equation for a massive field φ_ω (with an effective mass m^*) in a 3D Minkowski spacetime, in which one of the spatial coordinates $z = \tau$ behaves as a timelike variable. The metric coefficients g_{ik} of this flat 2 + 1 dimensional Minkowski spacetime may be defined as [3,13]:

$$g_{00} = -\varepsilon_1 \text{ and } g_{11} = g_{22} = -\varepsilon_2 \qquad (2)$$

As demonstrated in [13], a nonlinear optical Kerr effect may "bend" this 2 + 1 Minkowski spacetime, resulting in effective gravitational force between the extraordinary photons. It was also predicted that for the effective gravitational constant inside the metamaterial to be positive, negative self-defocusing Kerr medium must be used as a dielectric host of the metamaterial [13].

Artificial hyperbolic metamaterials described by $\varepsilon_x = \varepsilon_y = \varepsilon_1 > 0$ and $\varepsilon_z = \varepsilon_2 < 0$ are typically made out of metal wire array structures, as illustrated in Figure 1.

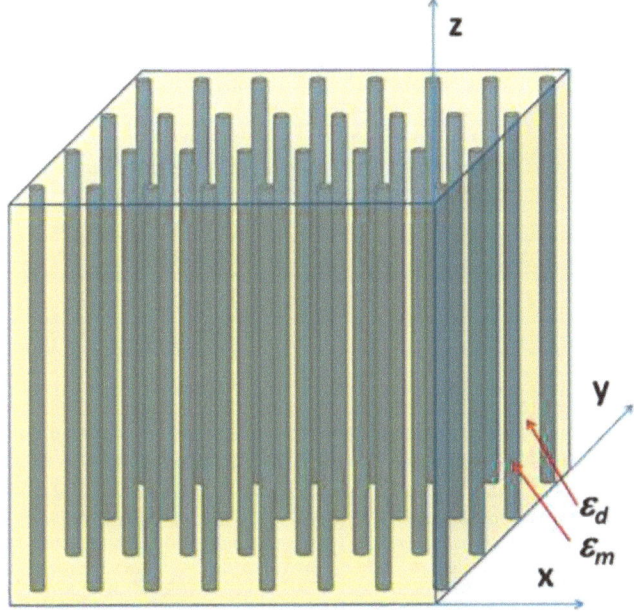

Figure 1. Typical geometry of a metal wire array hyperbolic metamaterial.

If the Maxwell–Garnett approximation is applied to such geometry [20], the diagonal components of the permittivity tensor of the metamaterial may be acquired as

$$\varepsilon_1 = \varepsilon_{x,y} = \frac{2f\varepsilon_m\varepsilon_d + (1-f)\varepsilon_d(\varepsilon_d + \varepsilon_m)}{(1-f)(\varepsilon_d + \varepsilon_m) + 2f\varepsilon_d} \approx \frac{1+f}{1-f}\varepsilon_d, \; \varepsilon_2 = \varepsilon_z = f\varepsilon_m + (1-f)\varepsilon_d \approx f\varepsilon_m \qquad (3)$$

where f is the metallic phase volume fraction, and $\varepsilon_m < 0$ and $\varepsilon_d > 0$ are the permittivities of the metal and the dielectric phase, respectively (note that $-\varepsilon_m \gg \varepsilon_d$ is typically assumed in the visible and infrared range). On the other hand, as mentioned above, several common natural materials, such as Al_2O_3, $ZrSiO_4$, TiO_2, etc., may also exhibit hyperbolic properties in the long wavelength infrared frequency range [21]. For example, Al_2O_3 is naturally hyperbolic in the 19.0–20.4 μm and 23–25 μm frequency bands. Such natural hyperbolic material examples look especially interesting near the phase transitions of the material, since they represent a natural physical situation in which the effective Minkowski spacetime may be "melted" as a function of material temperature T [22]. For example, in the melted state of Al_2O_3, the material becomes an isotropic liquid, and the effective Minkowski spacetime experienced by the extraordinary photons becomes an ordinary Euclidean space.

2. Methods: Acoustic Waves in Hyperbolic Materials as Analogues of Gravitational Waves

While the relationship between the dielectric permittivity and density of a material may be somewhat complicated, in general, the Clausius–Mossotti relation is typically used for this purpose [23]. In the case that the material consists of a mixture of two or more species, the molecular polarizability contribution from each species a, indexed by i, contributes to the overall dielectric permittivity as follows:

$$\frac{\varepsilon - 1}{\varepsilon + 2} = \sum_i \frac{N_i \alpha_i}{3\varepsilon_0} \qquad (4)$$

where N_i is the molecular concentration of the respective species and e_0 is the dielectric permittivity of the vacuum. As a result, for small deviations of e with respect to its average number, we may write

$$\Delta \varepsilon = \frac{(\varepsilon + 2)^2}{3} \sum_i \frac{\Delta N_i \alpha_i}{3\varepsilon_0} \qquad (5)$$

Equations (2), (3) and (5) clearly indicate that acoustic waves (the oscillations of molecular concentration DN) in a hyperbolic metamaterial act as classical gravitational waves, which perturb the effective metric of a flat 2 + 1 dimensional Minkowski spacetime (see Equation (2)) experienced by the photons propagating inside the metamaterial. In fact, these "gravitational waves" are known to be quite pronounced in such natural hyperbolic material as sapphire (Al_2O_3) due to its very strong piezoelectric behavior.

Furthermore, the sound waves in hyperbolic metamaterials may be quantized in a straightforward fashion, thus giving rise to the quantum mechanical description of sound waves in terms of phonons. The well-known Hamiltonian for this system is

$$H = \sum \frac{p_i^2}{2m} + \frac{1}{2} m \omega^2 \sum (x_i - x_j)^2 \qquad (6)$$

where m is the mass and w is the oscillation frequency of each atom (assuming for simplicity that they are all equal), and x_i and p_i are the position and momentum operators, respectively (the second sum is made over the nearest neighbors). The resulting quantization in momentum space is

$$k_n = \frac{2\pi n}{Na} \qquad (7)$$

where a is the interatomic distance. The harmonic oscillator eigenvalues or energy levels for the mode ω_k are:

$$E_n = \left(\frac{1}{2} + n\right) \hbar \omega_k \qquad (8)$$

While in the $k \to 0$ limit the dispersion relation of phonons is linear, this behavior changes near p/a. The picture of "acoustic" and "optical" phonons arises generically if different kinds of atoms are present in the crystalline lattice of the material. For example, if a one-dimensional lattice is made of two types of atoms of mass m_1 and m_2 connected by a chemical bond, which may be characterized by spring constant K, two phonon modes result [24]:

$$\omega_\pm = K\left(\frac{1}{m_1} + \frac{1}{m_2}\right) \pm \sqrt{\left(\frac{1}{m_1} + \frac{1}{m_2}\right)^2 - \frac{4 \sin^2 \frac{ka}{2}}{m_1 m_2}} \qquad (9)$$

The plus sign corresponds to the "optical mode" in which the two adjacent atoms move against each other, and the minus sign results in the acoustic mode in which they move together. These modes are shown schematically in Figure 2. We should also mention that similar to photons, the dispersion law of phonons in a metamaterial may also be made hyperbolic [25,26], so that both photons and phonons will live in an effective Minkowski spacetime. Quite obviously, the phonons in this picture become the analogues of gravitons. With this conclusion, our analogue model of quantum gravity is basically complete.

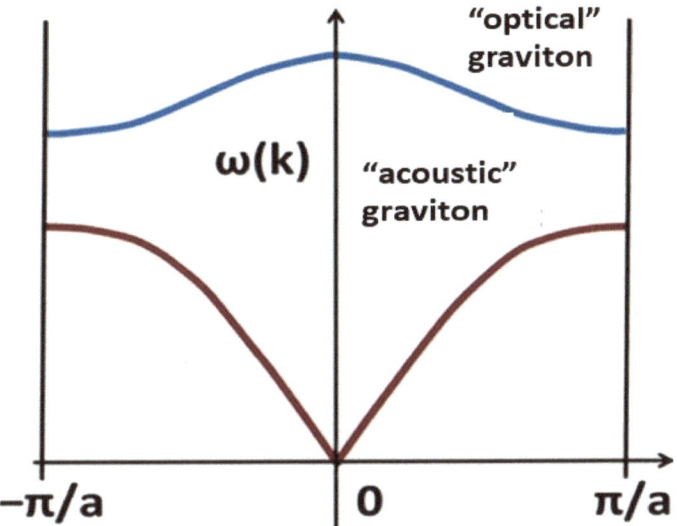

Figure 2. Dispersion curves of "acoustic" and "optical" gravitons.

3. Discussion: Analogue Quantum Gravity Effects in Hyperbolic Metamaterials

One of the main predictions of various versions of quantum gravity theories is the existence of the minimum length (which typically coincides with the Planck scale) [27]. In the hyperbolic (meta)material version of quantum gravity, this minimum length appears naturally due to the finite interatomic distance a. However, the minimum length physics actually look more complicated and quite interesting in such natural ferroelectric hyperbolic materials as Al_2O_3 and $BaTiO_3$. Based on Equations (1) and (2), the effective length element in a hyperbolic metamaterial equals

$$dl^2 = (-\varepsilon_2)\left(dx^2 + dy^2\right) \tag{10}$$

Therefore, the minimum length experienced by the extraordinary photons is

$$l_{min} = \sqrt{-\varepsilon_2}\, a \tag{11}$$

Since $-\varepsilon_2$ is temperature-dependent and may actually diverge near the critical temperature T_c of a ferroelectric phase transition, the analogue quantum gravity effects may become quite pronounced near T_c. Indeed, near the transition temperature, the dielectric susceptibility, χ, of these materials diverges following the Curie–Weiss law:

$$\chi = (\varepsilon - 1) = \frac{C}{T - T_c} \tag{12}$$

where C is the Curie–Weiss constant of the material.

On the other hand, at the "melting point" of the effective Minkowski spacetime (which was experimentally observed in a ferrofluid-based hyperbolic metamaterial [22] by tuning the f parameter in Equation (3)), $-\varepsilon_2$ changes sign, and therefore it transitions through the $\varepsilon_2 = 0$ point as a function of temperature. As a result, the effective "minimum length" inside the metamaterial becomes zero. The possibility of switching quantum gravity effects on and off as a function of metamaterial temperature looks very interesting and attractive. Moreover, since some naturally hyperbolic ferroelectric materials may exhibit quantum criticality [28], truly quantum effects associated with the emergence of the effective Minkowski spacetime at zero temperature may also be studied in the experiment (see also [12]).

Another important prediction of quantum gravity theories is the various effects related to particle creation by gravitational fields, such as Hawking radiation, Unruh effect, cosmological particle creation, etc. Let us discuss how these kinds of effects may be experimentally observed and studied within the scope of our model. Based on the analogy between sound and gravitational waves described above, sonoluminescence [29–35] appears to be the most natural choice to search for such effects. Moreover, Eberlein [36,37] already pointed out deep connections between sonoluminescence in liquids and the Unruh effect. According to Eberlein, sonoluminescence occurs when the rapidly moving surface of a microscopic bubble created by ultrasound converts virtual photons into real ones. Like in the Unruh effect, the resulting sonoluminescence spectrum appears to be similar to a black-body spectrum. For example, the radiated spectral density within the scope of this model is

$$P(\omega) = 1.16 \frac{(\varepsilon - 1)^2}{64\varepsilon} \frac{h}{c^4 \gamma} \left(R_0^2 - R_{\min}^2 \right)^2 \omega^3 e^{-2\gamma \omega} \tag{13}$$

where R_0 and $R_{\min}$ correspond to changes in bubble radius and γ describes the timescale of the bubble collapse (see Equations (4) and (12) from [36]). This expression is indeed proportional to the energy density of the thermal radiation given by the usual Planck expression

$$\frac{dU}{d\omega} = \frac{1}{\pi^2 c^3} \frac{\hbar \omega^3}{e^{\hbar \omega / kT} - 1} \tag{14}$$

if we assume that $T \sim 1/\gamma$.

As was recently predicted in [38], the Unruh effect is supposed to be strongly enhanced inside hyperbolic metamaterials, so that many orders of magnitude smaller accelerations may be used to observe the Unruh radiation. Following this prediction, we may demonstrate that sonoluminescence in ferrofluid-based hyperbolic metamaterials [14,22] will be strongly enhanced too. Let us consider a microscale sonoluminescent bubble inside the ferrofluid, as illustrated in Figure 3. The enhancement of the Unruh effect in hyperbolic metamaterials originates from the modification of the conventional Planck expression for the energy density of thermal radiation (Equation (14)) due to the huge enhancement of the photonic density of states $r(w)$ inside the metamaterial [3,7]. In particular, for the nanowire array metamaterial design shown in Figure 1, this enhancement factor equals

$$\frac{S_T}{S_T^{(0)}} \approx \frac{5}{16\pi^2} \left(\frac{k_{\max}^2}{k_T k_p} \right)^2 \tag{15}$$

where S_T is the energy flux along the symmetry axis of the metamaterial and $S_T^{(0)}$ is the usual Planck value for the energy flux (see Equation (9) from [38]). The characteristic k-vectors in Equation (15) are $k_{\max} \sim 1/a$ (defined by the structural parameter a of the metamaterial), $k_T = k_B T / \hbar c$ is the typical thermal momentum, and k_p is the typical "plasma momentum" [39] of the metamaterial, which is defined as

$$k_p = \sqrt{\frac{4\pi N}{m^*}} \frac{e}{c} \tag{16}$$

where N and m^* are the free charge carrier density in the metamaterial and their effective mass, respectively. Since, in a typical metamaterial, $k_{\max}$ is several orders of magnitude larger than k_T and k_p [40], the enhancement factor defined by Equation (15) may reach up to ten orders of magnitude (recall that $a \sim 1$ nm in natural hyperbolic materials). As a result, a similar strong enhancement may be expected for the radiated spectral density of sonoluminescence defined by Equation (13).

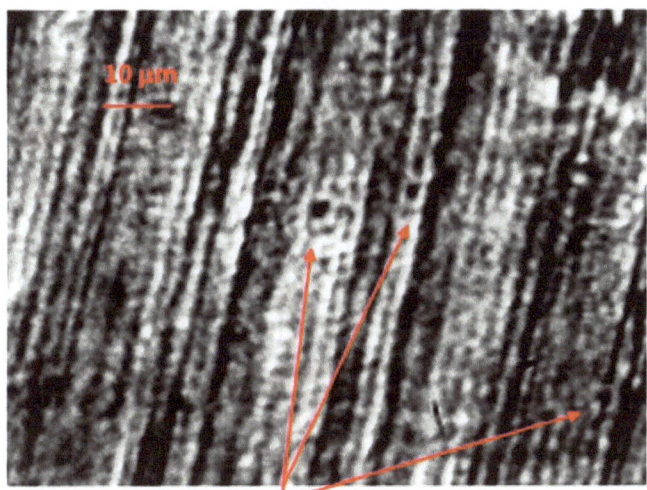

Figure 3. Photo of a ferrofluid-based self-assembled hyperbolic metamaterial [22]. The cobalt nanoparticle chains are formed inside the ferrofluid after application of external magnetic field. Several micron-scale bubbles inside the ferrofluid are indicated by arrows.

To summarize, it appears that several quantum gravity effects find interesting analogues in artificial and natural hyperbolic metamaterials. While the described analogies are obviously not perfect, the fact that these effects may be studied in the lab [14,22] make them a very useful tool to develop more intuition on the actual inner working of quantum gravity.

4. Conclusions

In conclusion, based on the fact that extraordinary photons in hyperbolic metamaterials [41–43] may be described as living in an effective Minkowski spacetime (which is defined by the peculiar form of the strongly anisotropic dielectric tensor in these metamaterials), we have demonstrated that sound waves in hyperbolic metamaterials look similar to gravitational waves [44–47]. As a result, within the scope of this model, the quantized sound waves (phonons) look similar to gravitons [48–50]. Such an analogue model of quantum gravity looks especially interesting near the phase transitions in hyperbolic metamaterials where it becomes possible to switch quantum gravity effects on and off at will as a function of metamaterial temperature. We also predicted strong enhancement of sonoluminescence in ferrofluid-based hyperbolic metamaterials, which looks analogous to such important quantum gravity effects as particle creation in gravitational fields [51–61].

Funding: This research received no external funding.

Institutional Review Board Statement: Not applicable.

Informed Consent Statement: Not applicable.

Acknowledgments: V.N.S. acknowledges SET/FCSM Grant at Towson University.

Conflicts of Interest: The author declares no conflict of interest.

References

1. Jacob, Z.; Alekseyev, L.V.; Narimanov, E. Optical hyperlens: Far-field imaging beyond the diffraction limit. *Optics Express* **2006**, *14*, 8247–8256. [CrossRef] [PubMed]
2. Smith, D.R.; Kolinko, P.; Schurig, D. Negative refraction in indefinite media. *JOSA B* **2004**, *21*, 1032–1043. [CrossRef]
3. Smolyaninov, I.I.; Narimanov, E.E. Metric signature transitions in optical metamaterials. *Phys. Rev. Lett.* **2010**, *105*, 067402. [CrossRef]
4. Krishnamoorthy, H.S.N.; Jacob, Z.; Narimanov, E.; Kretzschmarand, I.; Menon, V.M. Topological transitions in metamaterials. *Science* **2012**, *336*, 205–209. [CrossRef] [PubMed]
5. Smolyaninov, I.I.; Hung, Y.J.; Davis, C.C. Magnifying superlens in the visible frequency range. *Science* **2007**, *315*, 1699–1701. [CrossRef] [PubMed]
6. Liu, Z.; Lee, H.; Xiong, Y.; Sun, C.; Zhang, X. Far-field optical hyperlens magnifying sub-diffraction-limited objects. *Science* **2007**, *315*, 1686. [CrossRef]
7. Jacob, Z.; Smolyaninov, I.I.; Narimanov, E.E. Broadband Purcell effect: Radiative decay engineering with metamaterials. *Appl. Phys. Lett.* **2012**, *100*, 181105. [CrossRef]
8. Jacob, Z.; Kim, J.-Y.; Naik, G.V.; Boltasseva, A.; Narimanov, E.E.; Shalaev, V.M. Engineering photonic density of states using metamaterials. *App. Phys. B* **2010**, *100*, 215. [CrossRef]
9. Noginov, M.A.; Li, H.; Barnakov, Y.A.; Dryden, D.; Nataraj, G.; Zhu, G.; Bonner, C.E.; Mayy, M.; Jacob, Z.; Narimanov, E.E. Controlling spontaneous emission with metamaterials. *Opt. Lett.* **2010**, *35*, 1863. [CrossRef]
10. Narimanov, E.; Noginov, M.A.; Li, H.; Barnakov, Y. Darker than black: Radiation-absorbing metamaterial. In Proceedings of the Quantum Electronics and Laser Science Conference, San Jose, CA, USA, 16–21 May 2010. OSA Technical Digest (CD) (Optical Society of America, 2010), Paper QPDA6.
11. Narimanov, E.E.; Smolyaninov, I.I. Beyond Stefan-Boltzmann law: Thermal hyper-conductivity. *arXiv* **2011**, arXiv:1109.5444.
12. Smolyaninov, I.I. Quantum topological transition in hyperbolic metamaterials based on high Tc superconductors. *J. Phys. Condens. Matter* **2014**, *26*, 305701. [CrossRef] [PubMed]
13. Smolyaninov, I.I. Analogue gravity in hyperbolic metamaterials. *Phys. Rev. A* **2013**, *88*, 033843. [CrossRef]
14. Smolyaninov, I.I.; Yost, B.; Bates, E.; Smolyaninova, V.N. Experimental demonstration of metamaterial "multiverse" in a ferrofluid. *Opt. Express* **2013**, *21*, 14918–14925. [CrossRef] [PubMed]
15. Sedov, E.S.; Iorsh, I.V.; Arakelian, S.M.; Alodjants, A.P.; Kavokin, A. Hyperbolic metamaterials with Bragg Polaritons. *Phys. Rev. Lett.* **2015**, *114*, 237402. [CrossRef] [PubMed]
16. Tekin, B. Hyperbolic metamaterials and massive Klein-Gordon equation in (2 + 1)-dimensional de Sitter spacetime. *Phys. Rev. D* **2021**, *104*, 105004. [CrossRef]
17. Narimanov, E.E.; Kildishev, A.V. Metamaterials: Naturally hyperbolic. *Nat. Photonics* **2015**, *9*, 214–216. [CrossRef]
18. Smolyaninov, I.I. Vacuum in strong magnetic field as a hyperbolic metamaterial. *Phys. Rev. Lett.* **2011**, *107*, 253903. [CrossRef]
19. Chernodub, M.N. "Spontaneous electromagnetic superconductivity of vacuum in a strong magnetic field: Evidence from the Nambu–Jona-Lasinio model. *Phys. Rev. Lett.* **2011**, *106*, 142003. [CrossRef]
20. Wangberg, R.; Elser, J.; Narimanov, E.E.; Podolskiy, V.A. Nonmagnetic nanocomposites for optical and infrared negative-refractive-index media. *J. Opt. Soc. Am. B* **2006**, *23*, 498–505. [CrossRef]
21. Korzeb, K.; Gajc, M.; Pawlak, D.A. Compendium of natural hyperbolic materials. *Opt. Express* **2015**, *23*, 25406–25424. [CrossRef]
22. Smolyaninov, I.I.; Smolyaninova, V.N. Fine tuning and MOND in a metamaterial "multiverse". *Sci. Rep.* **2017**, *7*, 8023. [CrossRef] [PubMed]
23. Rysselberghe, P.V. Remarks concerning the Clausius–Mossotti law. *J. Phys. Chem.* **1932**, *36*, 1152–1155. [CrossRef]
24. Misra, P.K. §2.1.3 Normal modes of a one-dimensional chain with a basis. In *Physics of Condensed Matter*; Academic Press: Cambridge, MA, USA, 2010; p. 44.
25. Shen, C.; Xie, Y.; Sui, N.; Wang, W.; Cummer, S.A.; Jing, Y. Broadband acoustic hyperbolic metamaterial. *Phys. Rev. Lett.* **2015**, *115*, 254301. [CrossRef] [PubMed]
26. Smolyaninov, I.I.; Smolyaninova, V.N. Hybrid acousto-electromagnetic metamaterial superconductors. *Physica C* **2020**, *577*, 1353730. [CrossRef]
27. Garay, L.J. Quantum gravity and minimum length. *Int. J. Mod. Phys. A* **1995**, *10*, 145–166. [CrossRef]
28. Rowley, S.E.; Spalek, L.J.; Smith, R.P.; Dean, M.P.M.; Lonzarich, G.G.; Scott, J.F.; Saxena, S.S. Quantum criticality in ferroelectrics. *Nat. Phys.* **2014**, *10*, 367–372. [CrossRef]
29. Barber, B.P.; Putterman, S.J. Observation of synchronous picosecond sonoluminescence. *Nature* **1991**, *352*, 318. [CrossRef]
30. Brenner, M.P.; Hilgenfeldt, S.; Lohse, D. Single-bubble sonoluminescence. *Rev. Mod. Phys.* **2002**, *74*, 425. [CrossRef]
31. Barber, B.P.; Hiller, R.A.; Löfstedt, R.; Putterman, S.J.; Weninger, K.R. Defining the unknowns of sonoluminescence. *Phys. Rep.* **1997**, *281*, 65–143. [CrossRef]
32. Prosperetti, A.; Hao, Y. Modelling of spherical gas bubble oscillations and sonoluminescence. *Phil. Trans. Royal Soc. A* **1999**, *357*, 203–223. [CrossRef]
33. Wu, C.C.; Roberts, P.H. Shock-wave propagation in a sonoluminescing gas bubble. *Phys. Rev. Lett.* **1993**, *70*, 3424. [CrossRef] [PubMed]

34. Kwak, H.Y.; Na, J.H. Hydrodynamic solutions for a sonoluminescing gas bubble. *Phys. Rev. Lett.* **1996**, *77*, 4454. [CrossRef] [PubMed]
35. Greenspan, H.P.; Nadim, A. On sonoluminescence of an oscillating gas bubble. *Phys. Fluids A Fluid Dyn.* **1993**, *5*, 1065. [CrossRef]
36. Eberlein, C. Theory of quantum radiation observed as sonoluminescence. *Phys. Rev. A* **1996**, *53*, 2772–2787. [CrossRef] [PubMed]
37. Eberlein, C. Sonoluminescence as quantum vacuum radiation. *Phys. Rev. Lett.* **1996**, *76*, 3842. [CrossRef]
38. Smolyaninov, I.I. Enhancement of Unruh effect near hyperbolic metamaterials. *Euro. Phys. Lett.* **2021**, *133*, 18001. [CrossRef]
39. Pfender, E. Heat and momentum transfer to particles in thermal plasma flows. *Pure Appl. Chem.* **1985**, *7*, 1179–1195. [CrossRef]
40. Tumkur, T.U.; Gu, L.; Kitur, J.K.; Narimanov, E.E.; Noginov, M.A. Control of absorption with hyperbolic metamaterials. *Appl. Phys. Lett.* **2012**, *100*, 161103. [CrossRef]
41. Guo, Z.; Jiang, H.; Chen, H. Hyperbolic metamaterials: From dispersion manipulation to applications. *J. Appl. Phys.* **2020**, *127*, 071101. [CrossRef]
42. Poddubny, A.; Iorsh, I.; Kivshar, Y. Hyperbolic metamaterials. *Nat. Photonics* **2013**, *7*, 948–957. [CrossRef]
43. Ferrari, L.; Wu, C.; Lepage, D.; Zhang, X.; Liu, Z. Hyperbolic metamaterials and their applications. *Prog. Quantum Electron.* **2015**, *40*, 1–40. [CrossRef]
44. Einstein, A.; Rosen, N. On gravitational waves. *J. Frankl. Inst.* **1937**, *223*, 43–54. [CrossRef]
45. Landau, L.D.; Lifshitz, E.M. *The Classical Theory of Fields*; Pergamon Press: Oxford, UK, 1975; pp. 356–357.
46. Weber, J. Detection and generation of gravitational waves. *Phys. Rev.* **1960**, *117*, 306. [CrossRef]
47. Cervantes-Cota, J.L.; Galindo-Uribarri, S.; Smoot, G.F. A brief history of gravitational waves. *Universe* **2016**, *2*, 22. [CrossRef]
48. Goldhaber, A.S.; Nieto, M.M. Photon and graviton mass limits. *Rev. Mod. Phys.* **2010**, *82*, 939. [CrossRef]
49. Penrose, R. The nonlinear graviton. *Gen. Relativ. Gravit.* **1976**, *7*, 171–176. [CrossRef]
50. Gross, D.J.; Jackiw, R. Low-energy theorem for graviton scattering. *Phys. Rev.* **1968**, *166*, 1287. [CrossRef]
51. Zeldovich, Y.B.; Starobiskii, A.A. Rate of particle production in gravitational fields. *JETP Lett.* **1977**, *26*, 252–255.
52. Martin, J. Inflationary perturbations: The cosmological Schwinger effect. *Lect. Notes Phys.* **2008**, *738*, 193–241.
53. Unruh, W.G. Notes on black hole evaporation. *Phys. Rev. D* **1976**, *14*, 870–892. [CrossRef]
54. Wald, R.M. The back reaction effect in particle creation in curved spacetime. *Commun. Math. Phys.* **1977**, *54*, 1–19. [CrossRef]
55. Grib, A.A.; Levitskii, B.A.; Mostepanenko, V.M. Particle creation from vacuum by a nonstationary gravitational field in the canonical formalism. *Theoret. Math. Phys.* **1974**, *19*, 349–361. [CrossRef]
56. Parker, L. Particle creation in expanding universes. *Phys. Rev. Lett.* **1968**, *21*, 562. [CrossRef]
57. Ford, L.H. Gravitational particle creation and inflation. *Phys. Rev. D* **1987**, *35*, 2955. [CrossRef]
58. Sexl, R.U.; Urbantke, H.K. Production of particles by gravitational fields. *Phys. Rev.* **1969**, *179*, 1247. [CrossRef]
59. Woodhouse, N.M.J. Particle creation by gravitational fields. *Phys. Rev. Lett.* **1976**, *36*, 999. [CrossRef]
60. Mottola, E. Particle creation in de Sitter space. *Phys. Rev. D* **1985**, *31*, 754. [CrossRef]
61. Rubakov, V.A. Particle creation in a tunneling universe. *JETP Lett.* **1984**, *39*, 107–110.

Review

Maximal Kinematical Invariance Group of Fluid Dynamics and Applications

V. V. Sreedhar * and Amitabh Virmani

Chennai Mathematical Institute, H1 SIPCOT IT Park, Kelambakkam, Chennai 603103, India; avirmani@cmi.ac.in
* Correspondence: sreedhar@cmi.ac.in

Abstract: The maximal kinematical invariance group of the Euler equations of fluid dynamics for the standard polytropic exponent is larger than the Galilei group. Specifically, the inversion transformation ($\Sigma : t \to -1/t, \vec{x} \to \vec{x}/t$) leaves the Euler equation's invariant. This duality has been used to explain the striking similarities observed in simulations of the supernova explosions and laboratory implosions induced in plasma by intense lasers. The inversion symmetry extends to discontinuous fluid flows as well. In this contribution, we provide a concise review of these ideas and discuss some applications. We also explicitly work out the implosion dual of the Sedov's explosion solution.

Keywords: fluid dynamics; symmetries; shock conditions

1. Introduction

Surprises lurk in unexpected corners of physics. This review summarises a body of results that ensue from one such surprise, viz. the striking similarity between the earlier simulations of supernova explosions and the experimental evolution of an imploding plasma contained in a fusion capsule bombarded by high-intensity lasers [1]. It was hoped [1] that, over time, laser experiments would become more in line with actual supernovae behaviour. Hence, considerable efforts were devoted to simulating astrophysical systems in the laboratory. Modern supernova simulations have become much more complex; several new physical effects and numerical techniques are incorporated [2]. At the present stages of development, it is not clear how much one can learn about the astrophysical systems in the laboratory setting. However, the observations mentioned above have led to some intriguing theoretical developments. In this paper we concentrate on discussing the theoretical explanation [3–5] for the observed similarity [1], and some ramifications of the resulting analysis [6]. We limit our considerations mostly to references [3–6].

Earlier computational studies of the evolution of a supernova remnant (as cited in [1]) usually used initial conditions of dense pressure-free ejecta expanding ballistically outwards from the site of the explosion, taken for convenience to be the origin, and interacting with a stationary ambient medium of much lower density and negligible pressure. At early times, the bulk of the ejecta expands ballistically, except for a thin interaction region on the outside consisting of a forward shock running into the ambient medium, a zone of hot-shocked ambient medium, a contact discontinuity, a zone of shocked ejecta, and a reverse shock propagating into the ejecta. At later times, when the mass of the swept-up ambient medium becomes comparable with the ejecta mass, the reverse shock detaches itself from the contact discontinuity and implodes on the origin.

In the laboratory plasma, we have, initially, a stationary sphere of high density material surrounded by a low density converging flow. The inflowing gas has to decelerate at the shock front, building up pressure and driving a reverse shock which leads to an implosion. From an experimental point of view, a perfectly spherically symmetric explosion is not realistic. The ejecta emerging from a supernova explosion is also highly nonuniform on a wide range of scales making it computationally challenging to calculate its evolution.

From a theoretical point of view, it is interesting to study the underlying symmetry that enables one to map an explosion to an implosion. One can straightaway rule out time-reversal as an answer because the supernova explosions occur over astronomical time-scales, while the plasma implosions happen in a few nano-seconds.

Both an exploding star and an imploding plasma can be modelled by the equations of a perfect fluid, as we are taught in standard textbooks [7,8]. The explanation offered in Refs. [3,4] for the observed explosion–implosion duality stems from a hitherto unnoticed nonlinear symmetry of these equations, which we expand upon in the next section.

This analysis of ref. [4] highlights that the maximal kinematical invariance group of the Euler equations of fluid dynamics for the standard polytropic exponent is larger than the Galilei group. The techniques required to establish this find applications in other situations, viz. fluid flows in non-inertial frames. The Earth's oceans and atmosphere are important examples of fluid flows in non-inertial reference frames, where the Earth's rotation provides the underlying non-inertial frame. In order to describe oceanic and atmospheric fluid flows, it is natural to analyse fluid phenomena in the Earth's frame. This requires adding Coriolis forces to the right hand side of the fluid equations. The Coriolis force terms lead to surprising phenomenon: weather storms and ocean's currents. Going to the non-inertial reference frame allows us to separate out the rotational component of the fluid flows. We can then concentrate on the parts of flow patterns that matter the most.

Related situations arise when fluid flows are characterised by a large degree of expansion or contraction. Poludnenko and Khokhlov [6] considered the formulation of Euler equations of fluid dynamics in an expanding or contracting or possibly rotating reference frame. The motivation being that by going to an appropriately chosen frame we can discard the expanding or contracting or rotating nature of the fluid flows. The frame motion is adjusted to minimise the local fluid velocities. This method allows to accommodate efficiently large degrees of change in the flow extent, such as those encountered in astrophysical flows: supernovae, contracting stars, etc. Their work investigated numerical computations in such non-inertial reference frames.

As in the case of rigidly rotating reference frames, going to an expanding or contracting reference frame requires adjustments of the fluid flow equations. Ref. [6] argued that these adjustments do not come at any additional numerical cost: the new equations can be as easily implemented numerically using any of the standard numerical schemes. However, by separating out the global component of the fluid flow, it leads to significant improvement in the physical understanding of the fluid flows, which would be difficult to extract in inertial reference frame simulations. (More precisely, in numerical work it is important to work with smaller local fluid velocities. If a fluid flow is dominated by the global component associated with expansion or contraction or rotation, then it is inefficient to model such flows in inertial frames.) They carried out extensive numerical testing of the method for a variety of reference frames representative of realistic applications.

The rest of the article is organised as follows. In Section 2, we review the maximal kinematical invariance group of fluid dynamics, based on the original work of O' Raifeartaigh and one of the authors [4]. In Section 3, we discuss the symmetries of discontinuous flows, based on our original work with Oliver Jahn [5]. In Section 4, we review the work of Poludnenko and Khokhlov [6], who considered the formulation of Euler equations of fluid dynamics in non-inertial reference frames. In Section 5, we present the conclusions. Appendix A explicitly works out the implosion dual of Sedov's explosion solution. Throughout the review we will be concise, referring the reader to original references for further details.

2. The Fluid Equations

This section is based on the original work of O'Raifeartaigh and one of the authors [4]. The general fluid dynamic equations in n-dimensional space are (see, e.g., textbooks [7,8])

$$D\rho = -\rho \nabla \cdot \mathbf{u}, \tag{1}$$
$$\rho D\mathbf{u} = -\nabla p + \mathbf{V}, \tag{2}$$
$$D\epsilon = -(\epsilon + p)\nabla \cdot \mathbf{u}, \tag{3}$$

where the convective derivative D and the viscosity terms $\mathbf{V}$ are defined by

$$D = \frac{\partial}{\partial t} + \mathbf{u} \cdot \nabla, \tag{4}$$

$$V_i = \nabla_j\left(\eta(\nabla_j u_i + \nabla_i u_j - \frac{2}{n}\delta_{ij}\nabla_k u_k)\right) + \nabla_i(\zeta \nabla_k u_k), \tag{5}$$

respectively. In the above equations $\rho, \mathbf{u}, p, \epsilon$ stand for the density, the velocity vector field, the pressure, and the energy density respectively, and η and ζ are the bulk and shear viscosity fields. These partial differential equations are augmented by an algebraic condition called the equation of state that relates the pressure and energy density. According to the *polytropic* equation of state

$$p = (\gamma_0 - 1)\epsilon \quad \Longrightarrow \quad p + \epsilon = \gamma_0 \epsilon, \tag{6}$$

where the constant γ_0 is called the polytropic exponent. This equation can be used to eliminate p from the fluid equations. Further, by making the substitution

$$\epsilon = \chi \rho^{\gamma_0} \tag{7}$$

the equations can be rewritten in the form

$$D\rho = -\rho \nabla \cdot \mathbf{u}, \tag{8}$$
$$\rho D\mathbf{u} = -(\gamma_0 - 1)\nabla(\chi \rho^{\gamma_0}) + \mathbf{V}, \tag{9}$$
$$D\chi = 0. \tag{10}$$

2.1. Action Formulation

The fluid equations may be given an action formulation by switching off the viscosity fields, i.e., $\eta = \zeta = 0$. Such fluids are called inviscid, or perfect fluids, and the equations are called Euler's equations. We next set $\chi = 1$ without loss of consistency, representing the isentropicity condition. Further, the Clebsch parametrisation [9–11]

$$\mathbf{u} = \nabla \phi - \nu \nabla \theta, \tag{11}$$

allows us to isolate the irrotational parts by setting $\nu = \theta = 0$. The resulting action for inviscid, isentropic, irrotational flows in three dimensions is given by,

$$S = \int d^3x dt \left[\rho\left(\dot{\phi} - \frac{1}{2}(\nabla\phi)^2\right) - \rho^{\gamma_0}\right]. \tag{12}$$

It may be mentioned in passing that the terms contained in the parentheses represent the Hamilton–Jacobi function for a free particle.

The symmetries of the aforementioned special flows, represented by the action, follow from the requirement of its form-invariance. The transformation properties of the fields in the general fluid equations may be extracted from these transformation properties once again by requiring the equations to transform covariantly.

2.2. Symmetries

We begin a priori with the most general transformations involving the coordinates and fields. The transformed coordinates ξ and τ are defined by

$$\vec{\xi} = \vec{\xi}(\vec{x},t), \quad \tau = \tau(\vec{x},t), \quad \tilde{\phi} = \tilde{\phi}(\xi,\tau,\phi), \quad \tilde{\rho} = \tilde{\rho}(\xi,\tau,\rho). \tag{13}$$

Substituting the transformations and demanding the form-invariance of the action produces a set of equations which can be solved exactly to yield [4],

$$\vec{\xi} = f(t)\,(\mathbf{R}\vec{x} + \vec{a} + \vec{v}t), \tag{14}$$

$$\tilde{\rho} = f^{-3}(t)\rho, \tag{15}$$

$$\tilde{\phi} = \phi + \lambda(\xi,\tau), \tag{16}$$

where

$$\tau = \frac{\alpha t + \beta}{\gamma t + \delta}, \quad f(t) = \frac{1}{\gamma t + \delta}, \quad \alpha\delta - \beta\gamma = 1, \tag{17}$$

$$\frac{\partial \lambda}{\partial \tau} - \frac{1}{2}\left(\frac{\partial \lambda}{\partial \xi}\right)^2 = 0. \tag{18}$$

In the above, **R** represents the usual rotation matrix, $\vec{a}$ the translations, $\vec{v}$ the boosts, and $f(t)$, a time-dependent scale parameter. The $\alpha,\beta,\gamma,\delta$ represent parameters of the $SL(2,R)$ group, a non-relativistic remnant of the special conformal group, to be discussed in Section 2.4. For details on $\lambda(\xi,\tau)$, we refer the reader to the original reference [4]. We note that the following discrete symmetries are permitted:

$$(\alpha,\beta,\gamma,\delta) \sim (\alpha,-\beta,-\gamma,\delta) \sim (-\alpha,\beta,\gamma,-\delta) \sim (-\alpha,-\beta,-\gamma,-\delta). \tag{19}$$

2.3. Transformation Functions for General Flows

The transformation functions for general flows may be obtained by requiring the general fluid equations to transform covariantly. It is straightforward to see that for general non-isentropic flows, the equations transform covariantly if χ is a scalar under the coordinate transformations.

The velocity vector transforms inhomogeneously as,

$$\tilde{\vec{u}} = (\gamma t + \delta)\vec{u} - \gamma\vec{x} \tag{20}$$

We note that these transformations do not preserve the condition $\nabla \cdot \mathbf{u} = 0$, implying that unlike Galilean symmetry, the above symmetry is valid only when the fluid is compressible.

The viscosity fields transform similar to scalar densities [4]. This implies that the symmetry we are discussing is broken in the case of Navier–Stokes equations which approximate the viscosity to be constant.

2.4. The Maximal Invariance Group

Let g be a general element of the symmetry group G obtained by setting $\beta = \gamma = 0, \alpha = 1$. It follows

$$\vec{\xi} = \mathbf{R}\vec{x} + \vec{v}t + \vec{a}, \quad \tau = t \tag{21}$$

In this case,

$$\tilde{\rho} = \rho \quad \text{and} \quad \tilde{\vec{u}} = \vec{u} + \vec{v} \tag{22}$$

We notice that these correspond to the static Galilei transformations.

Let σ denote an element of the $SL(2, R)$ group obtained by setting $\vec{a} = \vec{v} = 0$. In this case,

$$\vec{\xi} = f(t)\vec{x}, \quad \tau = \frac{\alpha t + \beta}{\gamma t + \delta}, \tag{23}$$

$$\tilde{\rho} = (\gamma t + \delta)^3 \rho \quad \text{and} \quad \tilde{\vec{u}} = (\gamma t + \delta)\vec{u} - \gamma \vec{x}. \tag{24}$$

This represents a combination of dilatations and inversions, which are a nonrelativistic limit of the special conformal group.

It is a straightforward exercise to check that G is an invariant subgroup:

$$\sigma^{-1} \cdot g(\mathbf{R}, \vec{a}, \vec{v}) \cdot \sigma = g(\mathbf{R}, \vec{a}_\sigma, \vec{v}_\sigma), \tag{25}$$

where

$$\begin{pmatrix} \vec{a}_\sigma \\ \vec{v}_\sigma \end{pmatrix} = \begin{pmatrix} \delta & \beta \\ \gamma & \alpha \end{pmatrix} \begin{pmatrix} \vec{a} \\ \vec{v} \end{pmatrix}. \tag{26}$$

It therefore follows that the full group under which the fluid equations are invariant under the specified transformation properties for the coordinates and the fields is a semi-direct product

$$\mathcal{G} = SL(2,R) \wedge G.$$

A special element of the group viz. $\Sigma : (\alpha, \beta, \gamma, \delta) = (0, -1, 1, 0)$ corresponds to a composition of an inversion and reversal of time, and plays an important role in the explosion–implosion duality discovered by Drury and Mendonça in [3]. A curious result follows immediately: $\Sigma^2 = P$, where P is the parity operator.

Cosets defined using these elements, $g_\Sigma(\mathbf{R}, \vec{a}, \vec{v}) = \Sigma \cdot g(\mathbf{R}, \vec{a}, \vec{v})$ have the interesting property that they are fourth roots of Galilean transformations,

$$g_\Sigma^4(\mathbf{R}, \vec{a}, \vec{v}) = g\left(\mathbf{R}^4, (\mathbf{R}^2 - 1)(\mathbf{R}\vec{a} - \vec{v}), (\mathbf{R}^2 - 1)(\mathbf{R}\vec{v} + \vec{a})\right) \tag{27}$$

Since $(\mathbf{R}\vec{a} - \vec{v})$ and $(\mathbf{R}\vec{v} - \vec{a})$ are linearly independent, it follows that every Galilean transformation is a fourth power of a coset transformation [4].

2.5. Conservation Laws

Euler's equations for a perfect fluid can be expressed in the form of conservation laws for mass, momentum, and energy, as follows:

$$\frac{\partial \rho}{\partial t} = -\frac{\partial}{\partial x_j}(\rho u_j), \tag{28}$$

$$\frac{\partial}{\partial t}(\rho u_i) = -\frac{\partial}{\partial x_j}(\rho u_i u_j + \delta_{ij} p), \tag{29}$$

$$\frac{\partial}{\partial t}\left(\frac{1}{2}\rho \vec{u}^2 + \epsilon\right) = -\frac{\partial}{\partial x_j}\left[\left(\frac{1}{2}\rho \vec{u}^2 + \epsilon + p\right) u_j\right]. \tag{30}$$

These equations can be expressed succinctly as follows:

$$\partial_\mu J_\rho^\mu = \partial_\mu J_{\vec{p}}^\mu = \partial_\mu J_H^\mu = 0. \tag{31}$$

The zeroth components of the above currents namely, $\rho, \rho\vec{u}, \frac{1}{2}\rho\vec{u}^2 + \epsilon$, give the charge densities which, when integrated over all space, give the conserved charges. The corresponding current densities are

$$J_\rho^j = \rho u_j, \tag{32}$$

$$J_{P_i}^j = (\rho u_i u_j + \delta_{ij} p), \tag{33}$$

$$J_H^j = \left(\frac{1}{2}\rho \vec{u}^2 + \epsilon + p\right) u_j. \tag{34}$$

The conservation laws corresponding to rotations ($\delta x^i = \omega^{ij} x^j$), boosts ($\delta x^i = v^i t$), dilatations ($\delta x^i = \lambda x^i, \delta t = 2\lambda t$), and expansions ($\delta x^i = -\mu t x^i, \delta t = -\mu t^2$) can be stated similarly,

$$\partial_\mu J_L^\mu = \partial_\mu J_K^\mu = \partial_\mu J_D^\mu = \partial_\mu J_A^\mu = 0, \tag{35}$$

where the appropriate charge densities are [5],

$$\vec{L} = \vec{P} \times \vec{x}, \quad \vec{K} = \vec{P} t - \rho \vec{x}, \quad D = -2tH + \vec{x} \cdot \vec{P}, \quad A = t^2 H - t\vec{x} \cdot \vec{P} + \frac{\rho}{2}\vec{x}^2. \tag{36}$$

The corresponding current densities can also be written down in a straightforward manner

$$\vec{J}_{L_i} = \epsilon_{ikl} x_k \vec{J}_{P_l}, \tag{37}$$

$$\vec{J}_{K_i} = t \vec{J}_{P_i} - x_i \vec{J}_\rho, \tag{38}$$

$$\vec{J}_D = x_i \vec{J}_{P_i} - 2t \vec{J}_H, \tag{39}$$

$$\vec{J}_A = \frac{1}{2}\vec{x}^2 \vec{J}_\rho - t x_i \vec{J}_{P_i} + t^2 \vec{J}_H. \tag{40}$$

These laws follow as a direct consequence of Noether's theorem. They will be useful in studying flows with discontinuities, a topic to which we now pass.

3. Discontinuous Flows

This section is based on our original work with Oliver Jahn [5]. As long as the flows are smooth, i.e., the functions $\rho, \vec{u}, p, \epsilon$ are smooth functions of $\vec{x}, t$, Equations (1)–(3) and (31) are equivalent. Real flows, however, may develop discontinuities as they evolve. Such flows are described by *weak* solutions of differential Equations [12]. A weak solution is generally piecewise smooth. The smooth parts satisfy the differential equations in the usual *strong* form. The entire solution required to specify the course of motion of the initial conditions is obtained by supplementing the strong solutions by jump conditions. The jump conditions are derived from the conservation laws. We briefly review these concepts in the next two subsections. For pedagogical discussions on these topics we refer the reader to [7,8,13].

3.1. Weak Solutions and Jump Conditions

By definition, any, possibly non-smooth, function $J^\mu(\vec{x}, t)$ that satisfies

$$\int \partial_\mu \omega(\vec{x}, t) J^\mu(\vec{x}, t) d^3 x dt = 0, \tag{41}$$

for all test functions $\omega(\vec{x}, t)$ is called a weak solution of the differential equation $\partial_\mu J^\mu = 0$.

Suppose $J^\mu(\vec{x}, t)$ has a jump discontinuity across a hypersurface $\mathcal{S}$ in $\vec{x}, t$ space while otherwise being continuously differentiable in some neighbourhood $\mathcal{N}$ of $\mathcal{S}$; see Figure 1. Let $\omega(\vec{x}, t)$ be a test function with support in region $\mathcal{N}$. Let $\mathcal{R}$ be the part of the region $\mathcal{N}$ that lies on one side of $\mathcal{S}$, say to the right. We take $\omega(\vec{x}, t) = 0$ on the boundary of $\mathcal{R}$, except on $\mathcal{S}$. Then by Gauss's theorem,

$$\int_\mathcal{R} \partial_\mu \omega \, J^\mu d^3 x dt + \int_\mathcal{R} \omega \, \partial_\mu J^\mu d^3 x dt = \int_\mathcal{R} \partial_\mu (\omega J^\mu) d^3 x dt = \int_\mathcal{S} \omega n_\mu J^\mu d\mathcal{S}, \tag{42}$$

since $\omega(\vec{x}, t) = 0$ on the boundary of $\mathcal{R}$, except on $\mathcal{S}$. Here n^μ is the outward normal to the hypersurface $\mathcal{S}$. The second integral on the left hand side is zero because the conservation law holds in the strong sense in the interior of $\mathcal{R}$. If we integrate similarly over the left part of the support $\omega(\vec{x}, t)$, and add the two results we obtain for a weak solution

$$0 = \int_{\mathcal{S}} \omega\, n_\mu \Delta J^\mu dS, \qquad (43)$$

where Δf denotes the difference of the two limiting values of the the function f on the two sides of the hypersurface $\mathcal{S}$, i.e., the jump of the function. This result follows as the normal which points outwards by convention, flips its sign when we move from the right to the left side of the hypersurface. Since $\omega(\vec{x}, t)$ is an arbitrary function, it follows that

$$n_\mu \Delta J^\mu = 0 \quad \text{on } \mathcal{S}. \qquad (44)$$

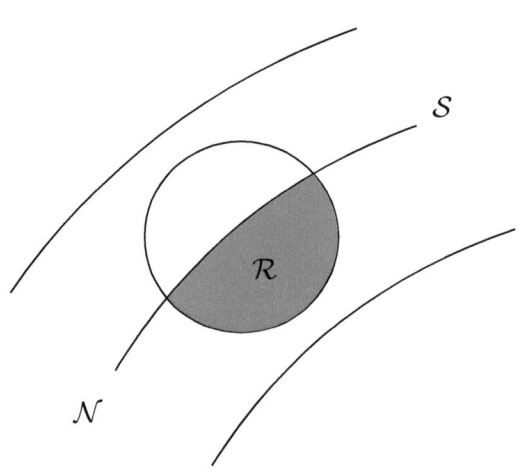

Figure 1. Diagram for the jump condition.

3.2. Rankine–Hugoniot Conditions

The general expression for the jump condition derived above can be applied to the conservation laws derived earlier. The conservation laws associated with mass, momentum, and energy yield

$$n_\mu \Delta J^\mu_{(\rho)} = 0, \qquad (45)$$

$$n_\mu \Delta J^\mu_{(\vec{P})} = 0, \qquad (46)$$

$$n_\mu \Delta J^\mu_{(H)} = 0, \qquad (47)$$

and are called the Rankine–Hugoniot conditions in the fluid dynamics literature [7,8,13,14]. Similar equations can be derived for the other conservation laws, viz.

$$n_\mu \Delta J^\mu_{(\vec{L})} = 0, \qquad (48)$$

$$n_\mu \Delta J^\mu_{(\vec{K})} = 0, \qquad (49)$$

$$n_\mu \Delta J^\mu_{(D)} = 0, \qquad (50)$$

$$n_\mu \Delta J^\mu_{(A)} = 0. \qquad (51)$$

These new jump conditions are associated with angular momentum, boosts, dilatations, and expansions.

3.3. Dual Rankine–Hugoniot Conditions

The new set of jump conditions are identically true because the current densities associated with angular momentum, boosts, dilatations, and expansions are linear combinations of the current densities associated with mass, momentum and energy conservation as shown in Equation (40). This suggests that the Rankine–Hugoniot conditions are invariant under the full kinematical invariance group of smooth flows including the $SL(2,R)$ part. We examine this point in what follows.

If an (abstract) symmetry generator T_r transforms under the $SL(2,R)$ transformation σ as

$$T'_r = \sigma^{-1} T_r \sigma = \sum_s M_{rs}(\sigma) T_s, \tag{52}$$

then the corresponding currents transform as [5]

$$J_r^{\mu'}(x') = \det\left(\frac{\partial x}{\partial x'}\right) \frac{\partial x^{\mu'}}{\partial x^\nu} \sum_s M_{rs}(\sigma) J_s^\nu(x). \tag{53}$$

The fact that the currents transform similar to vector densities can be appreciated by looking at the temporal components, which pick up the multiplicative factor $(\gamma t + \delta)^3$.

Arranging the currents in a column $J^\mu = \left(J^\mu_{(\rho)},\ J^\mu_{(\vec{K})},\ J^\mu_{(\vec{P})},\ J^\mu_{(A)},\ J^\mu_{(D)},\ J^\mu_{(H)}\right)^T$, one has the following transformation matrix

$$M = \begin{pmatrix} 1 & 0 & 0 & 0 & 0 & 0 \\ 0 & \alpha & \beta & 0 & 0 & 0 \\ 0 & \gamma & \delta & 0 & 0 & 0 \\ 0 & 0 & 0 & \alpha & -\alpha\beta & \beta^2 \\ 0 & 0 & 0 & -2\alpha\gamma & (\beta\gamma + \alpha\delta) & -2\beta\gamma \\ 0 & 0 & 0 & \gamma^2 & -\gamma\delta & \delta^2 \end{pmatrix}. \tag{54}$$

Using $\alpha\delta - \beta\gamma = 1$, it is easy to check that the matrix M has unit determinant. The temporal components transform according to the transformations,

$$\rho' = (\gamma t + \delta)^3 \rho, \tag{55}$$

$$\vec{P}' = (\gamma t + \delta)^3 (\delta \vec{P} + \gamma \vec{K}), \tag{56}$$

$$H' = (\gamma t + \delta)^3 (\gamma^2 A - \delta\gamma D + \delta^2 H). \tag{57}$$

Thus, ρ transforms under the singlet representation of $SL(2,R)$ as a scalar density. The translations and boosts constitute the doublet representation and transform similar to vector densities. Likewise the Hamiltonian, and the generators of dilatations and expansions transform similar to densities under the triplet (adjoint) representation of $SL(2,R)$. The transformation properties of the spatial components of the current can similarly be read off from the above matrix.

The dual Rankine-Hugoniot conditions are now easily obtained. The normal vector n_μ appearing in the jump condition (44) transforms like a covector

$$n'_\mu = \frac{\partial x^\nu}{\partial x^{\mu'}} n_\nu. \tag{58}$$

Thus, the transformed jump conditions for the currents are

$$n'_\mu \Delta J_r^{\mu'} \propto \det\left(\frac{\partial x}{\partial x'}\right) \sum_s M_{rs}(\sigma) n_\mu \Delta J_s^\mu(x) = 0 \quad \text{on } \mathcal{S}. \tag{59}$$

Since the determinant is smooth across the hypersurface $\mathcal{S}$, the factor in front can be omitted, and the transformed jump condition is a linear combination of the original jump conditions.

In particular, the transformed conditions for $J^\mu_{(\rho)}, J^\mu_{(\vec{P})}, J^\mu_{(H)}$ (the Rankine–Hugoniot conditions), become linear combinations of the jump conditions of $J^\mu_{(\rho)}, J^\mu_{(\vec{P})}, J^\mu_{(H)}, J^\mu_{(\vec{K})}, J^\mu_{(D)}$ and $J^\mu_{(A)}$. Specifically,

$$n'_\mu \Delta J^{\mu'}_{(\rho)} \propto n_\mu \Delta J^\mu_{(\rho)} \tag{60}$$

$$n'_\mu \Delta J^{\mu'}_{(\vec{P})} \propto n_\mu (\gamma \Delta J^\mu_{(\vec{K})} + \delta \Delta J^\mu_{(\vec{P})}) \tag{61}$$

$$n'_\mu \Delta J^{\mu'}_{(H)} \propto n_\mu (\delta^2 \Delta J^\mu_{(H)} - \gamma \delta \Delta J^\mu_{(D)} + \gamma^2 \Delta J^\mu_{(A)}) \tag{62}$$

The original Rankine-Hugoniot conditions, in conjunction with the new conditions (48)–(51), imply that the right hand sides of the above equations are identically zero, i.e., the Rankine-Hugoniot conditions are form-invariant [5].

In particular, this holds for the Drury–Mendonça transformation [3] $t \to -1/t$, $\vec{x} \to \vec{x}/t$ used to relate the explosion and implosion problems. This corresponds to the choice $(\alpha, \beta, \gamma, \delta) = (0, -1, 1, 0)$. We conclude that, if an explosion is described by the standard Rankine–Hugoniot conditions, the corresponding implosion is described by the dual Rankine–Hugoniot conditions.

3.4. Physical Conditions

As already mentioned, for a polytropic gas, $\epsilon = \chi \rho^{\gamma_0}$, which enables us to write the third of Euler's Equation (3) as $D\chi = 0$. χ transforms similar to a scalar. For a polytropic gas, it is well known [14] that χ is related to the specific entropy (entropy per unit mass) as follows:

$$S - S_0 = C_V \log[\chi(\rho, V)^{\gamma_0}], \tag{63}$$

where $C_V = R/(\gamma_0 - 1)$, R being the universal gas constant divided by the molecular weight, V, the volume, and S_0, an appropriate constant. Since χ transforms similar to a scalar, it follows that the specific entropy of a moving particle remains constant under an $SL(2, R)$ transformation. Hence a *physical* shock is mapped to a *physical* shock under the transformation.

The requirement that the transformation preserves the physicality of a shock puts a condition on the viscosity viz. that its positivity is preserved. As already pointed out, the viscosity fields transform as scalar densities, similar to ρ. It follows that the total viscosity, such as mass, is an invariant under the transformation.

4. Fluid Equations in Non-Inertial Frames

In this section, we review the work of Poludnenko and Khokhlov [6], who considered the formulation of Euler equations of fluid dynamics in non-inertial reference frames. We start with the inertial reference frame Euler Equations (1)–(3)

$$\partial_t \rho + \partial_i (\rho u_i) = 0, \tag{64}$$

$$\partial_t (\rho u_i) + \partial_j (\rho u_i u_j) + \partial_i p = 0, \tag{65}$$

$$\partial_t \epsilon + \partial_i ((\epsilon + p) u_i) = 0, \tag{66}$$

where ρ is the density, u_i the fluid velocity, p the pressure, and ϵ the total energy density. The total energy ϵ is related to the internal energy per unit mass e as,

$$\epsilon = e\rho + \frac{1}{2}\rho u^2. \tag{67}$$

Consider a non-inertial reference frame $\{\tilde{x}_i, \tau\}$ that expands or contracts with respect to the inertial frame $\{t, x_i\}$:

$$\tilde{x}_i = \frac{x_i}{a(t)}, \qquad \tau = \int_0^t \frac{dt}{a(t)^{\beta+1}}, \qquad (68)$$

where β is a constant and the scale factor $a(t)$ is a smooth non-vanishing function of time. (Poludnenko and Khokhlov [6] also considered additional rotational terms in transformation (68).) For simplicity we do not consider such terms; essential ideas are all captured by the simplified transformation). We use quantities with tilde signs to refer to quantities in the non-inertial reference frame $\{\tilde{x}_i, \tau\}$. Time t is the physical time, and τ is the computational time.

For the density, pressure, and energy density fields we introduce the scaling,

$$\tilde{\rho}(\tilde{x}, \tau) = a^\alpha \rho(x, t), \qquad (69)$$

$$\tilde{p}(\tilde{x}, \tau) = a^{\alpha + 2\beta} p(x, t), \qquad (70)$$

$$\tilde{e}(\tilde{x}, \tau) = a^{2\beta} e(x, t), \qquad (71)$$

where α and β are constant scaling exponents. A short calculation shows that

$$u_i = \frac{d}{dt} x_i(t) = a^{-\beta} \frac{d \ln a}{d\tau} \tilde{x}_i + a^{-\beta} \tilde{u}_i, \qquad (72)$$

and

$$\ddot{a} = \frac{1}{a^{2\beta+1}} \left[\frac{d^2 \ln a}{d\tau^2} - \beta \left(\frac{d \ln a}{d\tau} \right)^2 \right]. \qquad (73)$$

As a result, the mass conservation Equation (64) in the non-intertial frame (68) become,

$$\partial_\tau \tilde{\rho} + \tilde{\partial}_i (\tilde{\rho} \tilde{u}_i) = (\alpha - n) \frac{d \ln a}{d\tau} \tilde{\rho}, \qquad (74)$$

where $\tilde{\partial}_i$ are partial derivatives with respect to $\tilde{x}_i$ and n is the dimension of space. The momentum conservation Equation (65) becomes

$$\partial_\tau (\tilde{\rho} \tilde{u}_i) + \tilde{\partial}_j (\tilde{\rho} \tilde{u}_i \tilde{u}_j) + \tilde{\partial}_i \tilde{p} = (\alpha - n + \beta - 1) \frac{d \ln a}{d\tau} \tilde{\rho} \tilde{u}_i - a^{2\beta+1} \ddot{a} \tilde{\rho} \tilde{x}_i. \qquad (75)$$

The transformation of the energy Equation (66) to the non-intertial frame (68) is quite tedious. When the dust settles one finds,

$$\partial_\tau \tilde{e} + \tilde{\partial}_i ((\tilde{e} + \tilde{p}) \tilde{u}_i) = \frac{d \ln a}{d\tau} [(\alpha - n + 2\beta) \tilde{e} - n \tilde{p} - \tilde{\rho} \tilde{u}_i \tilde{u}_i] - a^{2\beta+1} \ddot{a} \tilde{\rho} \tilde{u}_i \tilde{x}_i. \qquad (76)$$

4.1. Conditions for Invariance of the Fluid Equations

There is subclass of transformations (68) that preserve the form of the Euler's equation. Let us look at this subclass in relation to the discussion of the previous sections. For the form invariance of mass conservation Equation (64) we require $\alpha = n$ from Equation (74). For the invariance of momentum and energy conservation (65) and (66), we require from Equations (75) and (76), $\ddot{a} = 0$, together with

$$\alpha = n, \qquad \beta = 1, \qquad (77)$$

and

$$\epsilon = \frac{n}{2} p + \frac{1}{2} \rho u^2. \qquad (78)$$

Recalling that the total energy ϵ is related to the internal energy e via (67), we have

$$e = \frac{np}{2\rho} \tag{79}$$

This is a restriction on the equation of state. For a polytropic gas $p = \chi \rho^{\gamma_0}$ we have the general relation,

$$e = \frac{p}{(\gamma_0 - 1)\rho}. \tag{80}$$

Thus, we conclude that the form invariance of Equations (74)–(76) singles out a special value of the polytropic index,

$$\gamma_0 = 1 + \frac{2}{n}. \tag{81}$$

For $n = 3$, $\gamma_0 = 5/3$. Low mass white-dwarf stars are well approximated by this polytropic index. These results are perfectly consistent with [4] reviewed in the previous sections. $\ddot{a} = 0$ implies,

$$a(t) = ct + d. \tag{82}$$

Thus, transformation (68) becomes

$$x_i \to \frac{x_i}{ct + d}, \qquad t \to -\frac{1}{c(\gamma t + d)}. \tag{83}$$

Comparing this with general $SL(2, R)$ transformations [4]

$$x_i \to \frac{x_i}{ct + d}, \qquad t \to \frac{at + b}{ct + d}, \qquad ad - bc = 1, \tag{84}$$

we have $a = 0, b = -1/c$. The scaling of density, pressure, and energy density (69)–(71) are also compatible with the scalings in [4], and so is the value of the polytropic index.

4.2. Non-Invariant Terms as Sources

Poludnenko and Khokhlov argue that the above formulation based on general scaling of the fluid variables provides a degree of flexibility, provided we treat the non-invariant terms as sources. They consider values of exponents other than in Equation (77) that do not leave the form of the equations invariant. For example, a set of exponents can be obtained by demanding the invariance of the first law of thermodynamics. For isentropic flows, the first law of thermodynamics in inertial frames take the form

$$ds = 0 \implies de = -pd\left(\frac{1}{\rho}\right), \tag{85}$$

which for fluid flows implies,

$$\partial_t e + u_i \partial_i e = \frac{p}{\rho^2}(\partial_t \rho + u_i \partial_i \rho). \tag{86}$$

In non-inertial frames (68), Equation (86) becomes

$$\partial_\tau \widetilde{e} + \widetilde{u}_i \widetilde{\partial}_i \widetilde{e} = \frac{\widetilde{p}}{\widetilde{\rho}^2}(\widetilde{\partial}_\tau \widetilde{\rho} + \widetilde{u}_i \widetilde{\partial}_i \widetilde{\rho}) - (\alpha \widetilde{p} - 2\beta \widetilde{e} \widetilde{\rho}) \frac{1}{\widetilde{\rho}} \frac{d \ln a}{d\tau}. \tag{87}$$

Using the polytropic equation of state (80), it simplifies to

$$\partial_\tau \widetilde{e} + \widetilde{u}_i \widetilde{\partial}_i \widetilde{e} = \frac{\widetilde{p}}{\widetilde{\rho}^2}(\widetilde{\partial}_\tau \widetilde{\rho} + \widetilde{u}_i \widetilde{\partial}_i \widetilde{\rho}) + (2\beta - \alpha(\gamma_0 - 1)) \frac{d \ln a}{d\tau} \widetilde{e}. \tag{88}$$

The choice of exponents

$$\alpha = n, \qquad \beta = \frac{n(\gamma_0 - 1)}{2}, \qquad (89)$$

ensures the invariance of the first law of thermodynamics together with the mass conservation for all values of the polytropic index. The momentum conservation equation no longer takes the conservative form, however. With this choice of exponents (89), the first of the source terms of the momentum conservation Equation (76),

$$(\alpha - n + 2\beta)\widetilde{\epsilon} - n\widetilde{p} - \widetilde{\rho}\,\widetilde{u}_i\,\widetilde{u}_i \qquad (90)$$

simplifies to

$$(\beta - 1)\widetilde{\rho}\,\widetilde{u}_i\,\widetilde{u}_i. \qquad (91)$$

This source term is proportional to the kinetic energy. This new set of exponents may be a preferred choice in numerical simulations if the thermal energy dominates the local kinetic energy in the non-inertial frame.

4.3. Primitive Fields as Simulation Variables

The key drawback in working with exponents (77) or (89) is the fact that they modify the primitive fields (69)–(71). From the general transformed equations (74)–(76), we immediately note that the homogeneous part of the equations is form-invariant. This allows for straightforward numerical implementations of the transformed equations for any value of the scaling exponents α and β treating the right hand side terms in Equations (74)–(76) as sources. The equations no longer take the form of conservation laws, but this is not a problem. In most practical situations this is a necessity. For example, if systems are governed by a non-polytropic equation of state, then we must work with source terms irrespective of the choice of the scaling exponents. We may as well work with the primitive fields as simulation variables, that is, we choose

$$\alpha = 0, \qquad \beta = 0. \qquad (92)$$

The use of primitive fields as simulation variables has the advantage of direct interpretation.

4.4. Numerical Results

In numerical work, source terms are frequently treated. Depending on the problem under consideration, source terms representing gravitational forces, electromagnetic forces, energy release due to radiation, etc are routinely added. Therefore, numerical computation in a moving frame can be performed at virtually no extra technical complication and at no extra computational cost. Poludnenko and Khokhlov mostly focus on tests of moving frame formulation of the fluid flow Equations (74)–(76) with zero exponents (92). They only briefly discuss other choices of scaling parameters. They perform their numerical simulations in a variety of frames for diverse physical problems. The details can be found in their paper. The key points are summarised as follows:

- They consider several types of non-inertial reference frames: accelerating, expanding, contracting, oscillating (sinusoidal) reference frames, etc.
- They treat in detail simulations of blast solutions (e.g., Sedov solution), converging shock solutions (e.g., Guderley blast wave solution), problems involving expansion of a gas sphere in vacuum, etc. They work in different reference frames best suited for the problem at hand.
- They note that the computation in moving frames does not introduce systematic errors. Numerical solutions properly converge to the exact ones when they are known, e.g., the Sedov solution.

- The method accuracy is valid even when solving the fluid equations for non-zero values of the exponents α and β.

There are some problems for which numerical simulations in non-inertial frames are not ideally suited. Such problems typically involve stationary regions of fluids in an inertial frame. The canonical example being the strong explosion in an otherwise stationary environment. (Expanding or collapsing environments where the ambient conditions are vacuous or dynamically unimportant can be optimally treated in non-inertial frames. In such problems, the ambient fluid can be set to be stationary in the computational–non-inertial–frame.) However, the demonstration in [6] that the numerical solution converges to the correct analytic Sedov solution is a crucial test of the method. The success of the simulation clearly shows that the non-conservative nature of the method does not introduce systematic errors and the Rankine–Hugoniot conditions are valid in the transformed reference frame too. The Rankine–Hugoniot conditions were shown to be valid in a subclass of transformed reference frames with scaling exponents Equation (77) in [5], as reviewed above. Further generalisation for different scaling exponents has not yet been carried out; however, given the numerical results it is likely that a useful formulation exists in more general situations.

5. Conclusions

In this article, we reviewed that the maximal kinematical invariance group of the Euler equations of fluid dynamics is larger than the Galilei group. Specifically, the inversion transformation ($\Sigma : t \to -1/t$, $\vec{x} \to \vec{x}/t$) leaves the Euler equations invariant. This duality has been used to explain the striking similarities observed in simulations of the supernova explosions and laboratory implosions induced in plasma by intense lasers. It is quite remarkable that the inversion symmetry extends to discontinuous fluid flows as well. We also reviewed how this comes about.

We summarised the work of Poludnenko and Khokhlov [6]. They presented methods for computation of fluid flows characterised by large degree of expansion or contraction. The key idea is the transformation to a non-inertial reference frame. The scaling transformation of the primitive fluid variables provides a degree of flexibility. The use of non-inertial frames often leads to non-conservative formulation of the fluid equations, however, this does not affect the accuracy of the numerical work. For many problems of astrophysical and geophysical interests, going to an appropriate non-inertial frame allows for a cleaner extraction of relevant physics. We focused on [6] because of its close connection to the invariance properties of fluid equations.

There are several other papers addressing these issues, see, for example [15,16] and references therein. Similar ideas are frequently used in simulations of galaxies and the large-scale structure in an expanding universe and in atmospheric simulations.

Author Contributions: All authors have contributed equally. All authors have read and agreed to the published version of the manuscript.

Funding: This research received no external funding.

Data Availability Statement: Not applicable.

Conflicts of Interest: The authors declare no conflict of interest.

Appendix A. Implosion Dual of the Sedov Explosion Solution

In this appendix we write down the explicit implosion solution dual of the Sedov explosion solution. To the best of our knowledge this has not been carried out before. (Drury and Mendonça in [3] have made some elementary remarks. They comment that Dwarkadas and Drury would publish details on the implosion solution dual to the Sedov solution in a separate paper. However, we are unable to locate the relevant references. Perhaps the results were not communicated to a journal. We will be glad if someone can point out relevant references to us.) We start with a brief review of the Sedov solution. We then apply the duality transformation to write the dual implosion solution. Some properties of the dual solution are presented. For the Sedov solution our presentation closely follows ([7], Section 99).

Appendix A.1. Sedov Solution

Sedov solution refers to an exact solution of compressible fluid dynamics equations in which spherical shock of great intensity propagates radially outwards as a result of a strong explosion. Strong explosion is characterised by the instantaneous release of energy E at the center. The equation of fluid dynamics in spherical symmetric situations take the form

$$\frac{\partial \rho}{\partial t} + \frac{\partial (\rho u)}{\partial r} + \frac{2\rho u}{r} = 0, \quad \frac{\partial u}{\partial t} + u \frac{\partial u}{\partial r} + \frac{1}{\rho}\frac{\partial p}{\partial r} = 0, \quad \frac{\partial s}{\partial t} + u \frac{\partial s}{\partial r} = 0, \quad (A1)$$

where $s = \ln(p\rho^{-\gamma})$. The last equation is the conservation of entropy. For the Sedov solution, the pressure discontinuity is very large: the pressure behind the shock is p_1 is much larger than the pressure in front of the shock p_0. Sedov solution neglects p_0 everywhere (more on this later). The flow pattern is completely determined by the energy released E and the ambient density ρ_0. The ratio of densities just behind and front of the shock is obtained by the Rankine–Hugoniot condition,

$$\rho_1 = \frac{\gamma+1}{\gamma-1}\rho_0, \quad (A2)$$

assuming $p_1 \gg p_0$. The shock front is defined by

$$R(t) = \xi_0 \left(\frac{E}{\rho_0}\right)^{1/5} t^{2/5}. \quad (A3)$$

The propagation velocity of the shock is

$$D = \frac{dR}{dt} = \frac{2}{5}\frac{R}{t}. \quad (A4)$$

Now, using the other two Rankine–Hugoniot conditions, which determine the gas velocity u_1 and pressure p_1 immediately behind the shock front, we obtain

$$p_1 = \frac{2}{\gamma+1}\rho_0 D^2, \quad u_1 = \frac{2}{\gamma+1} D. \quad (A5)$$

Note that as the shock expands p_1 and u_1 also change as a function of time. We define

$$\xi = \frac{r}{R(t)} \quad (A6)$$

and define

$$p(r,t) = \frac{8\rho_0}{25(\gamma+1)} \cdot \frac{r^2}{t^2} \cdot \check{p}(\xi), \quad (A7)$$

$$u(r,t) = \frac{4}{5(\gamma+1)} \cdot \frac{r}{t} \cdot \check{u}(\xi), \quad (A8)$$

$$\rho(r,t) = \rho_0 \frac{\gamma+1}{\gamma-1} \cdot \check{\rho}(\xi). \quad (A9)$$

Variables $\check{p}, \check{u}, \check{\rho}$ are dimensionless pressure, velocity, and density. These are functions of the dimensionless variable ξ. The Rankine–Hugoniot jumps conditions in terms of these functions become

$$\check{p} = \check{u} = \check{\rho} = 1 \quad \text{at} \quad \xi = \xi_0. \quad (A10)$$

These are the new boundary conditions. We warn the reader that there is a huge variation in the literature on the use of dimensionless variables for pressure, density, and velocity.

Using self-similarity of the solution one can argue that energy contained in a sphere of constant ζ remains constant in time. This gives an integral of motion. The argument proceeds as follows. For more details see [7]. Consider a spherical volume size r at constant ζ. It expands at the rate $\frac{2r}{5t}$. The energy exiting the sphere in time dt due to the motion of the fluid is

$$4\pi r^2 \cdot \rho v s. \left(h + \frac{1}{2}v^2\right) \cdot dt. \tag{A11}$$

This energy must be equal to the increase in the internal energy of the sphere in time dt due to its expansion

$$4\pi r^2 \cdot \rho \left(\epsilon + \frac{1}{2}v^2\right) \cdot \frac{2r}{5t} \cdot dt. \tag{A12}$$

Equating the two expressions give the first integral,

$$\frac{\check{p}(\zeta)}{\check{\rho}(\zeta)} = \frac{\gamma + 1 - 2\check{u}(\zeta)}{2\gamma \check{u}(\zeta) - \gamma - 1} \check{u}^2(\zeta). \tag{A13}$$

We obtain the remaining equations from the mass and entropy conservation equations. These equations simplify in the form

$$\frac{d\check{u}}{d\ln \zeta} + \left(\check{u} - \frac{\gamma + 1}{2}\right) \frac{d\ln \check{\rho}}{d\ln \zeta} = -3\check{u}, \tag{A14}$$

and

$$\frac{d}{d\ln \zeta}\left(\ln \frac{\check{p}}{\check{\rho}^\gamma}\right) = \frac{5(\gamma + 1) - 4\check{u}}{2\check{u} - (\gamma + 1)}. \tag{A15}$$

These equations can be integrated to give implicitly the functions $\check{p}(\zeta), \check{u}(\zeta), \check{\rho}(\zeta)$. They take the form

$$\left(\frac{\zeta_0}{\zeta}\right)^5 = \check{u}^2 \left(\frac{5(\gamma+1) - 2(3\gamma-1)\check{u}}{7-\gamma}\right)^{\nu_1} \left(\frac{2\gamma\check{u} - \gamma - 1}{\gamma - 1}\right)^{\nu_2}, \tag{A16}$$

$$\check{\rho} = \left(\frac{2\gamma\check{u} - \gamma - 1}{\gamma - 1}\right)^{\nu_3} \left(\frac{5(\gamma+1) - 2(3\gamma-1)\check{u}}{7-\gamma}\right)^{\nu_4} \left(\frac{\gamma + 1 - 2\check{u}}{\gamma - 1}\right)^{\nu_5}, \tag{A17}$$

where

$$\nu_1 = \frac{13\gamma^2 - 7\gamma + 12}{(3\gamma - 1)(2\gamma + 1)}, \tag{A18}$$

$$\nu_2 = -\frac{5(\gamma - 1)}{2\gamma + 1}, \tag{A19}$$

$$\nu_3 = \frac{3}{2\gamma + 1}, \tag{A20}$$

$$\nu_4 = \frac{13\gamma^2 - 7\gamma + 12}{(2 - \gamma)(3\gamma - 1)(2\gamma + 1)}, \tag{A21}$$

$$\nu_5 = \frac{2}{\gamma - 2}. \tag{A22}$$

Here we have corrected a few typos from [7] (the ν_5 there has a typo). The parameter ζ_0 is determined by the requirement that the total energy of the gas up to radius $R(t)$ is E. Other details of the solution can be found in [7,13]. Although reference [13] does not discuss the explicit solution, the discussion on the physical properties of the solution is very thorough and lucid.

Appendix A.2. Duality in Spherical Coordinates

In order to work out the implosion dual it is instructive to first work out the invariance of the simplified fluid Equation (A1) in spherical coordinates. To this end, we define the new time (now we use the notation $\tilde{t}$ for the new time as opposed to τ) and the new radial variable $\tilde{r}$

$$\tilde{r} = \frac{r}{a(t)}, \qquad \tilde{t} = \int_0^t \frac{dt}{a(t)^{\beta+1}}, \qquad \text{(A23)}$$

and rescale pressure and density as,

$$\tilde{\rho}(\tilde{r},\tilde{t}) = a^\alpha \rho(r,t), \qquad \text{(A24)}$$
$$\tilde{p}(\tilde{r},\tilde{t}) = a^\alpha + 2\beta p(r,t). \qquad \text{(A25)}$$

These transformations give

$$u = a^{-\beta}\left(\tilde{u} + \tilde{r}\frac{d\ln a}{d\tilde{t}}\right). \qquad \text{(A26)}$$

As for the derivatives we need to use

$$\partial_t X = \frac{\partial X}{\partial \tilde{t}} \cdot \frac{\partial \tilde{t}}{\partial t} + \frac{\partial X}{\partial \tilde{r}} \cdot \frac{\partial \tilde{r}}{\partial t} = a^{-\beta-1}\frac{\partial X}{\partial \tilde{t}} - \tilde{r}a^{-\beta-1}\frac{d\ln a}{d\tilde{t}}\frac{\partial X}{\partial \tilde{r}}, \qquad \text{(A27)}$$

and

$$\partial_r X = \frac{\partial X}{\partial \tilde{r}} \cdot \frac{\partial \tilde{r}}{\partial r} = a^{-1}\frac{\partial X}{\partial \tilde{r}}. \qquad \text{(A28)}$$

Now it is not difficult to verify that:

1. The continuity equation is form invariant for $\alpha = 3$.
2. The momentum equation is form invariant provided $\ddot{a} = 0$ and $\beta = 1$.
3. The entropy equation is form invariant for $\gamma = 5/3$.

Appendix A.3. Implosion Dual

We choose $a(t) = t$. Then,

$$\tilde{r} = \frac{r}{t}, \qquad \tilde{t} = -\frac{1}{t}. \qquad \text{(A29)}$$

Since the radial variable does not undergo an inversion, the interior Sedov solution is mapped to an interior solution, and the exterior solution is mapped to an exterior solution. Due to this, the dual solution does not satisfy any physically interesting boundary conditions, i.e., it cannot be compared with standard implosion solutions of the sort discussed in say, chapter XII of [13]. For a simple physically interesting laboratory realisable implosion solution, one would require the interior to be stationary at fixed density and negligible pressure. This is certainly not the case for the dual solution. On the contrary the exterior solution is at zero pressure (hence there the speed of sound is zero) and the fluid is moving.

We ask in what sense is the solution an implosion solution. Does it satisfy expected properties, specifically the Rankine–Hugoniot jumps conditions? We take $0 < \tilde{r} < \infty$ and $-\infty < \tilde{t} < 0$. For the initially stationary exterior region the velocity transformation gives

$$\tilde{u}(\tilde{r},\tilde{t}) = \frac{\tilde{r}}{\tilde{t}}. \qquad \text{(A30)}$$

Since $\widetilde{t}$ is negative, the velocity of the exterior fluid is directed inwards; an implosion. The location of the shock is

$$\widetilde{R}(\widetilde{t}) = \frac{R(t)}{t} = -\widetilde{t}\, R(t) = \xi_0 \left(\frac{E}{\rho_0}\right)^{1/5} (-\widetilde{t})^{3/5}. \tag{A31}$$

As $\widetilde{t}$ increases from negative value towards zero, $\widetilde{R}(\widetilde{t})$ decreases, i.e., it represents an implosion. The velocity of the shock surface is

$$\frac{dR(\widetilde{t})}{d\widetilde{t}} = \frac{3}{5} \frac{\widetilde{R}(\widetilde{t})}{\widetilde{t}}. \tag{A32}$$

Since the inward velocity of the shock surface is smaller than the inward velocity of ambient fluid just outside, the fluid is injected into the interior region through the shock surface. Thus, in this set-up the exterior fluid of low (zero) pressure is getting compressed at the shock surface into the interior region.

Let us now calculate the velocity of the fluid just behind the shock surface. Recall $\xi = \frac{r}{R(t)}$. It follows that,

$$\xi = \frac{\widetilde{r}}{\widetilde{R}(\widetilde{t})}, \tag{A33}$$

so the interpretation of ξ as a dimensionless variable remains the same. We have

$$\widetilde{u}(\widetilde{r},\widetilde{t}) = -\frac{4}{5(\gamma+1)} \cdot \frac{\widetilde{r}}{\widetilde{t}} \cdot \check{u}(\xi) + \frac{\widetilde{r}}{\widetilde{t}}, \tag{A34}$$

for the interior region. Similarly, other variable can be constructed. At the shock surface,

$$\widetilde{u} = -\frac{4}{5(\gamma+1)} \frac{\widetilde{R}}{\widetilde{t}} + \frac{\widetilde{R}}{\widetilde{t}} = \frac{1+5\gamma}{5(1+\gamma)} \frac{\widetilde{R}}{\widetilde{t}}. \tag{A35}$$

the frame of the shock, we can confirm that these velocities satisfy the Rankine–Hugoniot conditions. The other Rankine–Hugoniot conditions can also be checked similarly. The post-shock pressure increases as $(-\widetilde{t})^{-19/5}$. These results are all consistent with the comments in Drury and Mendonça in [3].

References

1. Remington, B. Supernova Hydrodynamics Upclose. Science and Technology Review. Lawrence Livermore Library. 2000. Available online: https://str.llnl.gov/content/pages/past-issues-pdfs/2000.01.pdf (accessed on 31 March 2022).
2. Available online: https://www.mpa-garching.mpg.de/84411/Core-collapse-supernovae (accessed on 31 March 2022).
3. Drury, L.O.; Mendonça, J.T. Explosion implosion duality and the laboratory simulation of astrophysical systems. *Phys. Plasmas* **2000**, *7*, 5148. [CrossRef]
4. O'Raifeartaigh, L.; Sreedhar, V.V. The Maximal kinematical invariance group of fluid dynamics and explosion–implosion duality. *Ann. Phys.* **2001**, *293*, 215–227. [CrossRef]
5. Jahn, O.; Sreedhar, V.V.; Virmani, A. Symmetries of discontinuous flows and the dual Rankine–Hugoniot conditions in fluid dynamics. *Ann. Phys.* **2005**, *316*, 30–43. [CrossRef]
6. Poludnenko, A.Y.; Khokhlov, A.M. Computation of fluid flows in non-inertial contracting, expanding, and rotating reference frames. *J. Comput. Phys.* **2007**, *220*, 678–711. [CrossRef]
7. Landau, L.; Lifshitz, E. Fluid Mechanics. In *Course of Theoretical Physics*; Pergamon Press: Oxford, UK, 2013; Volume 6.
8. Thorne, K.S.; Blandford, R.D. *Modern Classical Physics*; Princeton University Press: Princeton, NJ, USA, 2017.
9. Lamb, H. *Hydrodynamics*; Cambridge University Press: Cambridge, UK, 1942.
10. Deser, S.; Jackiw, R.; Polychronakos, A.P. Clebsch (string) decomposition in d=3 field theory. *Phys. Lett. A* **2001**, *279*, 151–153. [CrossRef]
11. Jackiw, R.; Nair, V.P.; Pi, S.Y. Chern-Simons reduction and nonAbelian fluid mechanics. *Phys. Rev. D* **2000**, *62*, 085018. [CrossRef]
12. Richtmyer, R.D. *Principles of Advanced Mathematical Physics*; Springer: New York, NY, USA, 1978; Volume 1.
13. Zel'dovich, Y.B.; Raizer, Y.P. *Physics of Shock Waves and High-Temperature Hydrodynamic Phenomena*; Hayes, W.D., Probstein, R.F., Eds.; Dover Publication: Mineola, NY, USA, 2002.
14. Courant, R.; Friedrichs, K.O. *Supersonic Flow and Shock Waves*; Interscience Publishers: New York, NY, USA, 1948.

15. Trac, H.; Pen, U.L. A moving frame algorithm for high Mach number hydrodynamics. *New Astron.* **2004**, *9*, 443–465. [CrossRef]
16. Gledhill, I.M.A.; Roohani, H.; Forsberg, K.; Eliasson, P.; Skews, B.W.; Nordström, J. Theoretical treatment of fluid flow for accelerating bodies. *Theor. Comput. Fluid Dyn.* **2016**, *30*, 449–467. [CrossRef]

Communication

Spin Distribution for the 't Hooft–Polyakov Monopole in the Geometric Theory of Defects

Mikhail O. Katanaev [1,2]

[1] Steklov Mathematical Institute, ul. Gubkina, 119991 Moscow, Russia; katanaev@mi-ras.ru
[2] N. I. Lobachevsky Institute of Mathematics and Mechanics, Kazan Federal University, ul. Ktremlevskaya 35, 420008 Kazan, Russia

Abstract: Recently the 't Hooft–Polyakov monopole solutions in Yang–Mills theory were given new physical interpretation in the geometric theory of defects describing the continuous distribution of dislocations and disclinations in elastic media. It means that the 't Hooft–Polyakov monopole can be seen, probably, in solids. To this end we need to compute the corresponding spin distribution on lattice sites of crystals. The paper describes one of the possible spin distributions. The Bogomol'nyi–Prasad–Sommerfield solution is considered as an example.

Keywords: 't Hooft–Polyakov monopole; geometric theory of defects; disclination

Citation: Katanaev, M.O. Spin Distribution for the 't Hooft–Polyakov Monopole in the Geometric Theory of Defects. *Universe* **2021**, *7*, 256. https://doi.org/10.3390/universe7080256

Academic Editors: Arundhati Dasgupta and Alfredo Iorio

Received: 8 June 2021
Accepted: 20 July 2021
Published: 21 July 2021

Publisher's Note: MDPI stays neutral with regard to jurisdictional claims in published maps and institutional affiliations.

Copyright: © 2021 by the authors. Licensee MDPI, Basel, Switzerland. This article is an open access article distributed under the terms and conditions of the Creative Commons Attribution (CC BY) license (https://creativecommons.org/licenses/by/4.0/).

1. Introduction

Many real solids possess a spin structure. For example, ferromagnets are characterized by the distribution of magnetic moments described by the unit vector field (n-field) in the continuum approximation. This unit vector field may have defects (singularities) that are called disclinations. Disclinations define many important properties of media and attract much interest in physics (see, e.g., [1,2]). Real solids may have many disclinations, and it is highly likely to have the continuous description for their distribution. The approach based on the n-field is applicable for single disclinations when we can write down equations for the n-field with suitable boundary conditions on the corresponding cuts in media. However this approach is not applicable for the continuous distribution of disclinations because, in this case, the n-field has singularities at each point and therefore does not exist at all.

The promising approach to this problem is the geometric theory of defects [3,4] which describes defects in elastic media within the Riemann–Cartan geometry. To avoid the problem with singularities of the n-filed we introduce new variable–the $\mathbb{SO}(3)$ connection– which is smooth even for continuous distribution of disclinations. The curvature tensor for this connection acquires the physical interpretation as the surface density of the Frank vector characterizing disclinations. The Frank vector for a single straight linear disclination equals, by definition, to the total rotational angle of the n-field when it goes around the disclination axis which is a multiple of 2π. If defects are absent in some domain of media, then the $\mathbb{SO}(3)$ connection is a pure gauge, and the n field can be reconstructed in this domain in full agreement with standard models.

We note that the geometric theory of defects is a more general model also describing the distribution of dislocations that are defects in elastic media itself. These defects correspond to nontrivial torsion, the latter having physical interpretation as the surface density of the Burgers vector of dislocations. The Burgers vector for a single straight linear dislocation is equal, by definition, to the jump of the displacement vector when it goes around the dislocation axis. In the present paper, we deal only with disclinations assuming that the metric is Euclidean which corresponds to the absence of elastic stresses but the $\mathbb{SO}(3)$ connection is nontrivial.

There is some interest in describing effects of dislocations on physical properties of solids within the geometric theory of defects (see, e.g., [5–14]). In the present paper, we deal only with another defects–disclinations. As far as we know, the first application of the geometric theory of defects to disclinations was given in [15–17] where straight linear disclinations were described in full agreement with the classical theory [1]. Several examples of point disclinations were given in [18]. A short review of disclinations in the geometric theory of defects can be found in [19].

Since the Lie algebras $\mathbb{SO}(3)$ and $\mathbb{SU}(2)$ are isomorphic, then the static solutions of $\mathbb{SU}(2)$ gauge models can be considered as describing some distribution of disclinations and, possibly, dislocations. In particular, it was noted that the 't Hooft–Polyakov monopole has straightforward physical interpretation in the geometric theory of defects describing media with continuous distribution of disclinations and dislocations [20]. It means that the 't Hooft–Polyakov monopole can, in principle, be observed in solids. There arises a problem. On the one hand, the solution describes the continuous distribution of disclinations, and the n-field does not exist in a strict sense. On the other hand, there is a spin in each site of the crystal lattice. The problem relates to what is the corresponding spin distribution for a given nontrivial $\mathbb{SO}(3)$ connection? The answer to this question relies on the definition of the presumable paths of parallel transport of spins and requires additional physical assumptions which are not known at present. In the present paper, we consider one of the possible and simplest way to compute spin distribution for the 't Hooft–Polyakov monopoles. As an example, we consider the Bogomolny–Prasad–Sommerfield solution.

2. The 't Hooft–Polyakov Monopole

The famous 't Hooft–Polyakov monopoles [21,22] are the exact solutions of the field equations in the $\mathbb{SU}(2)$ gauge theory interacting with the triplet of scalar fields in the adjoint representation and $\lambda \varphi^4$-type interaction (for review, see, e.g., [23–25]). The solutions are static and spherically symmetric. Therefore, the problem is reduced to minimization of the three-dimensional Euclidean energy functional which can be regarded as the free energy expression in solid state physics. We consider the $\mathbb{SU}(2)$ connection components as the $\mathbb{SO}(3)$ connection because their Lie algebras coincide, the triplet of scalar fields being the source of defects. Moreover, we assume that the $\mathbb{SO}(3)$ group acts not only in the isotopic space but also in the tangent space to the space manifold $\mathbb{R}^3$, the metric of the space being Euclidean. So the 't Hooft–Polyakov monopoles correspond to Euclidean metric (and triad) and nontrivial $\mathbb{SO}(3)$ connection which give rise to nontrivial Riemann–Cartan geometry of space.

Let us consider three-dimensional Euclidean space $\mathbb{R}^3$ with Cartesian coordinates x^μ and Euclidean metric $\delta_{\mu\nu}$, $\mu, \nu = 1, 2, 3$. The spherically symmetric $\mathbb{SU}(2)$ gauge fields $A_\mu{}^i(x)$, $i = 1, 2, 3$ interacting with the triplet of scalar fields $\varphi^i(x)$ in the adjoint representation minimize the three-dimensional energy [23–25]:

$$\varepsilon := \int d^3x \left(\frac{1}{4} F^{\mu\nu i} F_{\mu\nu i} + \frac{1}{2} \nabla^\mu \varphi^i \nabla_\mu \varphi_i + \frac{1}{4} \lambda \left(\varphi^2 - a^2 \right)^2 \right), \tag{1}$$

where indices are raised and lowered by Euclidean metrics $\delta_{\mu\nu}$ and δ_{ij},

$$\begin{aligned} F_{\mu\nu}{}^i &:= \partial_\mu A_\nu{}^i - \partial_\nu A_\mu{}^i + A_\mu{}^j A_\nu{}^k \varepsilon_{jk}{}^i, \\ \nabla_\mu \varphi^i &:= \partial_\mu \varphi^i + A_\mu{}^j \varphi^k \varepsilon_{jk}{}^i. \end{aligned} \tag{2}$$

— are the curvature tensor components for $\mathbb{SU}(2)$-connection and the covariant derivative of scalar fields; $\lambda > 0, a > 0$ – are coupling constants, ε_{ijk} is the totally antisymmetric tensor, $\varepsilon_{123} := 1$, and $\varphi^2 := \varphi^i \varphi_i$.

The spherically symmetric ansatz is:

$$A_\mu{}^i = \frac{\varepsilon_\mu{}^{ij} x_j (K-1)}{r^2}, \qquad \varphi^i = \frac{x^i H}{r^2}, \tag{3}$$

where $K(r)$ and $H(r)$ are some dimensionless functions on radius $r := \sqrt{x^2}$, $x^2 := x^\mu x_\mu$.

We assume that the $\mathbb{SO}(3)$ group acts simultaneously in the isotropic space and base manifold. Therefore the difference between Greek and Latin indices disappear, but we shall distinguish them as long as possible.

The Euler–Lagrange equations for functional (1) in the spherically symmetric case reduce to:
$$r^2 K'' = K(K^2 + H^2 - 1),$$
$$r^2 H'' = 2HK^2 + \lambda \left(H^2 - a^2 r^2\right) H. \tag{4}$$

At present we know only a few exact analytic solutions to this system of equations for $\lambda = 0$ [26–28] (see also [29–32]). In the following, we consider the Bogomol'nyi–Prasad–Sommerfield solution [26,27]:
$$K = \frac{lr}{\mathrm{sh}(lr)}, \qquad H = \frac{lr}{\tanh(lr)} - 1, \qquad l > 0. \tag{5}$$

It is easily checked that this solution has finite energy (1).

Solutions of Equation (4) describe 't Hooft–Polyakov monopoles in the $\mathbb{SU}(2)$ gauge model. In the geometric theory of defects, we consider functional (1) as the expression for the free energy describing static distribution of disclinations and dislocations in elastic media with defects, the triplet of scalar fields being the source of defects. This is possible because the Lie algebra $\mathbb{SU}(2)$ is isomorphic to $\mathbb{SO}(3)$, and scalar fields are real (in the fundamental representation of $\mathbb{SO}(3)$ group).

The Euclidean metric means that elastic stresses are absent in media. The Cartan variables (triad and $\mathbb{SO}(3)$ connection) for monopole solutions are:
$$e_\mu{}^i = \delta_\mu^i, \qquad \omega_\mu{}^{ij} = A_\mu{}^k \varepsilon_k{}^{ij} = (\delta_\mu^j x^i - \delta_\mu^i x^j) \frac{K-1}{r^2}, \tag{6}$$
where we use the spherically symmetric $\mathbb{SO}(3)$-connection (3). In the considered case, simple calculations yield the following expressions for curvature and torsion:
$$R_{\mu\nu}{}^k := \frac{1}{2} R_{\mu\nu}{}^{ij} \varepsilon_{ij}{}^k = F_{\mu\nu}{}^k = \varepsilon_{\mu\nu}{}^k \frac{K'}{r} - \frac{\varepsilon_{\mu\nu}{}^j x_j x^k}{r^3}\left(K' - \frac{K^2-1}{r}\right), \tag{7}$$
$$T_{\mu\nu}{}^k = \left(\delta_\mu^k x_\nu - \delta_\nu^k x_\mu\right) \frac{K-1}{r^2}. \tag{8}$$

In the geometric theory of defects, curvature (7) and torsion (8) have physical meaning of surface densities of Frank and Burgers vectors, respectively. That is they are equal to k-th components of respective vectors on surface element $dx^\mu \wedge dx^\nu$. If s^μ is normal to the surface element, then there are the following densities of Frank and Burgers vectors:
$$f_\mu{}^i := \frac{1}{2} \varepsilon_\mu{}^{\nu\rho} R_{\nu\rho}{}^i = \frac{1}{3r} \delta_\mu^i \left(2K' + \frac{K^2-1}{r}\right) - \frac{1}{r}\left(\hat{x}_\mu \hat{x}^i - \frac{1}{3}\delta_\mu^i\right)\left(K' - \frac{K^2-1}{r}\right), \tag{9}$$
$$b_\mu{}^i := \frac{1}{2} \varepsilon_\mu{}^{\nu\rho} T_{\nu\rho}{}^i = \varepsilon_\mu{}^{ij} \hat{x}_j \frac{K-1}{r}, \tag{10}$$

where $\hat{x}^\mu := x^\mu / r$ and tensor $f_\mu{}^i$ is decomposed into irreducible components.

We see that 't Hooft–Polyakov monopoles describe a continuous distribution of disclinations and dislocations in media. The problem lies in what is seen in real crystals. We do not observe directly the distribution of the Frank and Burgers vectors, but we observe instead the distribution of spins on lattice sites of a crystal. So, we have to compute it for a given $\mathbb{SO}(3)$ connection. In what follows, we concentrate our attention only on the distribution of disclinations as the first step in the analysis.

3. Distributions of Spins

To compute the distributions of spins, we follow the idealogy adopted in lattice gauge models [33] (for review, see, e.g., [34]). For simplicity, we consider the cubic lattice. There is a spin at each lattice site. We ascribe the rotational $SO(3)$ matrix to each link of the lattice. This is possible because we know the connection, and the rotational matrix is given by the path-ordered integral of the connection along the link which corresponds to the parallel transport of spins. Thus we obtain the unique rotational matrix at each link. The problem is that curvature of the connection is nontrivial, and the parallel transport of spins depends on the path. In particular, we must avoid closed paths because a spin cannot take different directions at a given site. Thus we fix the spin at one site, say, at the origin of the coordinate system, and parallelly transport it to the whole lattice. The paths have to reach all sites but without closed loops. It is clear that there are many possibilities. To choose the unique one, we must engage additional physical assumptions. At the moment, we do not know how to do this uniquely and therefore consider one of the simplest possibilities, because it is interesting to see how the 't Hooft–Polyakov monopole may look in a crystal.

The following calculations are performed in the continuous approximation.

Let the spin (unit vector) $n_0 := n(0) = (0,0,1)$ at the origin be directed along the $x^3 := z$ axis. We parallelly transport it to a point x_1 along a path $x(t)$, $t \in [0,1]$, $x(0) := 0$, $x(1) = x_1$. A path is assumed to have no intersection points. In our notation, we use the covector $n = (n_1, n_2, n_3)$, $n_i := n^j \delta_{ji}$, instead of vector components (sure, they coincide in the Cartesian coordinates). Then the result of parallel transport is given by the path-ordered exponent (see, e.g., [35], Section 14.5).

$$n_i(x_1) = S_i{}^j n_{0j} = P \exp\left(\int_0^1 dt\, \dot{x}^\mu \omega_\mu\right)_i^j n_{0j}, \qquad (11)$$

where $S_i{}^j(x_1, x_0)$ is the rotational matrix and the dot denotes differentiation with respect to the path parameter t.

First, we consider the parallel transport from the origin to a point x_1^μ along the ray $x(t) := (x_1^1 t, x_1^2 t, x_1^3 t)$. Then $\dot{x}^\mu = (x_1^1, x_1^2, x_1^3)$ and the integrand for connection (6) is zero:

$$\dot{x}_1^\mu \omega_\mu{}^{ij} = x_1^\mu (\delta_\mu^j x^i - \delta_\mu^i x^j) \frac{K-1}{er^2} = (x_1^j x^i - x_1^i x^j) \frac{K-1}{er^2} = 0,$$

Because $x(t)$ is proportional to x_1 along a ray, and the parallel transport does not change the vector.

We see that parallel transport along rays from the origin in $\mathbb{R}^3$ produces no effect on spins for the 't Hooft–Polyakov connection independently of function $K(r)$.

Let us consider another possibility for the parallel transport. First, we transport the vector n_0 from the origin along the x^3 axis. Previous calculations tell us that it is not changed. Then we parallelly transport it along the rays in the perpendicular planes x^1, x^2. Thus the path is $x(t) := (y_1 t, y_2 t, z = \text{const})$ and $\dot{x}^\mu = (y_1, y_2, 0)$, where $x_1 = (y_1, y_2, z)$ is the final point (now we write the coordinate indices of y at the bottom to distinguish them from exponents). The integrand in Equation (11) for connection (6) becomes:

$$\dot{x}^\mu \omega_\mu{}^{ij}(t) = B^k(t) \varepsilon_k{}^{ij}, \qquad (12)$$

where

$$B^1 = -y_2 z t \frac{K-1}{er^2}, \qquad B^2 = y_1 z t \frac{K-1}{er^2}, \qquad B^3 = 0.$$

Now one can easily check that the integrands commute:

$$[\dot{x}^\mu \omega_\mu(t_1), \dot{x}^\nu \omega_\nu(t_2)]_i{}^j = -B^k(t_1) B^l(t_2) \varepsilon_{kl}{}^m \varepsilon_{mi}{}^j = 0, \qquad \forall t_1, t_2.$$

Therefore the ordered exponent (11) coincides with the usual one and can be easily calculated:

$$C^k \varepsilon_k{}^{ij} := \int_0^1 dt\, \dot{x}^\mu \omega_\mu{}^{ij} = \varepsilon_k{}^{ij} \int_0^1 dt\, B^k, \qquad (13)$$

where

$$C^1 = -\frac{y_2 z}{e\rho^2} I, \quad C^2 = -\frac{y_1 z}{e\rho^2} I, \quad C^3 = 0, \quad \rho^2 := y_1^2 + y_2^2,$$

and

$$I := \rho^2 \int_0^1 dt\, t\frac{K-1}{r^2} = \int_z^{\sqrt{\rho^2+z^2}} dr \left(\frac{K}{r} - \frac{1}{r}\right),$$

$$r^2 = \rho^2 t^2 + z^2 \quad \Rightarrow \quad rdr = \rho^2 t dt.$$

For the Bogomol'nyi–Prasad–Sommerfield solution (5), the integral can be easily taken:

$$I = \ln\left[\frac{z}{\sqrt{\rho^2+z^2}} \frac{\tanh(l\sqrt{\rho^2+z^2})}{\tanh(lz)}\right], \quad l \to 2l, \tag{14}$$

where we rescaled the constant $l \to 2l$ for simplicity. The rotational matrix for the Bogomol'nyi–Prasad–Sommerfield solution is:

$$S_i^j = \exp(C^k \varepsilon_k)_i^j = \delta_i^j \cos C + \frac{C^k \varepsilon_{ki}{}^j}{C} \sin C + \frac{C_i C^j}{C^2}(1 - \cos C), \tag{15}$$

where

$$C^2 := C^k C_k = \frac{z^2}{e^2 \rho^2} I^2.$$

For the Bogomol'nyi–Prasad–Sommerfield solution, the rotational matrix is:

$$S = \begin{pmatrix} \frac{y_2^2}{\rho^2} + \frac{y_1^2}{\rho^2}\cos C & -\frac{y_1 y_2}{\rho^2}(1 - \cos C) & -\frac{y_1}{\rho}\sin C \\ -\frac{y_1 y_2}{\rho^2}(1 - \cos C) & \frac{y_1^2}{\rho^2} + \frac{y_2^2}{\rho^2}\cos C & -\frac{y_2}{\rho}\sin C \\ \frac{y_1}{\rho}\sin C & \frac{y_2}{\rho}\sin C & \cos C \end{pmatrix}. \tag{16}$$

One can easily check that $S = 1$ for $\rho = 0$ or $z = 0$. In addition, the asymptotics at infinities are the same:

$$S \to 1 \quad \text{for} \quad \rho \to \infty, \; z = \text{const} \neq 0 \quad \text{or} \quad |z| \to \infty, \; \rho > 0.$$

Now we can easily calculate spins components:

$$n_1 = \frac{y_1}{\rho}\sin C, \quad n_2 = \frac{y_2}{\rho}\sin C, \quad n_3 = \cos C, \tag{17}$$

where

$$C = \frac{z}{e\rho} I.$$

We see, that the spin distribution is invariant with respect to rotations around the z axis, as expected. Therefore, without loss of generality, we may put $y_2 = 0$ to visualize the distribution. Then:

$$n_1 = \sin C, \quad n_2 = 0, \quad n_3 = \cos C,$$

where

$$C = \frac{z}{ey_1} \ln\left[\frac{z}{\sqrt{y_1^2+z^2}} \frac{\tanh(l\sqrt{y_1^2+z^2})}{\tanh(lz)}\right].$$

It implies that spins are rotated on the same plane $y_2 = 0$. The corresponding spin distributions are shown in Figure 1 for a different range of coordinates.

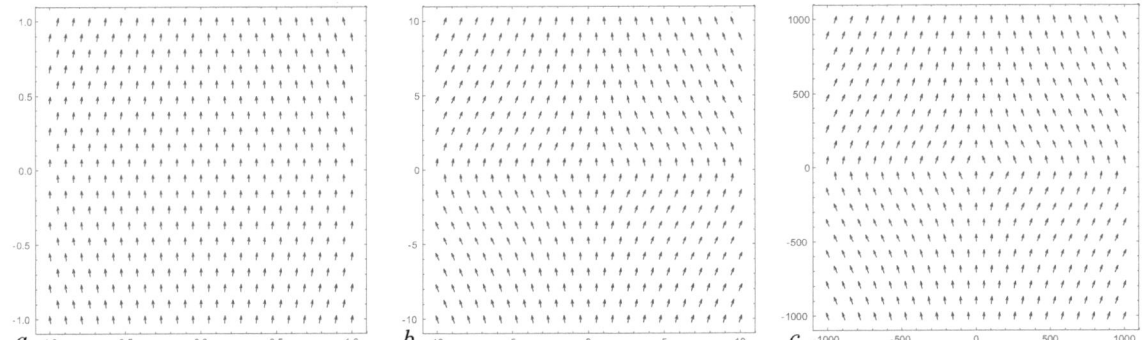

Figure 1. Spin distributions for the Bogomol'nyi–Prasad–Sommerfield solution for different ranges of coordinates: $y_1, z \in [-1, 1]$ (**a**); $y_1, z \in [-10, 10]$ (**b**); and $y_1, z \in [-1000, 1000]$ (**c**). The plots are drawn numerically for $e = 1$ and $l = 1$.

Notice that for large coordinate values, the function tanh goes exponentially to unity and:

$$C \approx \frac{z}{ey_1} \ln\left[\frac{z}{\sqrt{y_1^2 + z^2}}\right].$$

This function is homogeneous, and the spin distribution is self-similar. Therefore the Figure 1b,c are almost the same though the range of coordinates differs one hundred times. The difference appears only near the origin, as shown in Figure 1a.

4. Conclusions

The geometric theory of defects describes dislocations and disclinations in elastic media and crystals in the continuous approximation. It is well suited for the description of single defects as well as their continuous approximation. In the present paper, we consider media with Euclidean metric but nontrivial $\mathbb{SO}(3)$-connection. This assumption corresponds to the absence of elastic forces.

The famous 't Hooft–Polyakov monopoles are static spherically symmetric solutions of $\mathbb{SU}(2)$ Yang–Mills theory. The isomorphism of $\mathbb{SU}(2)$ and $\mathbb{SO}(3)$ Lie algebras implies that the 't Hooft–Polyakov monopoles have new physical interpretation in solid state physics. In contrast to the original model, the $\mathbb{SO}(3)$ group acts now not in the isotopic space but in the tangent space, giving rise to nontrivial torsion and curvature. These geometrical notions have physical interpretation as surface densities of Burgers and Frank vectors in the geometric theory of defects, respectively. To visualize these solutions in media, we have to compute the corresponding spin distributions. This is done in the present paper. We follow the prescription from lattice gauge theory. Spins are assumed to be located at sites of simple cubic lattice. Afterwards the rotational matrix for a given $\mathbb{SO}(3)$ connection is attributed to each link connecting neighboring sites. The spin is fixed at the origin and parallelly transported along links to all other sites along non-intersecting paths. There are many ways to do this, but we do not know the right prescription at present. Therefore we choose one of the simplest ways. First, we parallelly transport the spin along the z axis and then by radial rays in perpendicular planes. In this case, everything is computed analytically in the continuous approximation. The Bogomol'nyi–Prasad–Sommerfield solution is considered as an example for which we computed the spins distribution. The particular feature of the obtained distribution is its self-similarity at large scales. This is proved for the Bogomol'nyi–Prasad–Sommerfield solution, but seems to have place for other monopole solutions.

We leave interesting questions, such as what kind of media have to be chosen for experimental observation of monopoles and what is the right prescription for parallel transport of spins for reconciliation of the theory and practice, for future investigations. The present paper is a small step in this direction but it shows that this is possible.

Funding: The work was supported in part by the Russian Government Program of Competitive Growth of Kazan Federal University (Russian Academic Excellence Project "5–100").

Institutional Review Board Statement: Not applicable.

Informed Consent Statement: Not applicable.

Data Availability Statement: Not applicable.

Conflicts of Interest: The author declares no conflict of interest.

References

1. Landau, L.D.; Lifshits, E.M. *Theory of Elasticity*; Pergamon: Oxford, UK, 1970.
2. Kosevich, A.M. *Physical Mechanics of Real Crystals*; Naukova Dumka: Kiev, Ukraine, 1981. (In Russian)
3. Katanaev, M.O.; Volovich, I.V. Theory of defects in solids and three-dimensional gravity. *Ann. Phys.* **1992**, *216*, 1–28. [CrossRef]
4. Katanaev, M.O. Geometric theory of defects. *Phys. Uspekhi* **2005**, *48*, 675–701. [CrossRef]
5. Azevedo, S. Charged particle with magnetic moment in the background of line topological defect. *Phys. Lett.* **2003**, *A307*, 65–68. [CrossRef]
6. Furtado, C.; Moraes, F.; Carvalho, A.D.M. Geometric phases in graphitic cones. *Phys. Lett.* **2008**, *372*, 5368–5371. [CrossRef]
7. Katanaev, M.O. Torsion and Burgers vector of a tube dislocation. *Proc. Sci.* **2010**, *2010*, 1–7. [CrossRef]
8. Lazar, M.; Hehl, F. Cartan's spiral staircase in physics and, in particular, in the gauge theory of dislocations. *Found. Phys.* **2010**, *40*, 1298–1325. [CrossRef]
9. Randono, C.G.; Hughes, N.L. Torsional monopoles and torqued geometries in gravity and condensed matter. *Phys. Rev. Lett.* **2011**, *106*, 161102. [CrossRef]
10. Boehmer, C.G.; Obukhov, Y.N. A gauge theoretic approach to elasticity with microrotations. *Proc. R. Soc. Lond.* **2012**, *468*, 1391–1407.
11. Bakke, K.; Furtado, C. Abelian geometric phase due to the presence of an edge dislocation. *Phys. Rev.* **2013**, *87*, 012130. [CrossRef]
12. Katanaev, M.O. Rotational elastic waves in double wall tube. *Phys. Lett. A* **2015**, *379*, 1544–1548. doi:10.1016/j.physleta. [CrossRef]
13. Katanaev, M.O. Rotational elastic waves in cylindrical waveguide with wedge dislocation. *J. Phys. A* **2016**, *49*, 085202. [CrossRef]
14. Ciappina, M.F.; Iorio, A.; Pais, P.; Zampeli, A. Torsion in quantum field theory through time-loops on dirac materials. *Phys. Rev.* **2020**, *101*, 036021. [CrossRef]
15. Katanaev, M.O. Chern–Simons term in the geometric theory of defects. *Phys. Rev. D* **2017**, *96*, 84054, [CrossRef]
16. Katanaev, M.O. Description of disclinations and dislocations by the Chern–Simons action for SO(3) connection. *Phys. Part. Nucl.* **2018**, *49*, 890–893. [CrossRef]
17. Katanaev, M.O. Chern–Simons action and disclinations. *Proc. Steklov Inst. Math.* **2018**, *301*, 114–133. [CrossRef]
18. Katanaev, M.O.; Volkov, B.O. Point disclinations in the Chern–Simons geometric theory of defects. *Mod. Phys. Lett.* **2020**, *34*, 2150012. [CrossRef]
19. Katanaev, M.O. Disclinations in the geometric theory of defects. *Tr. Mat. Inst. Im. VA Steklova* **2021**, *313*, 87–108
20. Katanaev, M.O. The 't Hooft–Polyakov monopole in the geometric theory of defects. *Mod. Phys. Lett.* **2020**, *34*, 2050126, [CrossRef]
21. 't Hooft, G. Magnetic monopoles in unified gauge theories. *Nucl. Phys. B* **1974**, *79*, 276–284. [CrossRef]
22. Polyakov, A.M. Particle spectrum in the quantum field theory. *JETP Lett.* **1974**, *20*, 194–195.
23. Manton, N.; Sutcliffe, P. *Topological Solitons*; Cambridge University Press: Cambridge, MA, USA, 2004.
24. Rubakov, V.A. *Classical Theory of Gauge Fields*; Princeton University Press: Princeton, NJ, USA, 2002.
25. Shnir, Y. *Magnetic Monopoles*; Springer: Berlin/Heidelberg, Germany, 2005.
26. Prasad, M.K.; Sommerfield, C.H. Exact classical solution for the 't Hooft monopole and the Julia-Zee dyon. *Phys. Rev. Lett.* **1975**, *35*, 760–762. [CrossRef]
27. Bogomolny, E.B. The stability of classical solutions. *Sov. J. Nucl. Phys.* **1976**, *24*, 449–454.
28. Singleton, D. Exact Schwarzschild-like solution for Yang-Mills theories. *Phys. Rev. D* **1995**, *51*, 5911–5914. [CrossRef] [PubMed]
29. Lunev, F.A. Three-dimensional Yang-Mills-Higgs theory in gauge-invariant variables. *Phys. Lett. B* **1992**, *295*, 99–103. [CrossRef]
30. Lunev, F.A. Three-dimensional Yang-Mills-Higgs equations in gauge-invariant variables. *Theor. Math. Phys.* **1993**, *94*, 48–54. [CrossRef]
31. Lunev, F.A. Reformulation of QCD in the language of general relativity. *J. Math. Phys.* **1996**, *37*, 5351. [CrossRef]
32. Protogenov, A.P. Exact Classical Solutions of Yang-Mills Sourceless Equations. *Phys. Lett. B* **1977**, *67*, 62–64. [CrossRef]
33. Wilson, K.G. Confinement of quarks. *Phys. Rev. D* **1974**, *10*, 2445–2459. [CrossRef]
34. Makeenko, Y. *Methods of Contemporary Gauge Theory*; Cambridge University Press: Cambridge, MA, USA, 2002.
35. Katanaev, M.O. Geometric methods in mathematical physics. Ver. 4. *arXiv* **2020**, arXiv:1311.0733.

Review

Hunting Quantum Gravity with Analogs: The Case of High-Energy Particle Physics

Paolo Castorina [1,2,*], Alfredo Iorio [2] and Helmut Satz [3]

1. Istituto Nazionale di Fisica Nucleare, Sezione di Catania, I-95123 Catania, Italy
2. Faculty of Mathematics and Physics, Charles University, V Holešovičkách 2, 18000 Prague 8, Czech Republic
3. Fakultät für Physik, Universität Bielefeld, D-33501 Bielefeld, Germany
* Correspondence: paolo.castorina@ct.infn.it

Abstract: In this review, we collect, for the first time, old and new research results, and present future perspectives on how hadron production, in high-energy scattering processes, can experimentally probe fundamental questions of quantum gravity. The key observations that ignited the link between the two arenas are the so-called "color-event horizon" of quantum chromodynamics, and the (de)accelerations involved in such scattering processes. Both phenomena point to the Unruh (and related Hawking)-type effects. After the first pioneering investigations, such research studies continued, including studies of the horizon entropy and other "black-hole thermodynamical" behaviors, which incidentally are also part of the frontier of the analog gravity research itself. It has been stressed that the *trait d'union* between the two phenomenologies is that in both hadron physics and black hole physics, "thermal" behaviors are more easily understood, not as due to real thermalization processes (sometimes just impossible, given the small number of particles involved), but rather to a stochastic/quantum entanglement nature of such temperatures. Finally, other aspects, such as the self-critical organizations of hadronic matter and of black holes, have been recently investigated. The results of those investigations are also summarized and commented upon here. As a general remark, this research line shows that we can probe quantum gravity theoretical constructions with analog systems that are not confined to only the condensed matter arena.

Keywords: analogs; hadronic physics; quantum gravity

Citation: Castorina, P.; Iorio, A.; Satz, H. Hunting Quantum Gravity with Analogs: The Case of High-Energy Particle Physics. *Universe* **2022**, *8*, 482. https://doi.org/10.3390/universe8090482

Academic Editor: Jerzy Kowalski-Glikman

Received: 25 July 2022
Accepted: 7 September 2022
Published: 13 September 2022

Publisher's Note: MDPI stays neutral with regard to jurisdictional claims in published maps and institutional affiliations.

Copyright: © 2022 by the authors. Licensee MDPI, Basel, Switzerland. This article is an open access article distributed under the terms and conditions of the Creative Commons Attribution (CC BY) license (https://creativecommons.org/licenses/by/4.0/).

1. Introduction

Analogs have reached a level of maturity in theoretical modeling, e.g., [1], and experimental modeling, e.g., [2], which might bring them to the forefront in the experimental search for quantum gravity (QG) signatures, or, in general, in the theoretical research in fundamental high-energy physics, see, e.g., the contribution [3] to this Issue.

There are two obstacles. First, there is the skepticism of a large part of the theoretical community, which still do not trust analogs as a way to test the fundamental ideas; second, the need for a new era in the analog enterprise, namely to reach *dynamical* effects, rather than *kinematical* effects [1].

Here, we describe a line of research, initiated in [5,6], which addresses both problems. We focus on a specific high-energy scenario, where the effects of a large acceleration are evident; much of the subsequent work was carried out to understand the meaning of entropy in this context and its relation to BH entropy, which is a typical dynamical issue (e.g., recall that Wald's formula relates entropy to the action [7]).

The reproduction of aspects of gravitational physics, both classical and quantum, by means of analogs, is mainly based on condensed matter systems. Examples range from lasers [8–12] (see also the contribution [13] to this issue) to water-waves [14], and from Bose–Einstein condensates [2] to graphene [15–25], and more [1].

In particular, the detection of some form of the Unruh phenomenon [26–28] has been proposed in various set-ups [8–11,14,17,18,29–31]. However, in many of the proposed analog systems, the Unruh temperature

$$T_U = \frac{\hbar a}{2\pi c k_B} \qquad (1)$$

is still too small [5] for a direct experimental verification, as one sees that 1 m/s² →∼ 4 × 10⁻²¹ K. In (1), a is the uniform acceleration, and we explicitly kept the Planck constant, the speed of light, and the Boltzmann constant to ease the unit conversion. In the following, we shall set to one $\hbar$, c and k_B.

Some encouraging results come from femtosecond laser pulses that can produce an acceleration $a \simeq 10^{23}$ m/s² [11], with the associated Unruh temperature $T_U \sim 400$ K. On the other hand, the enormous accelerations (or decelerations) produced in relativistic heavy ion collisions, $a \simeq 4.6 \times 10^{32}$ m/s², have associated Unruh temperatures many orders of magnitude larger, $T_U \sim 1.85 \times 10^{12}$ K. A simple unit conversion shows that this is nothing else than the hadronization temperature T_h

$$T_U \sim 160 \text{ MeV} \sim T_h. \qquad (2)$$

This fact triggered the investigation of hadron production, in high-energy collisions, as a manifestation of the Unruh phenomenon in quantum chromodynamics (QCD) [5,6].

Of the latter we discuss it in this paper, by reviewing why such an interpretation is natural, commenting on the various ramifications, and speculating on the possible future directions. In other words, we elaborate on which aspects of this QCD phenomenology can be taken as viable analogs of specific aspects of QG.

The underlying idea behind the latter analogies is based on quark confinement as a phenomenon where a "horizon" (sometimes called "color horizon", see, e.g., [32,33]) hides those degrees of freedom to any observer, and only quantum (tunneling) effects could explain a radiation phenomenon [5]. This is a non-perturbative quantum phenomenon, related to the chromomagnetic properties of the QCD vacuum (see for example reference [34]), producing the squeezing of the chromoelectric field in quark–antiquark strings, with a constant energy density. Let us comment a bit more on this.

Quark confinement can be described as due to a potential that grows linearly at large distances, $V = \sigma r$. This corresponds to a constant acceleration; henceforth, the Rindler spacetime is the appropriate framework for this phenomenon. As well known, the Rindler metric is equivalent to the near-horizon approximation of the BH metric, with the acceleration equal to the surface gravity, k. Therefore, the *local* correspondence between a linear potential and the near-horizon dynamics of a BH is a strong analogy.

This is another perspective as to why quark confinement can be related to a "color horizon" [32,33], which hides the color degrees of freedom and a Rindler horizon, and is, in turn, associated with a specific BH (in [35], some proposals of specific BHs could account for this specific scenario). On the other hand, the Hawking radiation is a quantum phenomenon associated with tunneling and pair creation near the event horizon [36,37]. This is a clear dynamical correspondence to the string-breaking and quark–antiquark pair creation in the final process of the mechanism leading from the color degrees of freedom until the formation of hadrons.

Finally, another delicate dynamical issue involves the entropy associated with a "color event horizon". This is an entanglement entropy between the quantum field modes on the two sides of the horizon. As well known, such an entropy follows an area law [38–40], similar to the entropy of a BH [41,42], when logarithmic corrections are not included, or the entropy of a Rindler horizon [43]. Even though it is still an open question whether entanglement entropy alone could account for the whole BH entropy, this is yet another argument that strengthens the analogy between the two systems. Furthermore, in such QCD environments, the entropy is a quantity routinely considered, e.g., in (quantum) statistical models. Henceforth, we have measurable and natural candidates for quantities that can play the role of a BH entropy. As mentioned earlier, this is a very important milestone to move analogs to the next era; that is, the possibility to reproduce BH thermodynamics, with its intriguing fundamental open questions, such as the information paradox. For the sake of completeness, let us recall that the general Page approach (i.e., to the calculation of the entanglement entropy of an evaporating BH [44]) has been successfully applied to gluon shadowing in deep inelastic scattering [45], following the proposal in [46].

In this review paper, we collect, for the first time, the most important (old and new) results of this line of research; we comment on and discuss them. The paper is organized as follows. In Section 2, we recall the main features of the Unruh effect, and of the related BH physics, using the descriptions of the effects that make the link with hadron physics (that we want to disclose) easier; this Section is also important for setting the notation. In Section 3, we recollect three well-known aspects of the phenomenology of hadrons, which will be scrutinized using the analogies and links with gravitational physics in the Sections that follow. The hadronic phenomena described in Section 3.1 are reinterpreted as gravity analogs in Section 3.2; the hadronic phenomena described in Section 4 are reinterpreted as gravity analogs in Section 4.2; finally, the hadronic phenomena described in Section 5 are reinterpreted as gravity analogs in Section 5.2. We close the review with our conclusions in Section 6.

2. Accelerated Observers and near BH Horizon Observers

In this Section, we recall the main features of the Unruh effect, and the related BH physics, which will mostly be used in the realizations in hadronic physics, which we discuss later. In particular, we first discuss the interplay between pair production, tunneling, and the Unruh effect. We then mention the correspondence between the near-horizon BH metric and Rindler metric, and the area law obeyed by BH entropy.

Let us begin by discussing the Unruh effect and its relation to tunneling and pair production. For this part, we follow [5].

Consider the action, A, of a particle of mass m, subject to a constant force derived from a potential $\varphi(x)$:

$$A = -\int (m\,ds + \varphi\,dt). \tag{3}$$

For a constant force, the one-dimensional (1D) potential is $\varphi = -\sigma x$ modulo, an additive constant, and the equations of motion of the particle are

$$\frac{dp_x}{dt} = \sigma, \quad \frac{dp_\perp}{dt} = 0. \tag{4}$$

Using $ds^2 = (1 - v^2(t))\,dt^2$ and the equations of motion, one can evaluate action A [5]

$$\begin{aligned} A(\tau) &= \int^\tau dt\left(-m\sqrt{1-v(t)^2} + \sigma x(t)\right) \\ &= -\frac{m}{a}\mathrm{arcsinh}(a\tau) + \frac{\sigma}{2a^2}[a\tau(\sqrt{1+a^2\tau^2}-2) + \mathrm{arcsinh}(a\,\tau)] + \mathrm{const}. \end{aligned} \tag{5}$$

In quantum theory, the particle has a finite probability to be found under the potential barrier, σx, in the classically forbidden region. Mathematically, this comes about because action A, being an analytic function of τ, has an imaginary part

$$A(\tau) = \frac{m\pi}{a} - \frac{\sigma\pi}{2a^2} = \frac{\pi m^2}{2\sigma}, \tag{6}$$

which corresponds to the motion of a particle in Euclidean time, t_E, with the Euclidean trajectory

$$x(t_E) = a^{-1}\left(\sqrt{1 - a^2 t_E^2} - 1\right), \tag{7}$$

bouncing between the two identical points $x_a = -a^{-1}$ at $t_{E,a} = -a^{-1}$ and $x_b = -a^{-1}$ at $t_{E,b} = a^{-1}$, and the turning point $x_a = 0$ at $t_{E,a} = 0$.

In the quasi-classical approximation, the rate of tunneling under the potential barrier is given by

$$\Gamma_{vac \to m} \sim e^{-2\operatorname{Im}A} = e^{-\frac{\pi m^2}{\sigma}}, \qquad (8)$$

which gives the probability to produce a particle and its antiparticle (each of mass m) out of the vacuum, under the effects of a constant force σ. The ratio of the probabilities to produce states of masses M and m is then

$$\frac{\Gamma_{vac \to M}}{\Gamma_{vac \to m}} = e^{-\frac{\pi(M^2 - m^2)}{\sigma}}. \qquad (9)$$

The relation (9) had a double interpretation, in terms of both the Unruh and the Schwinger effects, see, e.g., [47–49] and references therein. Indeed, consider a detector with quantum levels m and M, moving with a constant acceleration. Each level is accelerated differently; however, if the splitting is not large, $M - m \ll m$, we can introduce the average acceleration of the detector

$$\bar{a} = \frac{2\sigma}{M + m}. \qquad (10)$$

Substituting (10) into (9), we arrive at

$$\frac{\Gamma_{vac \to M}}{\Gamma_{vac \to m}} = e^{\frac{2\pi(M - m)}{\bar{a}}}. \qquad (11)$$

This expression is reminiscent of the Boltzmann probabilistic weight in a heat bath, with an effective temperature, $T = \bar{a}/2\pi$. This is the Unruh effect.

A similar study of the Unruh radiation (tunneling through a barrier by WKB-like methods) was carried out in [50]. A more rigorous derivation of the Unruh effect can be given by recalling that the uniformly accelerated detector in the Minkowski space is equivalent to the inertial detector in the Rindler space. The vacuum in the Minkowski space is related to the vacuum in the Rindler space by a nontrivial Bogoliubov transformation, which shows that the Rindler vacuum is populated with thermal radiation of temperature $T = a/2\pi$ (for a review, see [28]).

We will now focus on another aspect of the Hawking–Unruh phenomenon that is crucial for the analogy between quark confinement and the physics of curved spacetime (which we shall discuss later)—the correspondence between the Rindler metric and the near-horizon approximation of a BH metric.

The Schwarzschild metric for a BH of mass M, in radial coordinates, is given by

$$ds^2 = f(r)dt^2 - f(r)^{-1}dr^2 - r^2[d\theta^2 + \sin^2\theta d\phi^2], \qquad (12)$$

with

$$f(r) = \left(1 - \frac{2GM}{r}\right). \qquad (13)$$

The equation $f(r) = 0$ sets the Schwarzschild radius, R_S, as the radius of the spherical event horizon

$$R_S = 2GM. \qquad (14)$$

This means that $M(R_S) = (2G)^{-1} R_S$, which is a linear law for the BH mass. This is particularly interesting if one notices that an analogous behavior is enjoyed by the confining potential of the strong interactions.

In Equation (12), the coordinate transformation [39] is

$$\eta = \frac{\sqrt{f(r)}}{\kappa}, \qquad (15)$$

where the surface gravity κ is given by

$$\kappa = \frac{1}{2}\left(\frac{\partial f}{\partial r}\right)_{r=R_S}, \tag{16}$$

one obtains, for $r \to R_S$, the BH metric in the near-horizon approximation

$$ds^2 = \eta^2\kappa^2 dt^2 - d\eta^2 - R^2(d\theta^2 + \sin^2\theta d\phi^2). \tag{17}$$

To compare the previous result with the Rindler metric of a constantly accelerated observer, let us recall the relations among Rindler coordinates, (ξ, τ) and the Minkowski coordinates (x, t)

$$x = \xi \cosh a\tau, \quad t = \xi \sinh a\tau, \tag{18}$$

where $a = \sigma/m$ denotes the acceleration in the instantaneous rest frame of m, and τ is the proper time. With these, the metrics of such an accelerating system (in spherical coordinates) are

$$ds^2 = \xi^2 a^2 d\tau^2 - d\xi^2 - \xi^2 \cosh^2 a\tau (d\theta^2 + \sin^2\theta d\phi^2). \tag{19}$$

Comparing Equations (19) and (17), it is evident that the system in uniform acceleration is the same as a system near a spherical BH horizon, provided we identify the surface gravity κ with the acceleration a.

The final topic of this Section will be entropy and its area law, a feature common to BH and constantly accelerated systems.

The gravitational entropy is related to the existence of a horizon, which forbids an observer to acquire knowledge (or what is happening beyond it). In a way, it could be seen as a measure of the ignorance of the fate of matter (and space) degrees of freedom that contribute to making the BH.

As well known, such entropy obeys the Bekenstein–Hawking area law [41,42]:

$$S_{\text{BH}} = \frac{1}{4}\frac{A}{\ell_P^2}, \tag{20}$$

where $\ell_P = \sqrt{G}$ is the Planck length, and, for Schwarzschild BH: $A = 4\pi R_S^2$. Once more the only parameter of interest is the mass of the BH: $M \sim R_S$.

On the other hand, it is also well known that access to the degrees of freedom describing an accelerated observer is also restricted by a horizon—the Rindler horizon. Therefore, the entropy of the so-called Rindler wedge was evaluated (similar to the BH). The computation was performed a long time ago [43], and it turns out that $S = (1/4)(A_a/\ell_P^2)$, where A_a is the area of a surface of the constant Rindler spatial coordinate, x, and the proper time, τ. If y and z are the Minkowski coordinates (we suppose that the acceleration is along the x-axis), the entropy is actually infinite; however, an entropy *density*, per unit area, can be defined for this spacetime.

Finally, we recall another well-known result, namely that the entanglement (hence, quantum) entropy of a bipartite system (which includes both the Rindler and the BH cases just discussed, due to their event horizons) also obeys an area law. This has been shown in various quantum field theoretical setups, see, e.g., [38–40].

In the following, we schematically recall those results of the phenomenological analysis of high-energy collision data, which will then be reconsidered in light of the gravity analog, in a separate dedicated Subsection.

3. Hadron Production in High-Energy Collisions

3.1. Statistical Hadronization Model

There is abundant multihadron production in high-energy collisions, starting from the electron-positron annihilation, and then in the proton–proton, proton–nucleus, and nucleus–nucleus scattering processes. The relative rates of the secondaries produced are well accounted for by ideal gases of all hadrons and hadronic resonances, at a fixed temperature T and baryochemical potential μ_B. This is known as the statistical hadronization model (SHM) [51–53]. There is one (well-known) caveat though. The strangeness production one finds is reduced with respect to the rates predicted by the SHM. However, this suppression can be taken into account by one further parameter, $0 < \gamma_s \leq 1$, if the predicted rate for a hadron species containing $\nu = 1, 2, 3$ strange quarks is suppressed by the factor γ_s^ν [54].

To describe such a resonance gas, the basic tool one needs is the grand-canonical partition function for an ideal gas at the temperature T in a spatial volume V

$$\ln Z(T) = V \sum_i \frac{d_i \gamma_s^{\nu_i}}{(2\pi)^3} \phi(m_i, T), \tag{21}$$

with d_i specifying the degeneracy (spin, isospin) of the species i, and m_i its mass. The sum runs over all species. For simplicity, we assume for the moment $\mu_B = 0$. Here,

$$\phi(m_i, T) = \int d^3p \, \exp\{\sqrt{p^2 + m_i^2}/T\} \sim \exp(-m_i/T), \tag{22}$$

is the Boltzmann factor for species i, so that the ratio of the production rates, N_i and N_j, for hadrons of species i and j, is given by

$$\frac{N_i}{N_j} = \frac{d_i \gamma_s^{\nu_i} \phi(m_i, T)}{d_j \gamma_s^{\nu_j} \phi(m_j, T)}, \tag{23}$$

where $\nu_i = 0, 1, 2, 3$ specifies the number of strange quarks in species i.

Both the temperature T and strangeness suppression factor γ_s were measured, at various collision energies, and for different collision configurations. The resulting temperature of the emerging resonance gas is found to have a universal value

$$T_c \simeq 160 \pm 10 \text{ MeV}, \tag{24}$$

for all (high) collision energies, where $\mu_B \simeq 0$ and all collision configurations, including hadron production in the e^+e^- annihilation.

Moreover, in heavy ion collisions at lower energy, the finite baryon density has a crucial role and the dynamics are dominated by Fermi statistics and baryon repulsion. In the $T - \mu_B$ plane, the dependence of the hadronization temperature on μ_B defines the chemical "freeze-out" curve, which can be described by specific (but poorly understood, see next Section) criteria [55–59].

Indeed, a fixed ratio between the entropy density, s, and the hadronization temperature, $s/T^3 \simeq 7$, or the average energy per particle, $<E>/N \simeq 1.08$ GeV reproduces the curve in the $T - \mu_B$ plane, as shown in Figure 1, where the percolation model result [55] is also plotted.

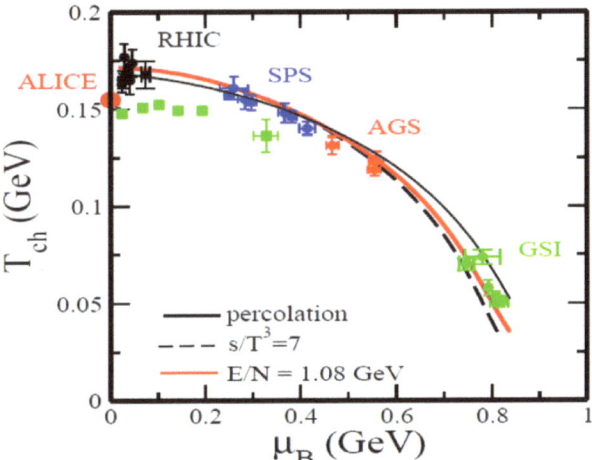

Figure 1. Freeze-out curve in the statistical hadronization model compared with the criteria discussed in the text. The green squares without error bars are the QCD lattice simulation data.

The agreement of the SHM with data on the abundances of different hadronic species, from e^+e^- annihilation to heavy ion collisions, is puzzling. In heavy ion collisions, it is possible to expect the emergence of statistical distributions as a result of intense reinteractions; however, this seems very implausible in the e^+e^- annihilation at high energies because the density of the produced hadron is small there.

Moreover, in e^+e^-, the jet structure, the angular distributions of the produced hadrons, and the inter–jet correlations, point to the important role of QCD dynamics of gluon radiation. Thus, the "phase space dominance" cannot be invoked. Indeed, in all high-energy collisions, for $\sqrt{(s)} \geq 20$ GeV, the hadronization temperature is essentially constant and independent from the initial configurations.

The previous aspects call for some *universal* mechanism at the root of hadron production, which has to be related to the way the QCD vacuum responds to color fields.

3.2. Analog Gravity Interpretation of the SHM and the QCD Hawking–Unruh Radiation

As mentioned in the introduction, the phenomenology of quark confinement can be seen as the effect of a Rindler force due to the string tension, σ. Let us now describe this phenomenon in more detail.

We recall in Section 2 that the basic mechanism of the Unruh radiation involves tunneling through the confining event horizon. This is most simply illustrated by hadron production through the e^+e^- annihilation into a q pair, see Figure 2.

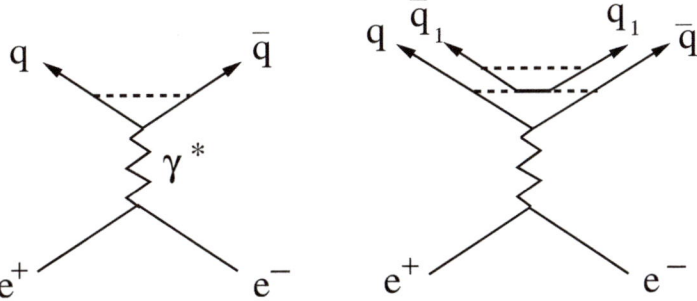

Figure 2. Quark formation in the e^+e^- annihilation

The first quark–antiquark pair, $q\bar{q}$, initially tries to separate. The attempt stops when both quarks hit the confinement horizon, i.e., when they both reach the end of the binding string, where their separation is R. At that point, the attempt to separate can only continue if a further quark–antiquark system is excited from the vacuum. Although the new pair, $q_1\bar{q}_1$, is at rest, in the overall center of mass system, each constituent has a transverse momentum k_T, determined by the uncertainty relation in terms of the transverse dimension of the string flux tube. The string theory [60] for the basic thickness gives

$$r_T = \sqrt{2/\pi\sigma}, \qquad (25)$$

leading to

$$k_T = \sqrt{\pi\sigma/2}. \qquad (26)$$

The maximum separation distance R is specified by

$$\sigma R = 2\sqrt{m_q^2 + k_T^2} = 2k_T, \qquad (27)$$

where we take $m_q = 0$ for the quark mass. From this, we obtain

$$R = \sqrt{2\pi/\sigma}, \qquad (28)$$

as the string-breaking distance. The departing quark q now pulls the newly formed $\bar{q}_1$ along, giving it an acceleration [6]

$$a = \sqrt{2\pi\sigma}. \qquad (29)$$

The $q_1\bar{q}_1$ pair eventually suffers the same fate as the q pair: it is separated to its confinement horizon, where it again excites a new pair, which is now emitted as the Unruh radiation of temperature

$$T_h = a/2\pi = \sqrt{\sigma/2\pi}, \qquad (30)$$

that is also the hadronization temperature, as we shall see in a moment. This process is sequentially repeated until the energies of the initial "driving" quarks q and $\bar{q}$ are exhausted.

The case of the e^+e^- annihilation corresponds to baryochemical potential, $\mu_B = 0$. Here, one finds the average value $\sigma \simeq 0.19 \pm 0.03$ GeV^2, see, e.g., [61], which with Equation (30) then leads to

$$T_h(\mu_B = 0) = \sqrt{\sigma/2\pi} \simeq 175 \pm 15 \text{ MeV}. \qquad (31)$$

for the freeze-out temperature at $\mu_B = 0$.

The fundamental mechanism in the Unruh scenario is quark (de)acceleration, leading to the string-breaking with the resulting pair production, as specified by Equation (27). As long as we assume a vanishing quark mass, the only dimensional parameter in the entire formalism is the string tension σ.

Therefore, the Unruh hadronization temperature is "universal"; this explains the observation of thermal hadron production in high-energy collisions in e^+e^- and pp interactions. In this respect, the emitted hadrons are "born in equilibrium" [62,63].

The previous analysis shows that the hadronization temperature corresponds to the Unruh temperature related to the string-breaking in high-energy collisions, where $\mu_B \simeq 0$.

As discussed, the dependence of the hadronization temperature on μ_B defines the chemical "freeze-out" curve, which turns out to be in agreement (see Figure 1) with a fixed ratio between the entropy density, s, the hadronization temperature, $s/T^3 \simeq 7$, and/or the average energy per particle, $<E>/N \simeq 1.08$ GeV, and/or $n \simeq 0.12$ fm^{-3}, where n is the number density.

Although the Unruh mechanism and the string-breaking provide theoretical bases for the production of newly formed hadrons in high-energy collisions, they do not address the roles of the nucleons already present in the initial state of the heavy ion collisions. However, the corresponding hadron formation gives clear meaning to the figures that characterize the whole freeze-out curve.

Indeed, as discussed in [64], the energy of the pair produced by string-breaking, i.e., of the newly formed hadron, is given by (cf Equations (26) and (27))

$$E_h = \sigma R = \sqrt{2\pi\sigma}. \tag{32}$$

In the central rapidity region of high-energy collisions, one has $\mu_B \simeq 0$, so that E_h is, in fact, the average energy $\langle E \rangle$ per hadron, with an average number $\langle N \rangle$ of newly produced hadrons. Hence, one obtains

$$\frac{\langle E \rangle}{\langle N \rangle} = \sqrt{2\pi\sigma} \simeq 1.09 \pm 0.08 \,\text{GeV}, \tag{33}$$

in accordance with the phenomenological fit obtained from the species abundances in high-energy collisions [56,58].

Next, we turn to the number density. For a single string-breaking, the number density is given by

$$n_{sb} \simeq \frac{1}{4\pi R^3/3}, \tag{34}$$

where R is the string-breaking distance, which turns out to be $R = 1/T_h$ for massless quarks. For $T_h \simeq 160$ MeV, consistent with our previous evaluation, one obtains $n_{sb} \simeq 0.129$ fm^{-3}.

Let us now consider the entropy. Since the event horizon is caused by color confinement, such entropy is an entanglement entropy of quantum field modes on both sides of the horizon (recall that, here, we have no real gravitational degrees of freedom). Its general form is [39,40]

$$S_{\text{ent}} = \alpha \frac{A}{r^2}, \tag{35}$$

where A is the area of the event horizon, r the scale of the characteristic quantum fluctuations, and α an undetermined numerical constant, which might as well be infinite. This expression shares the holographic structure (holography of entanglement entropy is a general result, see [38,65]) with the Bekenstein–Hawking entropy [41,42] for a BH given in (20)

$$S_{\text{BH}} = \frac{1}{4}\frac{A}{\ell_P^2}.$$

A relation similar to (20) also holds in the case of an accelerated observer [43]. Here, we take it to be valid in our case, where gravity is not involved and the entire entropy must be of the entanglement type. The scale of the characteristic quantum fluctuations is now given by the transverse string thickness in Equation (25), rather than the Planck length, ℓ_P, of the gravitational phenomena. One obtains

$$S_h = \frac{1}{4}\frac{A_h}{r_T^2} = \frac{1}{4}\frac{4\pi R^2}{r_T^2}, \tag{36}$$

for the entropy in the hadron production. The parameter R is given by Equation (28), and inserting these expressions into Equation (36) for the entropy associated with the hadron production gives

$$S_h = \pi^3, \tag{37}$$

and the entropy *density*, $s = S_h/V$ (here, $V = 4/3\pi R^3$), divided by T^3, turns out to be

$$\frac{s}{T^3} = \frac{S_h}{(4\pi/3)R^3 T^3} = \frac{3\pi^2}{4} \simeq 7.4, \tag{38}$$

as the freeze-out condition in terms of $s(T)$ and T. This result is in accordance with the value obtained for s/T^3 from species abundance analyses in terms of the ideal resonance gas model [58,59]. Moreover, within this picture, one can show [66] that QCD entropy, evaluated by lattice simulations in the region $T_c < T < 1.3T_c$, is in reasonable agreement with a melting color event horizon.

The analogy between the freeze-out temperature as a function of μ_B and the Hawking temperature for charged BH is discussed in [6]; another interesting aspect is that it can be translated to the temperature dependence on the collision energy $\sqrt{s}$, by considering $\mu_B(\sqrt{s})$ [67].

Since the Unruh temperature triggers the search for the gravitational BH, which in its near-horizon approximation better simulates the hadronization phenomenon, one can study which BH behind that Rindler horizon could reproduce the experimental behavior of $T(\sqrt{s})$. Although the complete hadronization process is in 4D spacetime, the hadronic Rindler spacetime should be better consider as the near-horizon approximation of the effective two-dimensional (2D) BH analog for the following two reasons

- New particle creation is effectively 2D because it can be described in terms of the evolution in time of the hadronic strings, which are one-dimensional objects [68].
- The near-horizon field dynamics are effectively 2D [36,69].

Provided certain natural assumptions hold, it has been shown [35] that the so-called exact string BH in 2D dilaton gravity [68] turns out to be the best candidate, as it fits the available data on $T(\sqrt{s})$, and that its limiting case, the Witten BH, is the unique candidate to explain the constant T for all elementary scattering processes at large energies.

To conclude this Section, we now turn to the strange quark mass and the interpretation *alla* Unruh of the strangeness enhancement.

At the beginning of this Section, we illustrated how the thermal hadron production process is a Hawking–Unruh mechanism. In doing so, we neglected the effects of the quark mass. If one includes them, the expression one obtains for acceleration is

$$a_q = \frac{\sigma}{w_q} = \frac{\sigma}{\sqrt{m_q^2 + k_q^2}}, \tag{39}$$

where $w_q = \sqrt{m_q^2 + k_q^2}$ is the effective mass of the produced quark, with m_q the bare quark mass, and k_q the quark momentum inside the hadronic system $q_1\bar{q}_1$ or $q_2\bar{q}_2$ (see Figure 3). Since the string breaks [6] when it reaches a separation distance

$$x_q \simeq \frac{2}{\sigma}\sqrt{m_q^2 + \frac{\pi\sigma}{2}}, \tag{40}$$

the uncertainty relation gives us $k_q \simeq 1/x_q$

$$w_q = \sqrt{m_q^2 + [\sigma^2/(4m_q^2 + 2\pi\sigma)]}, \tag{41}$$

for the effective mass of the quark. The resulting Unruh temperature depends on the quark mass; thus it is given by

$$T(q\bar{q}) \simeq \frac{\sigma}{2\pi w_q}. \tag{42}$$

Here, it is assumed that the quark masses for q_1 and q_2 are equal. For $m_q \simeq 0$, Equation (42) reduces to $T(00) \simeq \sqrt{\sigma/2\pi}$, as obtained in Equation (30).

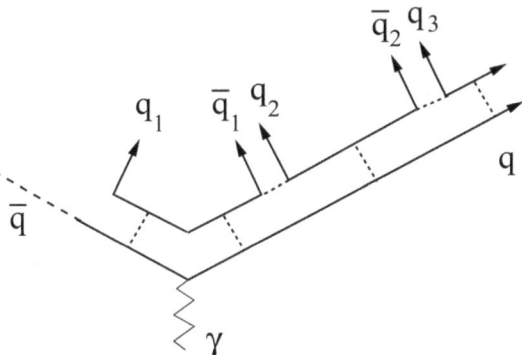

Figure 3. Sequential quark formation in the e^+e^- annihilation

If the produced hadron $\bar{q}_1 q_2$ consists of quarks of different masses, the resulting temperature has to be calculated as an average of the different accelerations involved. For one massless quark ($m_q \simeq 0$) and one of strange quark of mass m_s, the average acceleration becomes

$$\bar{a}_{0s} = \frac{w_0 a_0 + w_s a_s}{w_0 + w_s} = \frac{2\sigma}{w_0 + w_s}. \quad (43)$$

From this, the Unruh temperature of a strange meson is given by $T(0s) \simeq \sigma/\pi(w_0 + w_s)$ with $w_0 \simeq \sqrt{1/2\pi\sigma}$ and w_s is given by Equation (4) with $m_q = m_s$. Similarly, we obtain $T(ss) \simeq \sigma/2\pi w_s$ for the temperature of a meson consisting of a strange quark–antiquark pair (ϕ).

The scheme is readily generalized to baryons. The production pattern leads to an average of the accelerations of the quarks involved [70]. Thus, we have $T(000) = T(0) \simeq \sigma/2\pi w_0$ for nucleons, $T(00s) \simeq 3\sigma/2\pi(2w_0 + w_s)$ for Λ and Σ production, $T(0ss) \simeq 3\sigma/2\pi(w_0 + 2w_s)$ for Ξ production, and $T(sss) = T(ss) \simeq \sigma/2\pi w_s$ for that of Ωs.

Thus, we obtain a resonance gas picture with five different hadronization temperatures, as specified by the strangeness content of the hadron in question: $T(00) = T(000)$, $T(0s)$, $T(ss) = T(sss)$, $T(00s)$, and $T(0ss)$.

In other words, the event horizon of the color confinement leads to thermal behavior, but the resulting temperature depends on the strange quark content of the produced hadrons, causing a deviation from the full equilibrium and, hence, a suppression of strange particle production, without the introduction of the γ_s parameter. The resulting formalism was applied to the multihadron production in the e^+e^- annihilation over a wide range of energies to make a comprehensive analysis of the data, in the conventional (i.e., with γ_s) SHM and its modified Hawking–Unruh formulation [70,71]. The modified SHM, with the different Unruh temperature, gives a better fit with respect to the standard SHM formulation.

In the Hawking–Unruh formulation, the number of free parameters of the model does not increase since all previous temperatures were completely determined by the string tension and the strange quark mass. Apart from possible variations of the quantities of σ and m_s, the description is parameter-free.

In all cases, the temperature for a hadron carrying nonzero strangeness was lower than that of non-strange hadrons and this led to an overall strangeness suppression in elementary collisions, in good agreement with the data, without the introduction of the ad hoc parameter γ_s. Figure 4 reports the comparison between the SHM with one temperature and γ_s and the Hawking–Unruh-inspired approach.

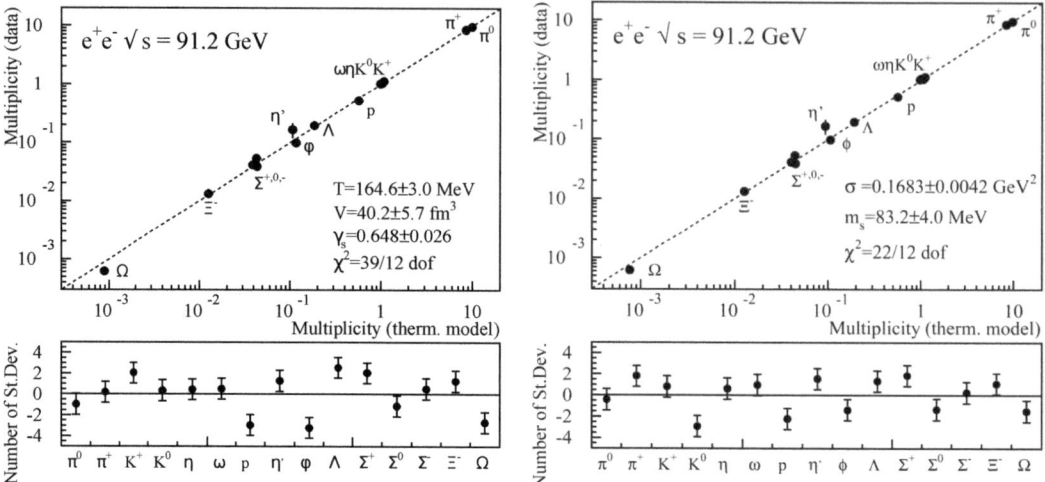

Figure 4. Comparison between the measured and fitted multiplicities of long-lived hadronic species in e^+e^- collisions at $\sqrt{s}$ = 91.25 GeV. (**Left**): statistical hadronization model with one temperature. (**Right**): Hawking–Unruh radiation model. See [70].

On the other hand, in nucleus–nucleus (AA, "large systems") collisions at $\sqrt{s} \geq 15$ GeV, the so-called strangeness enhancement with respect to e^+e^- and hadronic scattering (the "small" systems) has been observed, which in the standard SHM is described by the condition $\gamma_s = 1$ in AA with respect to $\gamma_s \simeq 0.5 - 0.6$ in small systems. Moreover, the same enhancement has been detected in proton–proton collisions at large energies and in large multiplicity events [72].

The translation of *alla* Unruh (of the strangeness enhancement) requires that the different temperature for various hadronic strangeness content disappear. Indeed, $T(00), T(0s), \ldots$ are derived from the breaking of a single string with the corresponding average acceleration and Unruh temperatures. On the other hand, as shown in reference [73], the universality among small and large systems is directly related to the initial parton density in the transverse plane.

If the initial setting is different but the collision energy and the large multiplicity cut produce initial states with similar entropy densities (i.e parton density in the transverse plane), the hadron production and other coarse-grain dynamical signatures are the same [73]. Therefore, for large parton density, there is a strong string overlap, as depicted in Figure 5.

Let us outline, in a simplified model, the mechanism that washes out the strangeness dependence of the Unruh temperature when, in a causally connected region, the parton density in the transverse plane is large.

Assume two species only: one scalar meson and one electrically neutral meson; that is, "pions" with mass m_π, and "kaons" with mass m_k and strangeness $s = 1$.

Let us consider a high-density system of quarks and antiquarks in a causally connected region for high-energy and high multiplicity events. Generalizing Equation (43), the average acceleration is given by

$$\bar{a} = \frac{N_l w_0 a_0 + N_s w_s a_s}{N_l w_0 + N_s w_s}, \tag{44}$$

where $N_l \gg 1$, $N_s \gg 1$ are, respectively, the number of light quarks and strange quarks.

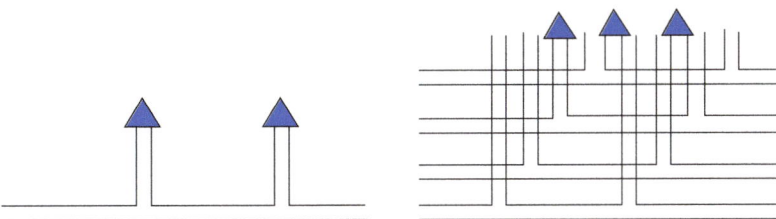

Figure 5. Left: Hadron production *alla* Unruh by a sequence of independent single string breakings. Right: Hadron production due to the overlap of different color event horizons for large parton density.

By assuming $N_l \gg N_s$, after simple algebra, the average temperature, $\bar{T} = \bar{a}/2\pi$, turns out to be

$$\bar{T} = T(00)\left[1 - \frac{N_s}{N_l}\frac{w_0 + w_s}{w_0}\left(1 - \frac{T(0s)}{T(00)}\right)\right] + O[(N_s/N_l)^2], \quad (45)$$

Now, in our "world of pions and kaons", one has $N_l = 2N_\pi + N_k$ and $N_s = N_k$ and, therefore,

$$\bar{T} = T(00)\left[1 - \frac{N_k}{2N_\pi}\frac{w_0 + w_s}{w_0}\left(1 - \frac{T(0s)}{T(00)}\right)\right] + O[(N_k/N_\pi)^2]. \quad (46)$$

On the other hand, in the Hawking–Unruh-based statistical calculation, the kaon–pion ratio, N_k/N_π, depends on the equilibrium (average) temperature $\bar{T}$; that is

$$N_k/N_\pi = \frac{m_k^2}{m_\pi^2}\frac{K_2(m_k/\bar{T})}{K_2(m_\pi/\bar{T})}, \quad (47)$$

where $K_2(x)$ denotes a Hankel function of a purely imaginary argument. Therefore, one has to determine the temperature $\bar{T}$ by self-consistency of Equation (46) with Equation (47). This condition implies

$$\frac{2[1 - \bar{T}/T(00)]w_0}{[1 - T(0s)/T(00)](w_s + w_0)} = \frac{m_k^2}{m_\pi^2}\frac{K_2(m_k/\bar{T})}{K_2(m_\pi/\bar{T})}, \quad (48)$$

which can be solved numerically. For $\sigma = 0.17$ Gev^2, $m_s = 0.083$ GeV (see Figure 4), the solution gives $\bar{T}/T(00) \simeq 0.97$.

In other words, this toy model shows that the non-equilibrium condition, with species-dependent temperatures, converges to an equilibrated system, with the average temperature, $\bar{T} \simeq T(00)$, for large parton density in a causally connected region.

4. Thermal Component in the Transverse Momentum Spectra

4.1. High-Energy Hadronic Processes

The transverse momentum, p_T, and spectra of hadrons produced in high-energy collisions, can be decomposed into two components: the exponential (or "soft") component and the power (or "hard") component. Their relative strengths, in deep inelastic scattering (DIS), depend drastically on the global structure of the event. Namely, the exponential component is absent in the diffractive events characterized by a rapidity gap [74,75].

The hard component is well understood, resulting from the high-momentum transfer scattering of quarks and gluons and their subsequent fragmentations. The "soft" component is ubiquitous in high-energy collisions and appears as a thermal spectrum. In nuclear collisions, given the high number of participants involved, one may expect thermalization to take place; however, it is hard to believe that this might occur in processes such as DIS or e^+e^- annihilation.

In [76], it was found that the following parametrization well describes the hadron transverse momentum distribution, both in hadronic collisions and in deep-inelastic scattering

$$\frac{d\sigma}{p_T dp_T} = A_{therm} e^{-m_T/T_{th}} + A_{hard}\left(1 + \frac{m_T^2}{n T_{th}^2}\right)^{-n}. \quad (49)$$

This clearly defines the soft/thermal components and the hard component parameterized by T_{th}. Here, $m_T = \sqrt{m^2 + p_T^2}$.

4.2. Analog Gravity Interpretation of the Origin of the Thermal Component in the Transverse Momentum Spectra

The strength of the chromoelectric field, in a single string-breaking, is determined by the string tension, and it describes the yields of the different hadronic species. However, to discuss the transverse momentum spectra of the produced hadrons (see Section 4), one has to take into account the increasing number of gluons in the wave functions of the colliding hadrons. This can be done by the parton saturation [77], or color glass condensate [78,79] picture. In this approach, the density of partons in the transverse plane is parameterized by the saturation momentum $Q_s(s,\eta)$, which depends on the c.m.s. collision energy-squared s and (pseudo-)rapidity η.

The temperature of the radiation from the resulting Rindler event horizon is given by [5]

$$T_U = T_{th} = c\,\frac{Q_s}{2\pi}, \quad (50)$$

where c is a constant [80]. T_{th} is related to the deceleration of partons in the transverse plane; moreover, $Q_s = T$ in the parametrization of the hard component in Equation (22) [74]. Therefore, one predicts a proportionality between the T_{th} and T, which has been verified [74,75].

The established proportionality of the parameters describing the thermal and hard components of the transverse momentum spectra supports the theoretical picture in which the soft hadron production is a consequence of the quantum evaporation from the event horizon formed by the deceleration in longitudinal color fields. The absence of the thermal component in diffractive interactions lends further support to this interpretation.

5. Self-Organization and Self-Similarity

5.1. Hadronic Spectrum

The typical illustration of self-organized criticality (SOC), proposed in the pioneering work [81], is the 'avalanche dynamics' of sandpiles. There, the number $N(s)$ of avalanches of size s observed over a long period was found to vary as a power of s, $N(s) = \alpha s^{-p}$. This means that the phenomenon is scale-free, so the same structure is found, again and again, at all scales. This phenomenon is often referred to as self-similarity: the system resembles itself at all scales.

Another example of self-similarity is found when partitioning naturals. Given a natural number, $N \in \mathbb{N}$, we can *decompose* it (in mathematical jargon) into the natural, whose sum gives $N = \sum_i N_i$, with no distinction of the order of N_is entering the sum, e.g., 3 = 2 + 1 and 3 = 1 + 2 would count the same as a decomposition of 3. On the other hand, we also have *compositions* of N, which are decompositions of N in which the *order* of the terms matters. In the following, according to the 'abuse' of language in the physics literature, we shall call the decompositions "unordered partitions of the integer" (UPIs) and the compositions "ordered partitions of the integer" (OPIs).

The number of OPIs of N, say $O(N)$, can be easily computed as

$$O(N) = 2^{N-1}. \quad (51)$$

In other words, the self-similarity pattern can be phrased as "large integers consist of smaller integers, which in turn consist of still smaller integers, and so on...".

Starting with the integer N, we need to know the number $n(N,k)$ that specifies how often a given integer k occurs in the set of all OPIs of N, e.g., considering $N = 3$, we have $n(3,3) = 1$, $n(3,2) = 2$, and $n(3,1) = 5$. To apply the formalism of SOC, we associate a weight $s(k)$ to each integer. The natural choice is $s(k) = O(k) = 2^{k-1}$ and the number $n(N,k)$ we are looking for, in a scale-free scenario, is given by

$$n(N,k) = \alpha(N)[s(k)]^{-p}. \tag{52}$$

For small values of N, $n(N,k)$ is readily obtained explicitly and one finds that the critical exponent is $p \simeq 1.26$.

The previous example is immediately reminiscent of the statistical bootstrap model of Hagedorn [62,82–84]. There, we have "fireballs composed of fireballs, which in turn are composed of smaller fireballs, and so on". Indeed, its general pattern is shown to be due to an underlying structure related to the OPIs [85].

More precisely, Hagedorn's bootstrap approach proposes that a hadronic colorless state, with overall mass m, can be partitioned into structurally similar colorless states. Then, those component colorless states can be partitioned into structurally similar colorless states, and so on. If the states were at rest, the situation would be identical to the OPI just discussed. Since the constituent fireballs, though, have intrinsic motions, the number of states, $\rho(m)$, corresponding to a given mass m, is determined by the bootstrap equation, which can be asymptotically solved [83]. This gives $\rho(m) \sim m^{-a} \exp(m/T_H)$, and T_H is the solution of

$$\left(\frac{2}{3\pi}\right)\left(\frac{T_H}{m_0}\right) K_2(m_0/T_H) = 2\ln 2 - 1, \tag{53}$$

with m_0 denoting the lowest possible mass and $K_2(x)$ denoting a Hankel function of pure imaginary argument. For $m_0 = m_\pi \simeq 130$ Mev, this leads to the Hagedorn temperature

$$T_H \simeq 150 \text{MeV}, \tag{54}$$

that is, approximately, the critical hadronization temperature found in statistical QCD. The cited solution gives $a = 3$, but other exponents could also be considered.

The previous expression of $\rho(m)$ is an asymptotic solution of the bootstrap equation, which diverges for $m \to 0$; hence, it cannot hold for small masses. Using for $\rho(m)$ a result similar to the one obtained in the dual resonance model, Hagedorn proposed

$$\rho(m) = \text{const.}(1 + (m/\mu_0))^{-a} \exp(m/T_H), \tag{55}$$

where $\mu_0 \simeq 1 - 2$ GeV is a normalization constant.

We should emphasize that the form of $\rho(m)$ is entirely due to the self-organized nature of the system. That is in no way a result of thermal behavior. We expressed the slope coefficient of m in terms of the Hagedorn "temperature" only because we have the analog gravity scenarios in mind, which will soon be discussed; however, by itself, this coefficient is exclusively of combinatorial origin.

5.2. Analog Gravity Interpretation of the Partitions of Integers for BH Self-Similarity

The celebrated *self-similarity* at work in the hadronic spectrum, recalled in Section 5, is typical of many physical setups that enjoy scale invariance, such as fractals, phase transitions at the critical point, etc. [86]. Among those, BH self-similarity [87–89] is surely one of the most interesting, if one wants to probe fundamental ideas of QG.

Some aspects of BH self-similarity are understood if one recalls that the Hawking temperature, T_H, of a Planck-sized BH ($T_H \approx l_P^{-1}$, where l_P is the Planck length) could be viewed as the Hagedorn temperature in string theory [90–92]. At that temperature, BH evaporation stops and a phase transition is expected to occur, in analogy to what happens at the phase transition between the hadrons and the quark–gluon–plasma phase [93]. Nonetheless, to properly speak of self-similarity, one would need to make sense of statements, such as, "large BHs could be viewed as formed by smaller BHs, formed in turn of even smaller BHs, and so on..."

In the work [94], some steps were moved in that direction, and a link was established, in simple terms, between the spaces of BH configurations and the OPI. This, in turn, shed new light on BH self-similarity, in the plain terms of the statement quoted above. In what follows, let us comment on this.

First, the model we refer to is the so-called "quantum BH" of Mukhanov and Bekenstein [95–97]. In that approach, the area of the BH event horizon is quantized

$$A = \alpha N l_P^2, \tag{56}$$

where $N \in \mathbb{N}$ and the "it from bit" [98] choice for the proportionality factor, $\alpha = 4\ln 2$, allows for a two-level spin-1/2 system description, $\uparrow$ or $\downarrow$, per given Planck cell. With these, BH entropy, S_{BH}, can be written as

$$S_{BH} = \frac{A}{4 l_P^2} = N \ln 2, \tag{57}$$

which is the entropy of a quantum system living in a Hilbert (configuration) space of dimension $\dim H = 2^N$, where each 2^N configuration has the same statistical weight, e.g., see [99] for this and other approaches.

Thus, on the one hand, the number of OPIs of N, $O(N) = 2^{N-1}$, whereas the number of configurations of the quantum BH is given by $C(N) = 2^N$. Therefore, if we want to relate the two ways of counting configurations, one needs to find a 2-to-1 map from the latter to the former.

In [94], this is achieved by distinguishing between BH configurations, differing not only by how many spins are up and how many are down, as in other approaches [99], but also by the *position* of the spin. The "spin-flip map", there introduced, does the job of halving the number of BH configurations in a consistent way (to associate spin states, on the one hand, and with the OPI of N, on the other hand).

The 2-to-1 map works as follows: For any one given OPI of N, it associates the two BH states that are obtained (one from the other) when all the spins that identify the given configuration are flipped, $\uparrow \leftrightarrow \downarrow$. Then, the rule that relates a given *pair* of BH (spin) configurations to a given OPI is the following (for details see [94]):

When a spin is next to an opposite spin, i.e., when $\uparrow$ is next to $\downarrow$ or when $\downarrow$ is next to $\uparrow$, in the OPI this corresponds to $1 + 1$, e.g., $(\uparrow, \downarrow, \uparrow, \ldots)$, and the spin-flips $(\downarrow, \uparrow, \downarrow, \ldots)$ both correspond in the OPI to the partition $1 + 1 + \ldots$. When the spin is likewise, it contributes with an integer that is the sum of how many times the spin does not flip, e.g., $(\uparrow, \uparrow, \downarrow, \ldots)$ and $(\downarrow, \downarrow, \uparrow, \ldots)$ correspond in the OPI to the partition $2 + \ldots$.

With these, one takes into account all possibilities; hence, the wanted 2-to-1 map from the BH configurations to the OPI (the "spin-flip map") is obtained. Having established that, we want to see how the self-similarity patterns of the OPI can be imported into the self-similarity of BHs.

To avoid overcounting some configurations or missing others, in [94], the authors constructed an operation, $\hat{+}$, which allowed obtaining the configuration space of the given BH only once, for any given partition. If we indicate with $\mathbf{N}$ such 2^N-dimensional configuration space, and $N_1 + N_2 + \cdots = N$ is a given OPI of N, such an operation must give $\mathbf{N_1} \hat{+} \mathbf{N_2} \hat{+} \cdots = \mathbf{N}$. Doing so, we establish a one-to-one correspondence between the OPI of N, and the way to combine the subspaces of $\mathbf{N}$, corresponding to the OPI. We report here the actual definition of such an operation:

Take each partition of N, say $N_1 + N_2 = N$, and write the spin configuration space associated with the first number of the sum, $\mathbf{N_1}$. Then, take the tensor product of each representative with all of the spin configurations of $\mathbf{N_2}$, explicitly including all spin-flipped configurations. The result of such an operation, $\mathbf{N_1}\hat{+}\mathbf{N_2}$, is all of the spin configurations of $\mathbf{N}$, with no redundant or missed configuration. The operation gives the same result for each OPI of N, including those with more than two terms. For the latter, one must start from the first term on the left, act with the second (as described), and the result needs to be acted upon with the next term, and so on, until the end.

The trivial example is $\mathbf{N} = \mathbf{N}$, where no composition is performed. The first non-trivial operation is $\mathbf{1}\hat{+}\mathbf{1}$, which originates from the partition $1 + 1 = 2$, so it must give $\mathbf{2}$:

$$\mathbf{1}\hat{+}\mathbf{1} = \uparrow \otimes \begin{matrix}\uparrow\\\downarrow\end{matrix} = \begin{matrix}\uparrow\uparrow\\\uparrow\downarrow\end{matrix} = \mathbf{2}. \tag{58}$$

Indeed, in the second-last term, the first line is one spin representative of $\mathbf{2}$, $(\uparrow,\uparrow)$, while the second line is one spin representative of $1 + 1$, $(\uparrow,\downarrow)$. The four-dimensional ($2^N = 2^2$), full configuration space, $\mathbf{2}$, is obtained when we spin-flip each final configuration: $(\uparrow,\uparrow),(\downarrow,\downarrow)$ and $(\uparrow,\downarrow),(\downarrow,\uparrow)$. Notice that this is a general feature of this operation: one can consider even just one single representative per each spin-flipped pair of the first term, perform the operation as described earlier, and then to obtain all configurations at the end of the procedure—apply the spin-flip.

We are now where we want to be. When $\mathbf{N}$ is the configuration space of a Mukhanov–Bekenstein quantum BH, we have found the BH self-similarity; in plain terms, we were searching for:

The configuration space, $\mathbf{N}$, of a BH is made of the configuration spaces of smaller BHs, which are made of configuration spaces of even smaller BHs, and again and again, until we reach N copies of $\mathbf{1}$, the configuration space of the tiniest (elementary) BH.

To any of the 2^{N-1} OPIs of N, we can associate one of the 2^{N-1} OPIs of $\mathbf{N}$

$$\sum_i N_i = N \to \hat{\sum}_i \mathbf{N_i} = \mathbf{N}, \ \sum_j M_j = N \to \hat{\sum}_j \mathbf{M_j} = \mathbf{N}, \ldots, \tag{59}$$

where $\hat{\sum}_i \mathbf{N_i} = \mathbf{N_1}\hat{+}\mathbf{N_2}\hat{+}\cdots$, whatever pattern we find in the OPI of N, it is found in the configuration space $\mathbf{N}$ of the BH, and then repeated for the smaller numbers, until we reach the "quantum" of the BH space, $\mathbf{1}$.

A suggestive pattern is given by

$$\mathbf{N} = \mathbf{1}\hat{+}(\mathbf{N-1}) = \mathbf{1}\hat{+}(\mathbf{1}\hat{+}(\mathbf{N-2})) = \mathbf{1}\hat{+}(\mathbf{1}\hat{+}(\mathbf{1}\hat{+}(\mathbf{N-3}))) = \cdots = \hat{\sum}_{i=1}^{N}\mathbf{1}. \tag{60}$$

Here, one can say that when the configuration space of the tiniest BH, $\mathbf{1}$, is isolated from the rest, this can be repeated until the complete splitting.

As wanted, in this picture, self-similarity does not require any change of description of the degrees of freedom (e.g., from the evaporating BH to the long string [92], see also [90]). What one does there is finds patterns within the configuration space of a given fixed BH. We are not considering either BH *evaporation* or BH *merging* [100].

Let us conclude this part by saying that the constructions of [94] may solve the problem we started with. On the other hand, they lack any dynamical consideration whatsoever, as only kinematics was the concern there. No configuration is preferred to any other, by virtue of the dynamical properties of the system. In other words, all configurations were treated equally and this can only give back the entropy of (57), which, with a strong abuse of the language, since we are in a quantum BH model, is sometimes referred to as "classical entropy".

This is likely something that will be fully amended only by the long-sought-for final QG theory, see, e.g., [101], which will tell us how these fundamental (fermionic) degrees of freedom (see, e.g., [102–104]) interact, and some, $O(\ln N)$, "quantum corrections" have been put forward based on perturbative quantum considerations [105–110].

On the other hand, the simple (simplistic) approach of [94] has two advantages. First, it is based on a non-interacting (free) spin model that some authors also consider to be a viable candidate [102–104]. Second, in order to use an information–theoretical approach, the selection of specific configurations over others is not appropriate. In fact, if a (quantum) BH has to be used as the ultimate (quantum) computer [111], then one expects all configurations to be treated equally. The actual evolution of the quantum states should not be fixed by a given spin model, but rather be governed by a specific Hamiltonian that "implements" the given "computation".

6. Conclusions

The interpretation of quark confinement as the effect of an (event) *horizon* for color degrees of freedom naturally leads to the view of hadronization as 'quantum tunneling' through such a horizon. With this view, hadron formation is the result of an Unruh phenomenon, related to the string-breaking/string formation mechanism. This is because the large-distance QCD potential generates a constant and large *acceleration*, $a \simeq 3.2 \times 10^{33}$ m/s^2, which is precisely what we need for a measurable Unruh effect, $T_U \sim 4 \times 10^{11}$K ~ 170 MeV.

This opens up the way for a clear explanation of the thermal behaviors of both arenas—hadron physics and BH physics. For instance, this immediately explains why the hadronization temperature, T_h, is universal when seen as a T_U. Indeed, T_h is found to be the same for small and large initial collision settings, whereas T_U is fixed, once and for all, by the value of the acceleration, a. This also explains why hadrons are born in equilibrium.

In fact, the Hawking–Unruh radiation is an example of a *stochastic* rather than *kinetic* equilibrium. The reason behind the randomization is not repeated (as well as casual collisions among particles), but rather the quantum entanglement between the degrees of freedom on the two sides of the barrier to the information transfer, which is the event horizon. The temperature is then determined by the strength of the "confining" field.

In the chromodynamics counterpart of this phenomenon, described in this review, the ensemble of all produced hadrons, averaged over all events, leads to the same equilibrium distribution as obtained in the hadronic matter by kinetic equilibrium. For a very high-energy collision, with a high average multiplicity, even one event alone can provide such equilibrium. The destruction of memory, which in kinetic equilibrium is achieved through many successive collisions, is here automatically provided by the tunneling process.

The above are the physical fundamental aspects common to both types of phenomena. On this, the analogy can be solidly established, and many results can be obtained, i.e., the string-breaking and BH entropy analogies, which reproduce the "magic numbers" characterizing the freeze-out curve; the strangeness production, at low parton density, which is due to different Unruh temperatures in the single string-breaking; at high-energy and multiplicity, the large parton density, in the transverse plane, which removes the different temperatures by string (or color event horizon) overlap, giving the strangeness enhancement; or self-similar behavior, characteristic of the hadronic production, which has driven research into the self-similarity of BH configurations.

Let us then close on an optimistic note, by stating that this new and original analog system of QG has many other results to grasp.

In particular, Unruh radiation should exhibit both spatial and temporal coherence, reflecting its quantum origin. In our case, the spatial coherence should be observable by probing the phase correlation between particle jets. This correlation exists in the gravitational case, although it cannot be detected since one particle of the pair remains trapped inside the event horizon. Indeed, in the condensed matter analog of the Unruh effect, this correlation has been observed [112].

Another interesting aspect concerns the relation between 'de-confinement' and restoration of the chiral symmetry. The Rindler metric corresponds to the near-horizon approximation of a black-hole metric. On the other hand, in the near-horizon approximation, the field theory becomes conformal and effectively two-dimensional. Therefore, there is no way, in the near-horizon approximation, to maintain a physical scale generated by symmetry breaking. From this point of view, the Unruh hadronization temperature and the critical temperature of the restoration of chiral symmetry are deeply related.

Funding: P.C. and A.I. were supported by the Charles University Research Center (UNCE/SCI/013).

Data Availability Statement: Not applicable.

Acknowledgments: P.C. gladly acknowledges the kind hospitality from the Institute of Particle and Nuclear Physics of Charles University. A. I. thanks the INFN, Sezione di Catania, for supporting a visit to Catania University, where this work was completed.

Conflicts of Interest: The authors declare no conflict of interest.

Note

1. In fact, different physical systems, governed by different Hamiltonians, Lagrangians, and equations of motion (dynamics) may exhibit analog features, such as the emergence of some sort of horizon, as with the vast majority of cases used to probe the Hawking–Unruh phenomenon [1]. This is similar to taking a snapshot of the evolution of the analog system, precisely when this "looks like" the target system (or, we believe it should "look like" the target system). With this, we can study the behavior of the target system using the analog system at that particular stage of the evolution. It is much more important though to be able to keep going, even just a little bit. Namely, it is important that the evolution of the analog system is similar to the one of the target system, at least in certain conditions and within a limited range. When this happens, we have a much better analog that can furnish much more information on the target system (these are the analogs introduced in the famous Feynman lecture of electrostatics [4]). This is particularly important when one wants to face issues, such as black-hole (BH) evaporation, which is a phenomenon intimately associated with the dynamics of the gravitational field and something impossible to capture in a single "snapshot".

References

1. Barceló, C.; Liberati, S.; Visser, M. Analogue Gravity. *Living Rev. Relativ.* **2005**, *8*, 12. [CrossRef] [PubMed]
2. Muñoz de Nova, J.R.; Golubkov, K.; Kolobov, V.I.; Steinhauer, J. Observation of thermal Hawking radiation and its temperature in an analogue black hole. *Nature* **2019**, *569*, 688–691. [CrossRef] [PubMed]
3. Acquaviva, G.; Iorio, A.; Pais, P.; Smaldone, L. Hunting Quantum Gravity with Analogs: The case of graphene. *Universe* **2022**, *8*, 455. [CrossRef]
4. Feynman, R.; Leighton, R.; Sands, M. *The Feynman Lectures on Physics*; Number v. 2 in Addison-Wesley World Student Series; Addison-Wesley: Reading, MA, USA, 1963.
5. Kharzeev, D.; Tuchin, K. From color glass condensate to quark–gluon plasma through the event horizon. *Nucl. Phys. A* **2005**, *753*, 316–334. [CrossRef]
6. Castorina, P.; Kharzeev, D.; Satz, H. Thermal hadronization and Hawking–Unruh radiation in QCD. *Eur. Phys. J. C* **2007**, *52*, 187. [CrossRef]
7. Wald, R.M. Black hole entropy is the Noether charge. *Phys. Rev. D* **1993**, *48*, R3427–R3431. [CrossRef] [PubMed]
8. Chen, P.; Tajima, T. Testing Unruh Radiation with Ultraintense Lasers. *Phys. Rev. Lett.* **1999**, *83*, 256–259. [CrossRef]
9. Schützhold, R.; Schaller, G.; Habs, D. Signatures of the Unruh Effect from Electrons Accelerated by Ultrastrong Laser Fields. *Phys. Rev. Lett.* **2006**, *97*, 121302. [CrossRef]
10. Schützhold, R.; Schaller, G.; Habs, D. Tabletop Creation of Entangled Multi-keV Photon Pairs and the Unruh Effect. *Phys. Rev. Lett.* **2008**, *100*, 091301. [CrossRef]
11. Kim, C.M.; Kim, S.P. Unruh effect and Schwinger pair creation under extreme acceleration by ultraintense lasers. *arXiv* **2017**. [CrossRef]
12. O'Raifeartaigh, L.; Sreedhar, V. The Maximal Kinematical Invariance Group of Fluid Dynamics and Explosion–Implosion Duality. *Ann. Phys.* **2012**, *293*, 215–227. [CrossRef]
13. Sreedhar, V.V.; Virmani, A. Maximal Kinematical Invariance Group of Fluid Dynamics and Applications. *Universe* **2022**, *8*, 319. [CrossRef]
14. Drori, J.; Rosenberg, Y.; Bermudez, D.; Silberberg, Y.; Leonhardt, U. Observation of Stimulated Hawking Radiation in an Optical Analogue. *Phys. Rev. Lett.* **2019**, *122*, 010404. [CrossRef]
15. Iorio, A. Weyl-gauge symmetry of graphene. *Ann. Phys.* **2011**, *326*, 1334–1353. [CrossRef]
16. Iorio, A. Using Weyl symmetry to make graphene a real lab for fundamental physics. *Eur. Phys. J. Plus* **2102**, *127*, 156. [CrossRef]
17. Iorio, A.; Lambiase, G. The Hawking–Unruh phenomenon on graphene. *Phys. Lett. B* **2012**, *716*, 334–337. [CrossRef]
18. Iorio, A.; Lambiase, G. Quantum field theory in curved graphene spacetimes, Lobachevsky geometry, Weyl symmetry, Hawking effect, and all that. *Phys. Rev. D* **2014**, *90*, 025006. [CrossRef]
19. Iorio, A. Curved spacetimes and curved graphene: A status report of the Weyl symmetry approach. *Int. J. Mod. Phys. D* **2015**, *24*, 1530013. [CrossRef]
20. Iorio, A.; Pais, P. Revisiting the gauge fields of strained graphene. *Phys. Rev. D* **2015**, *92*, 125005. [CrossRef]

21. Iorio, A.; Pais, P. (Anti-)de Sitter, Poincaré, Super symmetries, and the two Dirac points of graphene. *Ann. Phys.* **2018**, *398*, 265–286. [CrossRef]
22. Iorio, A.; Pais, P. Generalized uncertainty principle in graphene. *J. Phys. Conf. Ser.* **2019**, *1275*, 012061. [CrossRef]
23. Iorio, A.; Pais, P.; Elmashad, I.A.; Ali, A.F.; Faizal, M.; Abou-Salem, L.I. Generalized Dirac structure beyond the linear regime in graphene. *Int. J. Mod. Phys. D* **2018**, *27*, 1850080. [CrossRef]
24. Ciappina, M.F.; Iorio, A.; Pais, P.; Zampeli, A. Torsion in quantum field theory through time-loops on Dirac materials. *Phys. Rev. D* **2020**, *101*, 036021. [CrossRef]
25. Iorio, A.; Lambiase, G.; Pais, P.; Scardigli, F. Generalized uncertainty principle in three-dimensional gravity and the BTZ black hole. *Phys. Rev. D* **2020**, *101*, 105002. [CrossRef]
26. Hawking, S.W. Particle creation by black holes. *Commun. Math. Phys.* **1975**, *43*, 199–220. [CrossRef]
27. Unruh, W.G. Notes on black-hole evaporation. *Phys. Rev. D* **1976**, *14*, 870–892. [CrossRef]
28. Crispino, L.C.B.; Higuchi, A.; Matsas, G.E.A. The Unruh effect and its applications. *Rev. Mod. Phys.* **2008**, *80*, 787–838. [CrossRef]
29. Guedes, T.L.M.; Kizmann, M.; Seletskiy, D.V.; Leitenstorfer, A.; Burkard, G.; Moskalenko, A.S. Spectra of Ultrabroadband Squeezed Pulses and the Finite-Time Unruh-Davies Effect. *Phys. Rev. Lett.* **2019**, *122*, 053604. [CrossRef]
30. Smolyaninov, I.I. Giant Unruh effect in hyperbolic metamaterial waveguides. *Opt. Lett.* **2019**, *44*, 2224–2227. [CrossRef] [PubMed]
31. Kalinski, M. Hawking radiation from Trojan states in muonic Hydrogen in strong laser field. *Laser Phys.* **2005**, *15*, 1357–1361.
32. Recami, E.; Castorina, P. On quark confinement: Hadrons as «strong black holes». *Lett. Al Nuovo Cimento (1971–1985)* **1976**, *15*, 347–350. [CrossRef]
33. Salam, A.; Strathdee, J. Confinement through tensor gauge fields. *Phys. Rev. D* **1978**, *18*, 4596–4609. [CrossRef]
34. Di Giacomo, A. Understanding Color Confinement. *EPJ Web Conf.* **2014**, *70*, 00019. [CrossRef]
35. Castorina, P.; Grumiller, D.; Iorio, A. Exact string black hole behind the hadronic Rindler horizon? *Phys. Rev. D* **2008**, *77*, 124034. [CrossRef]
36. Parikh, M.K.; Wilczek, F. Hawking Radiation As Tunneling. *Phys. Rev. Lett.* **2000**, *85*, 5042–5045. [CrossRef] [PubMed]
37. Vanzo, L.; Acquaviva, G.; Criscienzo, R.D. Tunnelling methods and Hawking's radiation: Achievements and prospects. *Class. Quantum Gravity* **2011**, *28*, 183001. [CrossRef]
38. Srednicki, M. Entropy and area. *Phys. Rev. Lett.* **1993**, *71*, 666–669. [CrossRef]
39. Terashima, H. Entanglement entropy of the black hole horizon. *Phys. Rev. D* **2000**, *61*, 104016. [CrossRef]
40. Iorio, A.; Lambiase, G.; Vitiello, G. Entangled quantum fields near the event horizon and entropy. *Ann. Phys.* **2004**, *309*, 151–165. [CrossRef]
41. Bekenstein, J.D. Black Holes and Entropy. *Phys. Rev. D* **1973**, *7*, 2333–2346. [CrossRef]
42. Hawking, S.W. Black hole explosions? *Nature* **1974**, *248*, 30–31. [CrossRef]
43. Laflamme, R. Entropy of a Rindler wedge. *Phys. Lett. B* **1987**, *196*, 449–450. [CrossRef]
44. Page, D.N. Information in black hole radiation. *Phys. Rev. Lett.* **1993**, *71*, 3743–3746. [CrossRef]
45. Castorina, P.; Iorio, A.; Lanteri, D.; Lukeš, P. Gluon shadowing and nuclear entanglement entropy. *Int. J. Mod. Phys. E* **2021**, *30*, 2150010. [CrossRef]
46. Kharzeev, D.E.; Levin, E.M. Deep inelastic scattering as a probe of entanglement. *Phys. Rev. D* **2017**, *95*, 114008. [CrossRef]
47. Parentani, R.; Massar, S. Schwinger mechanism, Unruh effect, and production of accelerated black holes. *Phys. Rev. D* **1997**, *55*, 3603–3613. [CrossRef]
48. Gabriel, C.; Spindel, P. Quantum Charged Fields in (1+1) Rindler Space. *Ann. Phys.* **2000**, *284*, 263–335. [CrossRef]
49. Narozhny, N.; Mur, V.; Fedotov, A. Pair creation by homogeneous electric field from the point of view of an accelerated observer. *Phys. Lett. A* **2003**, *315*, 169–174. [CrossRef]
50. De Gill, A.; Singleton, D.; Akhmedova, V.; Pilling, T. A WKB-like approach to Unruh radiation. *Am. J. Phys.* **2010**, *78*, 685–691. [CrossRef]
51. Becattini, F. A thermodynamical approach to hadron production in e+ e- collisions. *Z. Für Phys. C Part. Fields* **1995**, *69*, 485–492. [CrossRef]
52. Cleymans, J.; Satz, H.; Suhonen, E.; Von Oertzen, D. Strangeness production in heavy ion collisions at finite baryon number density. *Phys. Lett. B* **1990**, *242*, 111–114. [CrossRef]
53. Andronic, A.; Braun-Munzinger, P.; Redlich, K.; Stachel, J. Decoding the phase structure of QCD via particle production at high energy. *Nature* **2018**, *561*, 321–330. [CrossRef]
54. Letessier, J.; Rafelski, J.; Tounsi, A. Gluon production, cooling, and entropy in nuclear collisions. *Phys. Rev. C* **1994**, *50*, 406–409. [CrossRef] [PubMed]
55. Magas, V.; Satz, H. Conditions for confinement and freeze-out. *Eur. Phys. J. C-Part. Fields* **2003**, *32*, 115–119. [CrossRef]
56. Cleymans, J.; Redlich, K. Unified Description of Freeze-Out Parameters in Relativistic Heavy Ion Collisions. *Phys. Rev. Lett.* **1998**, *81*, 5284–5286. [CrossRef]
57. Braun-Munzinger, P.; Stachel, J. Particle ratios, equilibration and the QCD phase boundary. *J. Phys. G Nucl. Part. Phys.* **2002**, *28*, 1971–1976. [CrossRef]
58. Cleymans, J.; Oeschler, H.; Redlich, K.; Wheaton, S. Transition from baryonic to mesonic freeze-out. *Phys. Lett. B* **2005**, *615*, 50–54. [CrossRef]

59. Tawfik, A. Influence of strange quarks on the QCD phase diagram and chemical freeze-out. *J. Phys. G Nucl. Part. Phys.* **2005**, *31*, S1105–S1110. [CrossRef]
60. Lüscher, M.; Münster, G.; Weisz, P. How thick are chromo-electric flux tubes? *Nucl. Phys. B* **1981**, *180*, 1–12. [CrossRef]
61. Yndurain, F. *The Theory of Quark and Gluon Interactions*; Theoretical and Mathematical Physics, Springer: Berlin/Heidelberg, Germany, 2013.
62. Hagedorn, R. Statistical thermodynamics of strong interactions at high energies. *Nuovo Cimento Suppl.* **1965**, *3*, 147–186.
63. Hagedorn, R. Thermodynamics of strong interactions. *Cargese Lect. Phys.* **1973**, *6*, 643–716.
64. Castorina, P.; Iorio, A.; Satz, H. Hadron freeze-out and Unruh radiation. *Int. J. Mod. Phys. E* **2015**, *24*, 1550056. [CrossRef]
65. Solodukhin, S.N. Entanglement Entropy of Black Holes. *Living Rev. Relativ.* **2011**, *14*, 8. [CrossRef] [PubMed]
66. Castorina, P.; Iorio, A. Confinement horizon and QCD entropy. *Int. J. Mod. Phys. A* **2018**, *33*, 1850211. [CrossRef]
67. Becattini, F.; Manninen, J.; Gaździcki, M. Energy and system size dependence of chemical freeze-out in relativistic nuclear collisions. *Phys. Rev. C* **2006**, *73*, 044905. [CrossRef]
68. Grumiller, D.; Kummer, W.; Vassilevich, D. Dilaton gravity in two dimensions. *Phys. Rep.* **2002**, *369*, 327–430. [CrossRef]
69. Birmingham, D.; Gupta, K.S.; Sen, S. Near-horizon conformal structure of black holes. *Phys. Lett. B* **2001**, *505*, 191–196. [CrossRef]
70. Becattini, F.; Castorina, P.; Manninen, J.; Satz, H. The thermal production of strange and nonstrange hadrons in e+ e- collisions. *Eur. Phys. J. C* **2008**, *56*, 493–510. [CrossRef]
71. Becattini, F.; Castorina, P.; Milov, A.; Satz, H. A comparative analysis of statistical hadron production. *Eur. Phys. J. C* **2010**, *66*, 377–386. [CrossRef]
72. Adam, J.; Adamová, D.; Aggarwal, M.M.; Rinella, G.A.; Agnello, M.; Agrawal, N.; Ahammed, Z.; Ahmad, S.; Ahn, S.U.; Aiola, S.; et al. Enhanced production of multi-strange hadrons in high-multiplicity proton–proton collisions. *Nat. Phys.* **2017**, *13*, 535–539. [CrossRef]
73. Castorina, P.; Iorio, A.; Lanteri, D.; Satz, H.; Spousta, M. Universality in hadronic and nuclear collisions at high energy. *Phys. Rev. C* **2020**, *101*, 054902. [CrossRef]
74. Bylinkin, A.A.; Kharzeev, D.E.; Rostovtsev, A.A. The origin of thermal component in the transverse momentum spectra in high energy hadronic processes. *Int. J. Mod. Phys. E* **2014**, *23*, 1450083. [CrossRef]
75. Baker, O.K.; Kharzeev, D.E. Thermal radiation and entanglement in proton-proton collisions at energies available at the CERN Large Hadron Collider. *Phys. Rev. D* **2018**, *98*, 054007. [CrossRef]
76. Bylinkin, A.A.; Rostovtsev, A.A. Parametrization of the shape of hadron-production spectra in high-energy particle interactions. *Phys. At. Nucl.* **2012**, *75*, 999–1005. [CrossRef]
77. Gribov, L.; Levin, E.; Ryskin, M. Semihard processes in QCD. *Phys. Rep.* **1983**, *100*, 1–150. [CrossRef]
78. McLerran, L.; Venugopalan, R. Computing quark and gluon distribution functions for very large nuclei. *Phys. Rev. D* **1994**, *49*, 2233–2241. [CrossRef]
79. McLerran, L.; Venugopalan, R. Gluon distribution functions for very large nuclei at small transverse momentum. *Phys. Rev. D* **1994**, *49*, 3352–3355. [CrossRef]
80. Kharzeev, D.; Levin, E.; Tuchin, K. Multiparticle production and thermalization in high-energy QCD. *Phys. Rev. C* **2007**, *75*, 044903. [CrossRef]
81. Bak, P.; Tang, C.; Wiesenfeld, K. Self-organized criticality: An explanation of the 1/f noise. *Phys. Rev. Lett.* **1987**, *59*, 381–384. [CrossRef]
82. Frautschi, S. Statistical Bootstrap Model of Hadrons. *Phys. Rev. D* **1971**, *3*, 2821–2834. [CrossRef]
83. Nahm, W. Analytical solution of the statistical bootstrap model. *Nucl. Phys. B* **1972**, *45*, 525–553. [CrossRef]
84. Hagedorn, R.; Rafelski, J. Analytic structure and explicit solution of an important implicit equation. *Commun. Math. Phys.* **1982**, *83*, 563–578. [CrossRef]
85. Blanchard, P.; Fortunato, S.; Satz, H. The Hagedorn temperature and partition thermodynamics. *Eur. Phys. J. C-Part. Fields* **2004**, *34*, 361–366. [CrossRef]
86. Zinn-Justin, J. *Quantum Field Theory and Critical Phenomena*; International series of monographs on physics; Clarendon Press: Oxford, UK, 2002.
87. Harms, B.; Leblanc, Y. Statistical mechanics of black holes. *Phys. Rev. D* **1992**, *46*, 2334–2340. [CrossRef] [PubMed]
88. Harms, B.; Leblanc, Y. Statistical mechanics of extended black objects. *Phys. Rev. D* **1993**, *47*, 2438–2445. [CrossRef] [PubMed]
89. Huang, W.H. Microcanonical statistics of black holes and the bootstrap condition. *Phys. Rev. D* **2000**, *62*, 043002. [CrossRef]
90. Veneziano, G. The Hagedorn Spectrum and the Dual Resonance Model: An Old Love Affair. In *Melting Hadrons, Boiling Quarks—From Hagedorn Temperature to Ultra-Relativistic Heavy-Ion Collisions at CERN: With a Tribute to Rolf Hagedorn*; Rafelski, J., Ed.; Springer International Publishing: Cham, Switzerland, 2016; pp. 69–74. [CrossRef]
91. Susskind, L.; Thorlacius, L.; Uglum, J. The stretched horizon and black hole complementarity. *Phys. Rev. D* **1993**, *48*, 3743–3761. [CrossRef]
92. Susskind, L.; Lindesay, J. *An Introduction to Black Holes, Information and the String Theory Revolution*; World Scientific: London, UK, 2004. [CrossRef]
93. Cabibbo, N.; Parisi, G. Exponential hadronic spectrum and quark liberation. *Phys. Lett. B* **1975**, *59*, 67–69. [CrossRef]
94. Castorina, P.; Iorio, A.; Smaldone, L. Quantum black holes, partition of integers and self-similarity. *Mod. Phys. Lett. A* **2022**. [CrossRef]

95. Mukhanov, V.F. Are black holes quantized? *Sov. J. Exp. Theor. Phys. Lett.* **1986**, *44*, 63–66.
96. Bekenstein, J.D.; Mukhanov, V. Spectroscopy of the quantum black hole. *Phys. Lett. B* **1995**, *360*, 7–12. [CrossRef]
97. Bekenstein, J.D. Quantum Black Holes as Atoms. In *Recent Developments in Theoretical and Experimental General Relativity, Gravitation, and Relativistic Field Theories*; World Scientific: London, UK, 1999; p. 92.
98. Wheeler, J. *Information, Physics, Quantum: The Search for Links*; Pamphlets on Physics, Physics Department, Princeton University: Princeton, NJ, USA, 1989.
99. Kiefer, C. Aspects of Quantum Black Holes. *J. Physics: Conf. Ser.* **2020**, *1612*, 012017. [CrossRef]
100. Abbott, B.P.; Abbott, R.; Abbott, T.D.; Abernathy, M.R.; Acernese, F.; Ackley, K.; Adams, C.; Adams, T.; Addesso, P.; Adhikari, R.X.; et al. Observation of Gravitational Waves from a Binary Black Hole Merger. *Phys. Rev. Lett.* **2016**, *116*, 061102. [CrossRef] [PubMed]
101. Rovelli, C.; Press, C.U.; Landshoff, P.; Nelson, D.; Sciama, D.; Weinberg, S. *Quantum Gravity*; Cambridge Monographs on Mathematical Physics, Cambridge University Press: Cambridge, UK, 2004.
102. Acquaviva, G.; Iorio, A.; Scholtz, M. On the implications of the Bekenstein bound for black hole evaporation. *Ann. Phys.* **2017**, *387*, 317–333. [CrossRef]
103. Iorio, A. Two arguments for more fundamental building blocks. *J. Phys. Conf. Ser.* **2019**, *1275*, 012013. [CrossRef]
104. Acquaviva, G.; Iorio, A.; Smaldone, L. Bekenstein bound from the Pauli principle. *Phys. Rev. D* **2020**, *102*, 106002. [CrossRef]
105. Kaul, R.K.; Majumdar, P. Logarithmic Correction to the Bekenstein-Hawking Entropy. *Phys. Rev. Lett.* **2000**, *84*, 5255–5257. [CrossRef]
106. Gupta, K.S.; Sen, S. Further evidence for the conformal structure of a Schwarzschild black hole in an algebraic approach. *Phys. Lett. B* **2002**, *526*, 121–126. [CrossRef]
107. Ghosh, A.; Mitra, P. Log correction to the black hole area law. *Phys. Rev. D* **2005**, *71*, 027502. [CrossRef]
108. Majhi, A.; Majumdar, P. 'Quantum hairs' and entropy of the quantum isolated horizon from Chern–Simons theory. *Class. Quantum Gravity* **2014**, *31*, 195003. [CrossRef]
109. Singleton, D.; Vagenas, E.C.; Zhu, T. Self-similarity, conservation of entropy/bits and the black hole information puzzle. *J. High Energy Phys.* **2014**, *2014*, 74. [CrossRef]
110. Ong, Y.C. GUP-corrected black hole thermodynamics and the maximum force conjecture. *Phys. Lett. B* **2018**, *785*, 217–220. [CrossRef]
111. Lloyd, S. Ultimate physical limits to computation. *Nature* **2000**, *406*, 1047–1054. [CrossRef] [PubMed]
112. Hu, J.; Feng, L.; Zhang, Z.; Chin, C. Quantum simulation of Unruh radiation. *Nat. Phys.* **2019**, *15*, 785–789. [CrossRef]

Review

Hunting Quantum Gravity with Analogs: The Case of Graphene [†]

Giovanni Acquaviva [1], Alfredo Iorio [2,*], Pablo Pais [2,3] and Luca Smaldone [4]

1 Arquimea Research Center, Camino de las Mantecas, 38320 Santa Cruz de Tenerife, Spain
2 Faculty of Mathematics and Physics, Charles University, V Holešovičkách 2, 18000 Prague 8, Czech Republic
3 Instituto de Ciencias Físicas y Matemáticas, Universidad Austral de Chile, Casilla 567, Valdivia 5090000, Chile
4 Faculty of Physics, Institute of Theoretical Physics, University of Warsaw, Ul. Pasteura 5, 02-093 Warsaw, Poland
* Correspondence: alfredo.iorio@mff.cuni.cz
† This paper is an extended version of our paper published in Alfredo Iorio. Analog hep-th, on Dirac Materials and in General. In Proceedings of the Corfu Summer Institute 2019 "School and Workshops on Elementary Particle Physics and Gravity" (CORFU2019), 31 August–25 September 2019.

Abstract: Analogs of fundamental physical phenomena can be used in two ways. One way consists in reproducing specific aspects of the classical or quantum gravity of quantum fields in curved space or of other high-energy scenarios on lower-energy corresponding systems. The "reverse way" consists in building fundamental physical theories, for instance, quantum gravity models, inspired by the lower-energy corresponding systems. Here, we present the case of graphene and other Dirac materials.

Keywords: analogs; Dirac materials; quantum gravity

1. Introduction

Richard Feynman wrote beautiful and visionary pages on analogs, in a famous lecture titled "Electrostatic Analogs", available in [1] (see also [2] for comments and discussions). There, he explains *how* it happens that different physical systems, among which a solid analogy can be established, are all described in a unified manner. In his now famous words, this happens because "the same equations have the same solutions". Therefore, if we have no access to certain regimes of system A, but they correspond to certain reachable regimes of the analogous system B, we can perform experiments on system B, and establish results that are valid for system A.

In the final, and less known, part of the lecture, he ventures into a visionary attempt to explain *why* this is so. This goes on until the thrilling hypothesis of the existence of *more elementary constituents* than the ones we deem to be fundamental. All those systems, including electrostatics itself, are just different coarse-grained versions of one dynamics, even more fundamental than quantum electrodynamics. Amazingly, Feynman realizes that the physical properties of space itself play a crucial role in the identification of such fundamental objects (that he calls "little Xons" [1]). It is precisely when space itself, besides matter, is included as part of the emergent phenomenon that these are also the conclusions of certain completely independent arguments of the contemporary quantum gravity (QG) [3–7].

As for the field of *gravity* analogs (see [8]), the seminal work is the 1981 paper of Bill Unruh [9], where he proposed to search for the experimental signatures of the Unruh effect [10] and of the Hawking effect [11], in a fluid-dynamical analog.

Due to our deeper theoretical understanding of these phenomena and to the higher experimental control of condensed matter systems, it is now becoming increasingly popular to reproduce that and other aspects of fundamental physics in analog systems. Examples include the Hawking phenomenon in Bose–Einstein condensates [12], the Weyl symmetry [13,14] and the related Hawking/Unruh phenomenon on graphene [15–17], gravitational and axial anomalies in Weyl semimetals [18], "Moiré gravity" in bilayer graphene [19,20]), and more.

Let us mention, for instance, the beautiful example of supernova explosions simulated in the laboratory by plasma implosions induced by intense lasers. Both systems are examples of fluid dynamics, and the Euler equations are invariant under an inversion transformation, which is an arbitrary uniform expansion or contraction of the system. This symmetry is studied in cosmology, and allows to map an explosion problem to a dual implosion problem. In principle, this duality allows the complete three-dimensional evolution of highly structured explosion ejecta to be modeled using a static target in an implosion facility. In [21], the maximal invariance group was determined to be the semi-direct product of the Galilei group with $SL(2, R)$, the latter containing time translations, dilations, and the inversion. Those results had an important impact on the field. More details are in the contribution [22] to this Issue.

Gravity analogs are not limited to condensed-matter systems, as shown by heavy-ion collisions at high energy [23–25]. In fact, hadron production in high-energy scattering processes is just a Unruh effect in quantum chromodynamics (QCD) [23], with the Unruh temperature, $T_U = (\hbar a)/(2\pi c k_B)$, given by the hadronization temperature, $T_h \simeq 160$ MeV, thus corresponding to an enormous acceleration, $a \simeq 4.6 \times 10^{32}$ m/s^2 that makes the effect very easy to detect. Here, we explicitly write the Planck constant, speed of light and Boltzmann constant. More details on this approach are in the contribution [26] to this Issue.

Here, we focus on the proposal of graphene as an analog of high-energy fundamental physics [13–17,27–29], based on the fact that its low-energy excitations [30] are massless Dirac pseudo-relativistic fermions (the matter fields ψ), propagating in a carbon two-dimensional honeycomb lattice. The emergent (long-wave limit) description of the latter is a surface (a portion of spacetime described by the "emergent" metric $g_{\mu\nu}$). Such a behavior is shared by a wide range of materials, ranging from silicene and germanene through d-wave superconductors to topological insulators [31]. Each of those materials has its own peculiarities, which allow for further extensions of results obtained with graphene, and hence permit to explore a wider range of the high-energy target systems. Let us now give some details.

Despite those impressive advances in the highly active area of analog physics, there are still two milestones to reach. One is to fully understand the epistemic role of analogs in fundamental high energy physics, as not all theorists would agree that analogs are much more than mere *divertissements*. In fact, experimental results obtained with analogs are not used as feedback for the target theories that they are analogs of (see, for example, [2,32]). Another milestone would be a reliable definition of an analog black-hole (BH) entropy, or at least, of a quantum field theory (QFT)-like entanglement entropy that, in the presence of horizons, might serve the scope of setting-up some form of the second principle of BH thermodynamics.

Any progress in this direction would be truly important for the QG research. Having some results there, we could eventually be able to address the so-called *information paradox*, i.e., the apparent loss of information during BH evaporation, a question that, most probably, cannot be entirely solved via theoretical reasonings. See, for example, [33–37] for different points of view.

In fact, there are plenty of unreachable regimes in fundamental physics, starting from BHs, that we do know to exist but that are not (easily or at all) reproducible in a laboratory. It is then of tremendous interest to establish solid criteria for such systems to correspond to other systems, within our reach, and to perform experiments on the latter to know of the former. On the other hand, when such correspondences are solidly established, why not infer from the analog system the most intimate nature of the target system? For instance, if QG behaves like graphene (under certain conditions for graphene and for certain specific regimes of QG), and since we still do not know how QG really is, why not trying to guess the whole QG picture from what we learned of the partial overlap between the two systems?

This is a less beaten track, but not a completely empty one. For instance, inspired by the findings of [13–17,27–29], in Ref. [6], the authors propose the existence of fundamental, high-energy constituents underlying both matter and space, and that these, at our low energies, exist in an entangled state. This entanglement is there because both matter and space emerge from the dynamics of the same more fundamental objects, whose existence can be inferred from the celebrated upper bound on the entropy of any system, conjectured by Bekenstein [3]. Quoting Feynman and paraphrasing Bekenstein, those objects are called "Xons" [2]. If such a view is correct, even matter that we deem to be fundamental, i.e., elementary, is in fact "quasi-matter", just like the massless quasi-particles, ψ, of graphene [30] that owe their properties to the interaction with the lattice[1]. The most noticeable result of this "quasi-particle picture" [6] is that the evaporation of a BH inevitably leads to an information loss, in the sense that, in general, there is a nonzero entanglement entropy associated to the final products of the evaporation. On the other hand, within the same picture, in [40], the authors describe BH evaporation from the point of view of the Xons. They see there that the Bekenstein bound [3–5] can be an effect of the Pauli exclusion principle, and that a full unitary picture, leading to a complete recovering of the initial information, is only possible if one could track the evolution of those fundamental constituents.

The paper is organized in two large Sections and some concluding remarks. Section 2 is dedicated to graphene and Dirac materials (DMs) as analogs of high-energy fundamental physics. Section 3 is dedicated to the QG that the latter research has inspired. Each large Section has many Subsections. As for Section 2, Sections 2.1–2.3 explain the main reasons why graphene is good at reproducing scenarios of fundamental physics; then Section 2.4 tells about old, new and future developments of this line of research, dedicating to each topic a brief Subsubsection; finally, Section 2.5 comments on the experimental search. As for Section 3, Section 3.1 introduces the quasi-particle picture in the QG context; Sections 3.2 and 3.3 deal with BH evaporation as seen from the quasi-particles and as seen from the Xons, respectively; the last Section 3.4 comments on recent work on how (classical) space emerges from the underlying (quantum) dynamics of Xons during BH evaporation. Section 4 is dedicated to our concluding remarks, which are a chance to point to future developments of the whole analog enterprise, in general, and those based on graphene, in particular.

2. Analog Gravity on Graphene

Graphene is an allotrope of carbon. It is one-atom-thick; hence it is the closest to a two-dimensional object in nature. It was theoretically speculated [41,42] and, decades later, it was experimentally found [43]. Its honeycomb lattice is made of two intertwined triangular sub-lattices L_A and L_B; see Figure 1. As is now well known, this structure is behind a natural description of the electronic properties of π electrons[2] in terms of massless, $(2+1)$-dimensional, Dirac (hence, relativistic-like) quasi-particles.

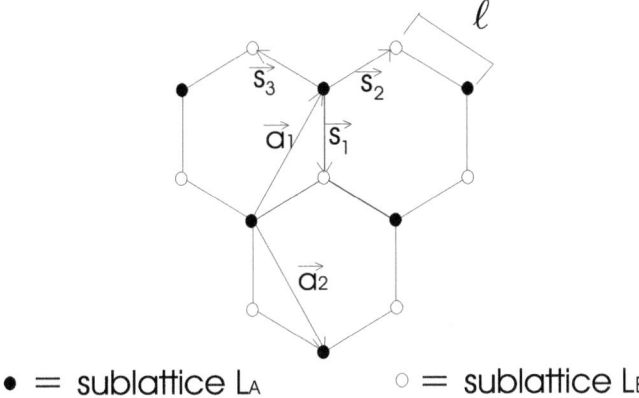

Figure 1. The honeycomb lattice of graphene, and its two triangular sublattices L_A and L_B. The choice of the basis vectors, $(\vec{a}_1, \vec{a}_2)$ and $(\vec{s}_1, \vec{s}_2, \vec{s}_3)$, is, of course, not unique. Here we indicate the one used in [17]. Figure taken from [16].

2.1. First Scale, $E < E_\ell$: From the Tight-Binding to the Dirac Hamiltonian

Such electrons, in the tight-binding low-energy approximation, are customarily described by the Hamiltonian (as here we use natural units, the reduced Planck constant is $\hbar = 1$)

$$H = -\eta \sum_{\vec{r} \in L_A} \sum_{i=1}^{3} \left(a^\dagger(\vec{r}) b(\vec{r} + \vec{s}_i) + b^\dagger(\vec{r} + \vec{s}_i) a(\vec{r}) \right), \quad (1)$$

where the nearest-neighbor hopping energy is $\eta \simeq 2.8 \, \text{eV}$, and a, $a^\dagger$ (b, $b^\dagger$) are the anti-commuting annihilation and creation operators, respectively, for the planar π electrons in the sub-lattice L_A (L_B); see Figure 1. All the vectors are bi-dimensional, $\vec{r} = (x, y)$, and, for the choice of basis vectors made in Figure 1, if we Fourier transform, $a(\vec{r}) = \sum_{\vec{k}} a(\vec{k}) e^{i \vec{k} \cdot \vec{r}}$, etc., then $H = \sum_{\vec{k}} (f(\vec{k}) a^\dagger(\vec{k}) b(\vec{k}) + \text{h.c.})$, with

$$f(\vec{k}) = -\eta \, e^{-i \ell k_y} \left(1 + 2 \, e^{i 3 \ell k_y / 2} \cos(\sqrt{3} \ell k_x / 2) \right), \quad (2)$$

where $\ell \simeq 1.4 \, \text{Å}$ is the graphene honeycomb lattice length (see Figure 1). Solving $E(\vec{k}) = \pm |f(\vec{k})| \equiv 0$ tells us if, in the first Brillouin zone (FBZ), conductivity and valence bands touch and where. Indeed, this does happen for graphene, pointing to a gapless spectrum, for which we expect massless excitations to emerge. Furthermore, the solution is not a Fermi *line* (the $(2+1)$-dimensional version of the Fermi surface of the $(3+1)$ dimensions), but instead, they are two Fermi *points*, $\vec{k}_\pm^D = \left(\pm \frac{4\pi}{3\sqrt{3}\ell}, 0 \right)$. Even if the mathematical solution to $|f(\vec{k})| = 0$ has six points, only the two indicated are unequivalent [30].

The label "D" on the Fermi points stands for "Dirac". This refers to the all-important fact that, near those points, the spectrum is *linear*, as can be seen from Figure 2, $E_\pm \simeq \pm v_F |\vec{k}|$, where $v_F = 3\eta\ell/2 \sim c/300$ is the *Fermi velocity*. This behavior is expected in a relativistic theory, whereas, in a non-relativistic system, the dispersion relations are usually quadratic.

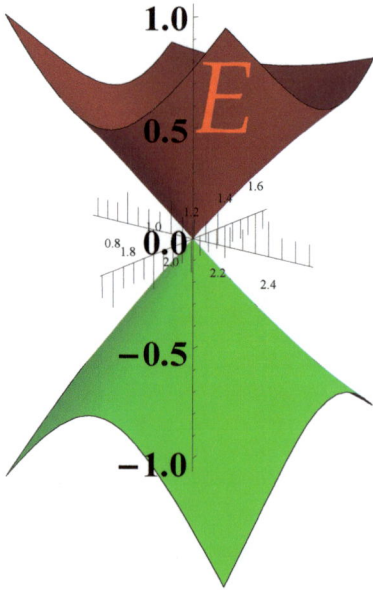

Figure 2. The linear dispersion relations near one of the Dirac points, showing the typical behavior of a relativistic-like system (the "v_F-light-cone" in k-space). Figure taken from [17].

If one linearizes around $\vec{k}_\pm^D$, $\vec{k}_\pm \simeq \vec{k}_\pm^D + \vec{p}$, then $f_+(\vec{p}) \equiv f(\vec{k}_+) = v_F(p_x + ip_y)$, $f_-(\vec{p}) \equiv f(\vec{k}_-) = -v_F(p_x - ip_y)$, and $a_\pm(\vec{p}) \equiv a(\vec{k}_\pm)$, $b_\pm(\vec{p}) \equiv b(\vec{k}_\pm)$. Therefore, the Hamiltonian (1) becomes

$$H|_{\vec{k}_\pm} \simeq v_F \sum_{\vec{p}} \left(\psi_+^\dagger \vec{\sigma} \cdot \vec{p}\, \psi_+ - \psi_-^\dagger \vec{\sigma}^* \cdot \vec{p}\, \psi_- \right) \qquad (3)$$

where $\psi_\pm \equiv \begin{pmatrix} b_\pm \\ a_\pm \end{pmatrix}$ are two-component Dirac spinors, and $\vec{\sigma} \equiv (\sigma_1, \sigma_2)$, $\vec{\sigma}^* \equiv (\sigma_1, -\sigma_2)$, with σ_i the Pauli matrices. Notice that here the 1/2-spinor description emerges from the two sublattice honeycomb structure instead of the intrinsic spin of the π electron.

Hence, if one considers the linear/relativistic-like regime only, the first scale is

$$E_\ell \sim v_F/\ell \sim 4.2\,\text{eV}. \qquad (4)$$

Notice that $E_\ell \sim 1.5\eta$, and that the associated wavelength, $\lambda = 2\pi/|\vec{p}| \simeq 2\pi v_F/E$, is $2\pi\ell$. The electrons' wavelength, at energies below E_ℓ, is large compared to the lattice length, $\lambda > 2\pi\ell$. Those electrons see the graphene sheet as a continuum.

The two spinors are connected by the inversion of the full momentum $\vec{k}_+^D + \vec{p} \to -\vec{k}_+^D - \vec{p} \equiv \vec{k}_-^D - \vec{p}$. Whether one needs one or both such spinors to describe the physics strongly depends on the given set-up. For instance, when only strain is present, one Dirac point is enough (see, for example, [27]), similar (see below here) to when certain approximations on the curvature are valid [15–17]. The importance and relevance of the two Dirac points for emergent descriptions of scenarios of the high-energy theoretical research were discussed at length in [28], where the role of grain boundaries and the related necessity for two Dirac points were explained in terms of a relation to spacetime torsion; see below. The full focus on torsion, though, is in [44].

When only one Dirac point is necessary over the whole linear regime, the following Hamiltonian well captures the physics of undeformed (planar and unstrained) graphene

$$H = -i v_F \int d^2x\, \psi^\dagger \vec{\sigma} \cdot \vec{\partial}\, \psi, \qquad (5)$$

where the two-component spinor is, for example, $\psi \equiv \psi_+$, we moved back to configuration space, $\vec{p} \to -i \vec{\partial}$, and sums turned into integrals because of the continuum limit. In various papers, this regime was exploited to a great extent until the inclusion of curvature and torsion in the geometric background. On the other hand, the regimes beyond the linear one were also investigated. There, granular effects associated with the lattice structure emerge; see [45] and also the related [46].

When both Dirac points are necessary, one needs to consider four-component spinors in a reducible representation [17,47,48] $\Psi \equiv \begin{pmatrix} \psi_+ \\ \psi_- \end{pmatrix}$, and 4×4 Dirac matrices $\alpha^i = \begin{pmatrix} \sigma^i & 0 \\ 0 & -\sigma^{*i} \end{pmatrix}$, $\beta = \begin{pmatrix} \sigma^3 & 0 \\ 0 & \sigma^3 \end{pmatrix}$, $i = 1, 2$. These matrices satisfy all the standard properties, see, e.g., [17,28].

With these, the Hamiltonian is

$$H = -i v_F \int d^2x \left(\psi_+^\dagger \vec{\sigma} \cdot \vec{\partial}\, \psi_+ - \psi_-^\dagger \vec{\sigma}^* \cdot \vec{\partial}\, \psi_- \right) = -i v_F \int d^2x\, \bar{\Psi} \vec{\gamma} \cdot \vec{\partial}\, \Psi. \qquad (6)$$

2.2. Second Scale, $E < E_r < E_\ell$: From the Flat Space to Curved Space Dirac Hamiltonian

In [13], the goal was to identify the conditions for graphene to get as close as possible to a full-power QFT in curved spacetime. Therefore, key issues had to be faced, such as the proper inclusion of the time variable in a relativistic-like description and the role of the nontrivial vacua and their relation to different quantization schemes for different observers. All this finds its synthesis in the Unruh or the Hawking effects, the clearest and unmistakable signatures of QFT in curved spacetime. Therefore, starting from [13,14], this road was pursued in [15,16]. Let us explain here the main issues and the approximations made there.

Besides the scale (4), when we introduce curvature, we also have a second scale. When this happens, E_ℓ is our "high energy regime", as we ask the curvature to be small compared to a maximal limiting curvature, $1/\ell^2$, otherwise: i) it would make no sense to consider a smooth metric, and ii) $r < \ell$ (where $1/r^2$ measures the intrinsic curvature), means that we should bend the very strong σ-bonds, an instance that does not occur. Therefore, our second scale is

$$E_r \sim v_F/r, \tag{7}$$

with $E_r = \ell/r \, E_\ell < E_\ell$. To have a quantitative handle on these scales, let us take, e.g., $r \simeq 10\ell$ as a small radius of curvature (high intrinsic curvature). To this corresponds an energy $E_r \sim 0.4\,\text{eV}$, whereas to $r \sim 1\,\text{mm} \sim 10^6\,\ell$, corresponds $E_r \sim 4\,\mu\text{eV}$. The "high energy" to compare with is $E_\ell \sim 4\,\text{eV}$.

When energies are within E_r (wavelengths comparable to $2\pi r$), the electrons experience the global effects of curvature. That is to say, at those wavelengths, they can distinguish between a flat and curved surface and, in particular, between, for example, a sphere and a pseudosphere. Therefore, whichever curvature $r > \ell$ we consider, the curvature effects are felt until the wavelength becomes comparable to $2\pi\ell$. The formalism we have used, though, considers all deformations of the geometric kind, except for torsion. Hence, this includes the intrinsic curvature and elastic strain of the membrane (on the latter, see [27]). However, the power stops before E_ℓ because there, local effects (such as the actual structure of the defects) play a role that must be taken into account in a QG-type theory. On the latter, the first steps were moved in [45] and also in [46] and in the forthcoming [49].

The intrinsic curvature is taken here as produced by disclination defects, that are customarily described in elasticity theory (see, for example, [50]), by the (smooth) derivative of the (non-continuous) $SO(2)$-valued rotational angle $\partial_i \omega \equiv \omega_i$, where $i = 1, 2$ is a "curved" spatial index[3]. The corresponding (spatial) Riemann curvature tensor is easily obtained

$$R^{ij}{}_{kl} = \epsilon^{ij}\epsilon_{kl}\epsilon^{mn}\partial_m\omega_n = \epsilon^{ij}\epsilon_{lk}2\mathcal{K}. \tag{8}$$

where $\mathcal{K}$ is the Gaussian (intrinsic) curvature of the surface. In this approach, we have included time, although the metric we adopted is

$$g_{\mu\nu}^{\text{graphene}} = \begin{pmatrix} 1 & 0 & 0 \\ 0 & & \\ 0 & & -g_{ij} \end{pmatrix}, \tag{9}$$

i.e., the curvature is all in the spatial part, and $\partial_t g_{ij} = 0$. Since the time dimension is included, the $SO(2)$-valued disclination field has to be lifted up to a $SO(1,2)$-valued (non-abelian) disclination field[4], $\omega_\mu{}^a$, $a = 0, 1, 2$, with $\omega_\mu^a = e_\mu^b \omega_b^a$, and the expression

$$\omega_a^d = \frac{1}{2}\epsilon^{bcd}\left(e_{\mu a}\partial_b E_c^\mu + e_{\mu b}\partial_a E_c^\mu + e_{\mu c}\partial_b E_a^\mu\right), \tag{10}$$

gives the relation between the disclination field and the metric (dreibein). All the information about intrinsic curvature does not change. For instance, the Riemann curvature tensor, $R^\lambda{}_{\mu\nu\rho}$, has only one independent component, proportional to $\mathcal{K}$, just like in (8) (see [13]).

With all of the above in mind, the hypothesis is that, when only curvature is important, the long wavelength/small energy electronic properties of graphene are well described by the following action

$$\mathcal{A} = i v_F \int d^3x \sqrt{g} \, \bar{\Psi}\gamma^\mu(\partial_\mu + \Omega_\mu)\Psi, \tag{11}$$

where $\Omega_\mu \equiv \omega_\mu{}^a J_a$, and J_a are the generators of $SO(1,2)$, the local Lorentz transformations in this lower-dimensional setting. Notice that J_a can never take into account the mixing of the $\psi_\pm$ because they are of the form $J^a = \begin{pmatrix} j_+^a & 0 \\ 0 & j_-^a \end{pmatrix}$, whereas what is necessary are generators of the form $K^a = \begin{pmatrix} 0 & k_+^a \\ k_-^a & 0 \end{pmatrix}$. This point was discussed at length in [28], within the Witten approach [51]. In that approach, the most general gauge field that takes into account curvature (intrinsic and extrinsic) and torsion has the following structure $A_\mu = \Omega_\mu + K_\mu$, where $K_\mu \equiv e^a{}_\mu K_a$, hence a Poincaré ($ISO(2,1)$) or (A)dS type of gauge theory, depending on the role played in here by the cosmological constant (on this see [15,16], and the review [17]). The matter, though, might be faced by taking an alternative view, for which the gauge fields are internal rather than spatiotemporal. In this case, a link with the supersymmetry (SUSY) introduced in [52] (that is a SUSY without superpartners, often referred to as *unconventional* SUSY (USUSY)) can be established, as is shown in [28] and in [53–55], as is briefly discussed in Section 2.4.2.

Let us clarify here an important point. Within this scenario, a nontrivial g_{00} in (9), and hence a clean nontrivial general relativistic effect (recall that $g_{00} \sim V_{\text{grav}}$) can only happen if specific symmetries and set-ups map the lab system into the wanted one. A lot of work went into it, e.g., [15,16], and went as far as producing measurable predictions of a Hawking/Unruh effect, for certain specific shapes. Let us recall here the main ideas behind this approach, which we may call the "Weyl symmetry approach" [17].

2.3. The Importance of Weyl Symmetry

First of all, one notices that the action (11) enjoys local Weyl symmetry

$$g_{\mu\nu} \to e^{2\sigma(x)} g_{\mu\nu} \quad \text{and} \quad \Psi \to e^{-\sigma(x)} \Psi, \tag{12}$$

that is an enormous symmetry among fields/spacetimes [56]. As explained in [13,14], to make the most of the Weyl symmetry of (11), we better focus on conformally flat metrics. The simplest metric to obtain in a laboratory is of the kind (9). For this metric, the Ricci tensor is $R_\mu{}^\nu = \text{diag}(0, \mathcal{K}, \mathcal{K})$. This gives as the only nonzero components of the Cotton tensor, $C^{\mu\nu} = \left(\epsilon^{\mu\sigma\rho} \nabla_\sigma R_\rho{}^\nu + \mu \leftrightarrow \nu\right)$, the result $C^{0x} = -\partial_y \mathcal{K} = C^{x0}$ and $C^{0y} = \partial_x \mathcal{K} = C^{y0}$. Since conformal flatness in $(2+1)$ dimensions amounts to $C^{\mu\nu} = 0$, this shows that all surfaces of constant $\mathcal{K}$ give rise in (9) to conformally flat $(2+1)$-dimensional spacetimes. This points the light-spot to surfaces of constant Gaussian curvature.

The result $C^{\mu\nu} = 0$ is intrinsic (it is a tensorial equation, true in any frame), but to exploit Weyl symmetry to extract non-perturbative exact results, we need to find the coordinate frame, say $Q^\mu \equiv (T, X, Y)$, where

$$g_{\mu\nu}^{\text{graphene}}(Q) = \phi^2(Q) g_{\mu\nu}^{\text{flat}}(Q). \tag{13}$$

Besides the technical problem of finding these coordinates, the issue to solve is the physical meaning of the coordinates Q^μ, and their practical feasibility. See [17,57].

Tightly related to the previous point is the conformal factor that makes the model *globally predictive, over the whole surface/spacetime*. The simplest possible solution would be a single-valued, and time independent $\phi(q)$, already in the original coordinates frame, $q^\mu \equiv (t, u, v)$, where t is the laboratory time, and, for example, u, v are the meridian and parallel coordinates of the surface.

Here, we are dealing with a spacetime that is embedded into the flat $(3+1)$-dimensional Minkowski. Although, as said, the focus is on intrinsic curvature effects, just like in a general relativistic context, issues related to the embedding, even just for the spatial part, are important. For instance, when the surface has negative curvature, one needs to move from the abstract objects of non-Euclidean geometry, to objects measurable in a Euclidean real laboratory. This involves the last point above about global predictability, and, in the case of negative curvature, necessarily leads to singular boundaries for the surfaces, as proved in a theorem by Hilbert, see, for example, [17,58]. Even the latter fact is, once more, a coordinates effect, due to our insisting in embedding a negative curvature surface in $\mathbb{R}^3$, and clarifies the hybrid nature of these emergent relativistic settings. The quantum vacuum of the field that properly takes into account the measurements processes, as for any QFT on a curved spacetime, was identified, including how the graphene hybrid situation can realize that [15,16]. As well known, this is crucial in QFT, in general, and on curved space, in particular.

The above leads us to propose a variety of set-ups, the most promising being the one obtained by shaping graphene as a Beltrami pseudosphere [15–17], a configuration that can be put into contact with three key spacetimes with horizon: the Rindler, the de Sitter and the Bañados–Teitelboim–Zanelli (BTZ) BH [59]. The predicted impact on measurable quantities is reported in the first papers, and then explored in the subsequent efforts of computer-based simulations.

2.4. Ramifications

Many other high energy scenarios can be reached with graphene and related systems that go under the name of Dirac materials (DMs) [31]. Here, we list some such directions.

2.4.1. Generalized Uncertainty Principles on DMs

In Ref. [45] (see also [46]), the realization in DMs of specific generalized uncertainty principles (GUPs) associated with the existence of a fundamental length scale was studied. The scenarios that one wants to reproduce there is that for which the commutation relations are modified by quantum gravity effects to be (see, for example, [60–69] and references therein)

$$[x_i, p_j] = i\hbar \left(\delta_{ij} - A \left(|\vec{p}|\delta_{ij} + \frac{p_i p_j}{|\vec{p}|} \right) + A^2 \left(|\vec{p}|^2 \delta_{ij} + 3 p_i p_j \right) \right), \qquad (14)$$

where $A = \tilde{A} \ell_P / \hbar$, with $\tilde{A}$ a phenomenological dimensionless parameter and $\ell_P \sim 10^{-35}$ m the Planck length.

In Ref. [45], it is shown that a generalized Dirac structure survives beyond the linear regime of the low-energy dispersion relations. Additionally, a GUP of the kind compatible with (14) related to QG scenarios with a fundamental minimal length (there, the graphene lattice spacing) and Lorentz violation (there, the particle/hole asymmetry, the trigonal warping, etc.) is naturally obtained. It is then shown that the corresponding emergent field theory is a table-top realization of such scenarios by explicitly computing the third-order Hamiltonian and giving the general recipe for any order. Remarkably, these results imply that going beyond the low-energy approximation does not spoil the well-known correspondence with analog massless quantum electrodynamics phenomena (as usually believed). Instead, it is a way to obtain the experimental signatures of quantum-gravity-like corrections to such phenomena.

In Ref. [46], the authors investigated the structure of the gravity-induced GUP in $(2+1)$-dimensions. They showed that the event horizon of the $M \neq 0$ BTZ micro-blackhole furnishes the most consistent limiting "gravitational radius" R_g (that is, the fundamental minimal length induced by gravitational effects). A suitable formula for the GUP and estimate the corrections induced by the latter on the Hawking temperature and Bekenstein entropy could be obtained. As for the role of graphene, it is shown that the extremal $M = 0$ case, and its natural unit of length introduced by the cosmological constant, $\ell = 1/\sqrt{-\Lambda}$, is a possible alternative to R_g, and DMs when shaped as hyperbolic pseudospheres represent condensed matter analog realizations of this scenario with $\ell = \ell_{DM}$. Due to the peculiarities of three-dimensional gravity [70], this configuration can still be regarded as a BH, even though $M = 0$; on this, see, for example, [71–73].

More work in this QG phenomenology direction is forthcoming [49]. There, it is found that even more GUPs are at work at different energy scales, and a link is established between the abstract coordinates satisfying the GUPs and the coordinates one measures in the lab.

With this in mind, one sees that our scales here are much more within reach than those of (14). Indeed, ℓ_P needs to be traded for the lattice spacing ℓ, that, for example, for graphene is $\ell_{graphene} \sim 1.4 \times 10^{-10}$ m. Therefore, we have much more hope to see in DMs the effects of the modifications to $[x_i, p_j] = i\hbar\,\delta_{ij}$ compared with the direct effects of $O(\ell_P)$.

2.4.2. Grain Boundaries on DMs and Two Scenarios: Witten 3D Gravity, and Ususy

In Ref. [28], two different high-energy-theory correspondences on DMs associated with *grain boundaries* (GBs) are proposed. We recall here that a GB can be realized as a line of disclinations of opposite curvature, for instance, pentagons and heptagons, arranged so that two regions (grains) of the membrane match. These grains have different relative orientations, given by the so-called misorientation angle θ, which characterizes the GB defect. Each side of the GB corresponds to one of the Dirac points (and the other is related by a parity transformation, see Appendix B of [28] for details) in the continuous π electron description. Therefore, the continuous limit description of the π electrons living in a honeycomb with GB needs the two inequivalent Dirac points. Even more, as the θ angle is related to a non-zero Burgers vector $\vec{b}$ through the Frank formula, and a non-zero $\vec{b}$ implies non-zero torsion in the continuous limit[5], such a description should take into account torsion.

The first correspondence points to a $(3+1)$-dimensional theory, with spatiotemporal gauge group $SO(3,1)$, with nonzero torsion, locally isomorphic to the Lorentz group in $(3+1)$ dimensions, or the de Sitter group in $(2+1)$ dimensions, in the spirit of $(2+1)$-dimensional gravity à la Witten [51]. The other correspondence treats the two Dirac fields as an internal symmetry doublet, and it is linked there with USUSY [52] with $SU(2)$ internal symmetry [53]. One of the properties of USUSY is the absence of gravitini, although it includes gravity and supersymmetry. Even if in $(2+1)$ dimensions it is constructed from a Chern–Simons connection containing fermion fields, the only propagating local degrees of freedom are the fermions [75]. Notice that in USUSY, the torsion of geometric backgrounds appears naturally, and its fully antisymmetric part is coupled with fermions.

Those results pave the way for the inclusion of GB in the emergent field theory picture associated with these materials, whereas disclinations and dislocations have already been well explored.

2.4.3. Particle–Hole Pairs in Graphene to Spot Spatiotemporal Torsion

In Ref. [44], assuming that dislocations could be meaningfully described by torsion, a scenario is proposed based on the role of time in the low-energy regime of two-dimensional DMs, for which coupling of the fully antisymmetric component of the torsion with the emergent spinor is not necessarily zero. That approach is based on the realization of an exotic *time loop*, that could be seen as oscillating particle–hole pairs. Although that is a theoretical paper, the first steps were moved toward testing the laboratory realization of these scenarios by envisaging *Gedankenexperiments* on the interplay between an external electromagnetic field (to excite the particle–hole pair and realize the time loops) and a suitable distribution of dislocations described as torsion (responsible for the measurable holonomy in the time loop, hence a current). The general analysis establishes that we need to move to a nonlinear response regime. Then the authors conclude by pointing to recent results from the interaction of laser–graphene that could be used to look for manifestations of the torsion-induced holonomy of the time loop, e.g., as specific patterns of suppression/generation of higher harmonics. As said before, USUSY takes into account torsion and couples its fully antisymmetric component with fermions in a very natural way. Therefore, it could play a significant role also in this exotic time loop [76].

2.4.4. Vortex Solutions of Liouville Equation and Quasi-Spherical Surfaces

In Ref. [57], the authors identified the two-dimensional surfaces corresponding to specific solutions of the Liouville equation of importance for mathematical physics, the non-topological Chern–Simons (or Jackiw–Pi [77,78]) vortex solutions, characterized by an integer [79] $N \geq 1$. Such surfaces, called $S^2(N)$, have positive constant Gaussian curvature, K, but are spheres only when $N = 1$. They have edges and, for any fixed K, have a maximal radius c that is found there to be $c = N/\sqrt{K}$. If such surfaces are constructed in a laboratory using DMs, these findings could be of interest to realize table-top Dirac massless excitations on nontrivial backgrounds. Then the types of three-dimensional spacetimes obtained as the product $S^2(N) \times \mathbb{R}$ are also briefly discussed.

2.5. Realization in the Labs

Besides the theoretical work just outlined, one should always aim at the actual realization of the necessary structures in real laboratories. See, for example, the work [58], where Lobachevsky geometry was realized via simulations by producing a carbon-based mechanically stable molecular structure arranged in the shape of a Beltrami pseudosphere. It was found there that this structure (i) corresponds to a non-Euclidean crystallographic group, namely a loxodromic subgroup of $SL(2, \mathbb{Z})$, and (ii) has an unavoidable singular boundary that is fully taken into account. That approach, substantiated by extensive numerical simulations of Beltrami pseudospheres of different sizes, might be applied to other surfaces of constant negative Gaussian curvature, and points to a general procedure to generate them. Such results pave the way for future experiments. More work is currently undergoing.

3. Graphene-Inspired Quantum Gravity: The Quasiparticle Picture

If the entropy of any physical system of volume V, including the entropy associated to space itself, is never bigger than the entropy of the BH whose event horizon coincides with the boundary of V [3]

$$S \leq S_{BH}, \tag{15}$$

this means that the associated Hilbert space, H, has finite dimension, $\dim(H) \sim e^{S_{BH}}$. This simple consideration poses serious questions.

In fact, at our energy scales, the world is well described by fields (matter) and the space they live in. Quantum fields, as we know them, act on infinite-dimensional Hilbert spaces, to which one should add the degrees of freedom surely carried by (the quanta of) space itself. How can then be that the *ultimate* Hilbert space, which must include all degrees of freedom, is not only separable, like for a single harmonic oscillator, but is actually finite-dimensional?

This logic points to the existence of something more fundamental, making both matter and space. Hence, the *elementary* particles of the standard model (leptons, quarks, etc.) would be, in fact, quantum *quasi-particles*, whose physical properties (spin, mass, etc.) are the effect of the interaction with a *lattice* whose emergent picture is, in turn, (classical) space. Inspired by Feynman [1] (see the Introduction here) these objects were called *Xons* [2]. To access the Xons, one needs resolutions of the order of the Planck length, which might not only be technically unfeasible, but actually impossible; see, for example, [80].

In Ref. [6], and later in [40], general arguments are provided regarding the connection between our low-energy quantum-matter-on-classical-space description and an hypothetical fundamental theory of the Xons. The reshuffling of the fundamental degrees of freedom during the unitary evolution then leads to an entanglement between space and matter. The consequences of such a scenario are considered in the context of BH evaporation (see, for example, [81–83]) and the related information loss: a simple toy model is provided in which an average loss of information is obtained as a consequence of the entanglement between matter and space. Pivotal for the previous study is the work of [84], where the Hawking–Unruh phenomenon is studied within an entropy–operator approach, à la thermo-field dynamics (TFD) [85,86] that discloses the thermal properties of BHs.

3.1. The Universal Quasiparticle Picture

Emergent, nonequivalent descriptions of the same underlying dynamics are ubiquitous in QFT [87], as, in general, the vacuum has a nontrivial structure with nonequivalent[6] "phases" [86]. That is, for a given basic dynamics (governed by an Hamiltonian or a

Lagrangian), one should expect several different Hilbert spaces, representing different "phases" of the system with distinct physical properties. Distinct excitations play the role of the elementary excitations for the given "phase", but their general character is that of the quasiparticles of condensed matter [85,86].

What it is added here to that QFT picture is the following:

- The degrees of freedom are finite, hence fields are necessarily emergent;
- Spacetime is also emergent.

Taking this view, the continuum of fields and space is then only the result of an approximation, of a limiting process. In general, there must be (many) microscopic configurations of the Xons giving rise to the *same* emergent space but to *different/non-equivalent* fields.

With this in mind (for details, see [6]), the generic state $|\psi\rangle \in H$ can be written as

$$|\psi\rangle = \bigoplus_{i=1}^{N_T} \sum_{I=1}^{p_i} \sum_{n=0}^{q_i-1} c_{In}^{(i)} |I_i\rangle \otimes |n_i\rangle, \qquad (16)$$

where the vectors $|I_i\rangle$ and $|n_i\rangle$ form a basis of $H_G^{p_i}$ and $H_F^{q_i}$, that are the Hilbert space of the "spatial degrees of freedom" (geometry) of dimension p_i and of the Hilbert space of the "matter degrees of freedom" (fields) of dimension q_i, respectively, and $c_{In}^{(i)}$ are numerical coefficients. Notice that N_T is the number of specific rearrangements (topologies) of the degrees of freedom.

By denoting with $P_{(i)} : H \mapsto T_{(i)}$ a projector onto $T_{(i)}$, a subspace with a given "topology", the associated density matrix, representing the state of the field, is

$$\rho_{(i)} = \text{Tr}_{H_G^{p_i}} |\psi\rangle_i \langle\psi|_i, \qquad (17)$$

where $|\psi\rangle_i \sim P_{(i)}|\psi\rangle$, and we trace away the degrees of freedom of the gravitational field. Correspondingly, the entropy of entanglement between matter and space, for a given topology of the lattice, is the usual expression[7]

$$S_{(i)} = -\text{Tr}_{H_F^{q_i}} \rho_{(i)} \ln \rho_{(i)}. \qquad (18)$$

This picture needs to be compared to the standard QFT picture, recalled earlier, of the non-equivalent field configurations, or "phases" *à la* TFD [85,86] where the mirror degrees of freedom, that characterize TFD (often called there the *tilde* degrees of freedom), model the degrees of freedom of the geometry. These degrees of freedom are then traced away, leaving us with quantities all referring solely to matter (fields). Indeed, the vacuum of TFD can be written as [85]

$$|0(\theta)\rangle = \sum_n \sqrt{w_n(\theta)} |n, \tilde{n}\rangle, \qquad (19)$$

where θ is a physical parameter labeling the different "'phases", w_n are probabilities such that $\sum_n w_n = 1$, and the states $|n, \tilde{n}\rangle$ (infinite in number) are the components of the condensate, each made of pairs of n quanta and their n mirror counterparts ($\tilde{n}$). Therefore, such a vacuum is clearly an entangled state. Notice that [85]

$$\langle 0(\theta)|0(\theta')\rangle \to 0, \qquad (20)$$

in the field limit, which formalizes the inequivalence we discussed. Notice also that, if one fixes θ, there is no unitary evolution to disentangle the vacuum, as the interaction with the environment and non-unitarity are the basis for the generation and the stability of such an entanglement [84].

The expected value of *field's* observables, O, are obtained by tracing away the *mirror* modes, $\tilde{n}$. In the TFD formalism, this corresponds to taking the vacuum expectation value over the vacuum (19)

$$\langle O \rangle \equiv \langle 0(\theta)|O|0(\theta)\rangle = \sum_n w_n(\theta) \langle n|O|n \rangle. \tag{21}$$

In particular, there is always an *entanglement entropy* associated to any field, given by, for example,

$$\langle S \rangle = \sum_k [n_k \ln n_k + (1-n_k)\ln(1-n_k)]. \tag{22}$$

where $n_k = \langle N_k \rangle$ is the expected value of the number operator for the given (fermionic, in this example) mode k. The analogy of (22) with (18) is stronger, if we think that in TFD the process of taking statistical averages through tracing is replaced, by construction [85], by taking vacuum expectation values (vevs) over the vacuum (19). Furthermore, as is well known, in the basis where the density matrix in the entropy (18) is diagonal, the entropy can be written as

$$\langle S \rangle = -\sum_n w_n(\theta) \ln w_n(\theta), \tag{23}$$

as shown, for example, in [86].

In this comparison, the mirror (tilde) image of the field mimics the effects of the entanglement with space where the field lives, even when the space is flat. This happens on a level that is both emergent and effective. This would have far reaching consequences, surely worth a serious exploration. For instance, the entanglement entropy associated to any field, would never be zero. Furthermore, this would explain why the attempt to quantize gravity as we quantize the matter fields, cannot make much sense.

To compare TFD entropies and the entropies obtained in the quasi-particle picture, a different point of view is taken in [40]. There, the authors focus on BH evaporation as seen from the point of view of the fundamental *X*ons, and were able to establish formulae and structures indeed similar to those of TFD. The main difference with TFD is that, at the level of the discrete structures related to *X*ons, the quantum field theoretical considerations illustrated above are only an approximation. In Section 3.3, we recall those results. Before doing so, let us focus on BH evaporation as seen from the point of view of the emergent quantum fields and emergent space.

3.2. Effects of the Quasiparticle Picture on Black Hole Evaporation

When applied to BH evaporation, the immediate consequence of the above is that it is impossible after the evaporation to retrieve the very same "phase" we had before the BH was formed. Hence, the information associated to the quantum fields before the formation of the BH is, in general, lost after the BH has evaporated, due to the entanglement between matter and space.

Even when the emergent spaces, before the formation and after the evaporation, are the same (say they are both Minkowski spacetimes), the emergent fields belong, in general, to non-equivalent Hilbert spaces. Therefore, even assuming unitary evolution at the X level, the initial and final Hilbert spaces of fields cannot be the same. There is always a *relic* matter–space entanglement entropy.

Looking at Equation (16), it is clear that the Hilbert space H can be written as

$$\mathsf{H} = \bigoplus_{i=1}^{N_T} \mathsf{H}_G^{p_i} \otimes \mathsf{H}_F^{q_i} \tag{24}$$

where we can now introduce measures, R_Fs, R_Gs, of the "degeneracies", $p_i = N_G R_G^i$, with the N_G classical geometries available (they represent the BH with mass $M^{(a)} = a\varepsilon$, where $a = 0, 1, \ldots N_G - 1$), and each classical geometry can be realized by R_G^i microstates.

On the other hand, $q_i = N_F R_F^i$, that is, each emergent field state can be realized by R_F^i indistinguishable microstates.

The analytic computations of the entanglement entropy demand a heavy toll, so in [6], the authors proceeded numerically. The case we present here is for the following choice of $N_G = 30$, $N_T = 2$, and $R_F^i = 1$, for each topology. The plots in Figure 3 show the entanglement entropies, corresponding to the three sets of values given in the box, as functions of the discrete evolution parameter k.

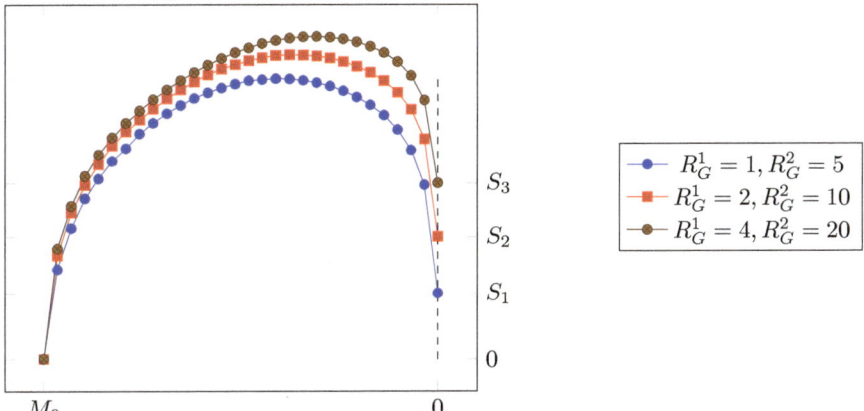

Figure 3. Entropy of the entanglement between matter and space, as a function of the decreasing mass of the evaporating BH. The initial and final points of this curve are in exact correspondence with the initial and final points of the Page curve. The plot here is for two topologies and three cases. The more microscopic realizations of macroscopic classical geometries are allowed, the higher the residual entropies. Here, $S_1 = 0.77$, $S_2 = 1.43$, $S_3 = 2.06$. Figure taken from [6].

As can be seen from the figure, the residual entropies are never zero, and are given by

$$S_1 = 0.77, \quad S_2 = 1.43, \quad S_3 = 2.06, \tag{25}$$

corresponding to the set of values in the box going from the top to the bottom, respectively. The more microscopic realizations of the same macroscopic geometry (i.e., the bigger the degeneracy R_G), the higher the relic entanglement entropy. This is as it must be.

The fact that, at the end of the evaporation, the entanglement entropy remains finite signals a dramatic departure from the *information conservation* scenario of the famous Page curve [82], presented here in Figure 4. There, the total Hilbert space has the dimension mn, and consists of two subsystems: the BH subsystem, of dimension $n \sim e^{A/4}$, where A is the area of the event horizon, and the radiation subsystem, of dimension $m \sim e^{s_{th}}$, where s_{th} is the thermodynamic radiation entropy. In Page's picture, there is no explicit mention of the degrees of freedom of space, and the evolution is taken to be unitary. Thus, in that picture, one sees that, when the BH is formed, there is no Hawking radiation outside; hence, $m = 1$ and $n = \dim \mathcal{H}$. The BH-radiation entanglement entropy, $S_{m,n}$ is trivially zero. As the BH evaporates, m increases, while n decreases, keeping $m \, n$ constant. Since the emitted photons are entangled with the particles under the horizon, $S_{m,n}$ increases, but only up to, approximately, half of the evaporation process. There, the information stored below the horizon starts to leak from the BH, so that $S_{m,n}$ decreases until full evaporation; hence $n = 1$ and $m = \dim \mathcal{H}$ and $S_{m,n}$ returns to zero.

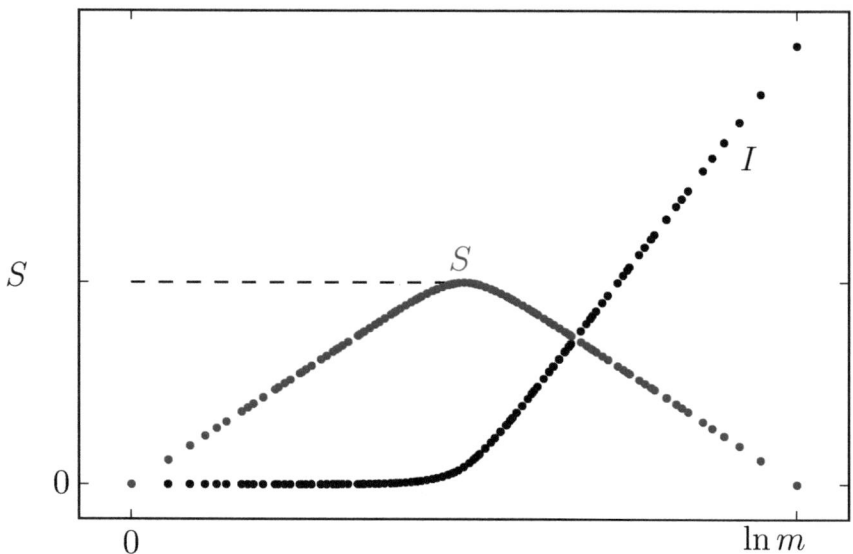

Figure 4. Page curve, representing the entanglement between matter modes inside the BH and matter modes of the radiation leaving the BH (in this picture there is no explicit reference to the degrees of freedom related to space) vs. the log of the dimension m of the Hilbert space of the radiation, obtained in [81,82]. The point $\ln m = 0$ corresponds to the initial mass of the BH, $M = M_0$. Indeed, $m = 1$ means that only the vacuum state populates the radiation subsystem of the Hilbert space, at the start of the evaporation. On the other hand, m_{max} corresponds a fully evaporated BH, $M = 0$. Figure taken from [6].

From the point of view of the quasiparticle picture, we may say that, even if one takes a conservative view for which the Xons evolve unitarily, nonunitarity is unavoidable:

- The unitary evolution may as well be only formally possible, but physically impossible to measure, for some form of a generalized uncertainty forbidding the necessary Planck scale localization/resolution (see, for example, [80]).
- The emergent description of the evolution is that of the combined system gravity + matter, and hence there is inevitably information loss, due to the relic entanglement of the matter field with the space.
- This description should apply also to standard nonunitary features of QFT, and we evoke here the possibility that the tilde degrees of freedom of TFD could be interpreted as "how the emergent fields see the degrees of freedom of space with which they are entangled".

Notice that this description does allow for an arbitrary number of different fields, and hence naturally includes the possibility of yet unknown ("dark") kinds of matter.

3.3. BH Evaporation as Seen from the Xons and the Unification of the Entropies

In Ref. [40], the authors describe BH evaporation from the point of view of the fundamental constituents, assuming they are fermions, so that only one excitation per quantum level is permitted. Because the Xons must be responsible for the formation of both matter and space, no geometric notions can be used. For example, it is assumed that only a finite number N of quantum states/slots are available to the system. This last condition is a non-geometric way of requiring that the system is localized in space. Moreover, it is not meaningful to refer to the *interior* and the *exterior* of a BH. Instead, the authors there distinguish between *free* and *interacting* Xons, respectively: BH evaporation is the process in which the number of *free* Xons decreases, $N \to (N-1) \to (N-2) \to \cdots$, while interacting Xons form matter (quasi-particles) and space (geometry), i.e., the environment.

The Hilbert space of *physical states* H is the subspace of a larger *kinematical* Hilbert space $K \equiv H_I \otimes H_{II}$, and it has dimension $\Sigma \equiv \dim H = 2^N$. Here I and II refer to BH and environment, respectively, in the sense explained above.

The state of the system $|\Psi(\sigma)\rangle \in H$ is [40]

$$|\Psi(\sigma)\rangle = \prod_{i=1}^{N} \sum_{n_i=0,1} C_i(\sigma) \left(a_i^\dagger\right)^{n_i} \left(b_i^\dagger\right)^{1-n_i} |0\rangle_I \otimes |0\rangle_{II}, \qquad (26)$$

where a and b are environment and BH ladder operators, respectively, and

$$C_i(\sigma) = (\sin\sigma)^{n_i} (\cos\sigma)^{1-n_i}. \qquad (27)$$

σ is an interpolating parameter going from 0 to $\pi/2$. We can also define TFD-like entropy operators

$$S_I(\sigma) = -\sum_{n=1}^{N} \left(a_n^\dagger a_n \ln\sin^2\sigma + a_n a_n^\dagger \ln\cos^2\sigma\right). \qquad (28)$$
$$S_{II}(\sigma) = -\sum_{n=1}^{N} \left(b_n^\dagger b_n \ln\cos^2\sigma + b_n b_n^\dagger \ln\sin^2\sigma\right). \qquad (29)$$

so that their averages on $|\Psi(\sigma)\rangle$ give the von-Neumann entropy of the two subsystems:

$$\mathcal{S}_I(\sigma) = \mathcal{S}_{II}(\sigma) = -N\left(\sin^2\sigma \ln\sin^2\sigma + \cos^2\sigma \ln\cos^2\sigma\right). \qquad (30)$$

Such entropy quantifies the entanglement between the environment and the BH. As for the original Page result, the entropy (30) shows that the BH evaporation at such fundamental level is a unitary process, with $\mathcal{S}(0) = \mathcal{S}(\pi/2) = 0$ and a maximum value $\mathcal{S}_{max} = N\ln 2 = \ln \Sigma$, so that $\Sigma = e^{\mathcal{S}_{max}}$. $\mathcal{S}_{max}$ quantifies the maximum information necessary in order to describe the BH and should be identified with the BH entropy before the evaporation. When the BH evaporates the mean number $N_{II}(\sigma) = N\cos^2\sigma$ of free Xons decreases, while the mean number of interacting Xons $N_I(\sigma) = N\sin^2\sigma$ increases. Then, BH and environment entropy should be

$$\mathcal{S}_{BH} = N\ln 2\cos^2\sigma, \qquad \mathcal{S}_{env} = N\ln 2\sin^2\sigma. \qquad (31)$$

Moreover, σ finds a natural explanation as a discrete parameter in the interval $[0, \pi/2]$, essentially counting the diminishing number of free Xons.

The entanglement, BH and environment entropy satisfy $\mathcal{S}_I \leq \mathcal{S}_{BH} + \mathcal{S}_{env} = \mathcal{S}_{max}$: the entropy of both BH and environment is bounded from above, in accordance with the Bekenstein bound. In Figure 5, these three entropies are plotted as functions of the discrete parameter σ.

For a full identification of $\mathcal{S}_{max}$ with the entropy $\mathcal{S}_{BH}$ of the initial BH, one would have (for a non-rotating, uncharged black hole)

$$N = \frac{4\pi M_0^2}{l_P^2 \ln 2}, \qquad (32)$$

where M_0 is the initial BH mass. Finally, identifying the N quantum levels with the quanta of area, in the spirit of Refs. [89–92], one obtains

$$A = 4Nl_P^2 \ln 2, \qquad (33)$$

for the BH horizon area. Notice that the value of $\alpha \equiv A/(Nl_P^2)$ could be inferred from measurement on BH quasinormal modes [93].

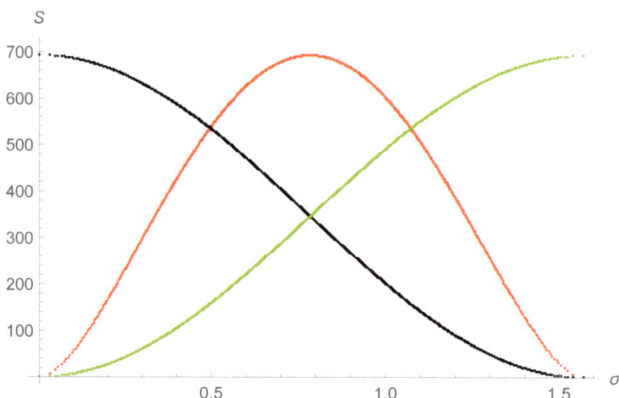

Figure 5. Plot of S_{BH} (black), S_{env} (green) and S_I (red), as function of σ, for $N = 1000$. Figure taken from [40].

3.4. Topological Phases and the Emergence of Space from Evaporating BHs

How is the existence of different phases of matter compatible with the finiteness of degrees of freedom? Such an issue is closely related with the evasion of the Stone–von Neumann theorem [94–96]. In fact, it is known that in quantum mechanics, all continuous, irreducible representations of Weyl–Heisenberg (for bosons) or Clifford (for fermions) algebra, are unitarily equivalent. However, as it was previously noted, such a theorem does not apply to QFT, where systems with an infinite number of degrees of freedom are studied [97–100]. The existence of unitarily inequivalent representations of canonical (anti)commutation relations permits to describe transitions among disjoint phases of the same system, in the QFT framework.

However, it is known that it is also possible to evade the Stone–von Neumann theorem by relaxing the continuity hypothesis [101]. This has been shown in quantum mechanical systems with a multiple-connected configuration space [102–104] or in polymer quantum mechanics [105–107].

In Ref. [108], the authors studied an example where both *thermodynamical* and *topological* disjoint phases are realized: a vortex solution in a QFT with a spontaneously broken $U(1)$ symmetry was analyzed by means of the *boson transformation method* [109–111]. Such an idea, firstly developed by H. Umezawa and collaborators, permits to describe classical extended objects emerging from an underlying QFT, by means of a canonical transformation performed on bosonic quasi-particle fields, which induces an inhomogeneous condensate on vacuum. Then, the authors showed that spontaneous symmetry breaking (SSB) is indeed possible, even when the volume of the system stays finite [108]. This represents a first step to understand the emergence of different phases in the Xons model.

The above method also permits to shed some light on the mechanism of the formation of space and quasi-particles from an underlying Xons dynamics. In Ref. [112], the authors face the delicate and fascinating issue of how space itself might be viewed as a classical extended object stemming from the SSB of underlying quantum dynamics, with the associated Goldstone bosons. In that case, discrete Xons Ψ_j are approximated by a field $\Psi(x)$, and the space structure and geometric tensors (metric, curvature, torsion) emerge as a result of the condensation of Goldstone bosons, while quasi-particles are described by fields on a classical (curved) space.

4. Concluding Remarks and Future Perspectives of the Graphene Analog Enterprise

QG and other fundamental scenarios can be tested also with analog experiments. In fact, the exciting and rapidly evolving field of analog physics is facing a new era. The interest is shifting from the reproduction of the *kinematical* aspects of the Hawking/Unruh phenomenon that has reached a climax of precision and accuracy, to the realization of some form of BH (thermo)dynamics. The latter is a challenging problem, but given its importance, even a partial solution is surely worth the effort.

The primary goals of the research in this field should then be to search for realizations of such dynamical aspects, and to learn from the above on QG. Here we have described the results found following the road of graphene. Let us now collect the many directions we see departing from there.

4.1. Hunting for Analog BH (Thermo)Dynamics

A conservative approach to BH evaporation [81,82] assumes that the evolution of the collapsing matter to produce a BH and its subsequent evaporation is a unitary process. This is what we would like to test in our analog systems. Indeed, current ongoing work [112] primarily focuses on the emergence of space in a QG scenario, as described in Section 3.4; henceforth, from there we are on the hunt for a BH dynamics on graphene and other DMs. In fact, the results of that general work will help us construct an experimentally sound geometry/gravity theory that describes the dynamics of the elastic DM membrane and explore the relations to existing gravity models. Having an action, we would be able to compute the Wald entropy in the usual way [113].

With this in mind, we are studying the realization of BHs on DM, based on the discoveries we summarized in Section 2.4. One important case under scrutiny is that of the BTZ BH realized using hyperbolic pseudospheres [46]. We shall operate through theoretical investigations but will interact more and more with the experimentalists to test the formulae obtained in [15,16] (or variants, obtained by refining earlier computations in the light of the new results; see, for example, [28]), and we shall produce more predictions of this kind for different samples' morphologies and various graphene observables.

In the "time-wise" approach, the focus will be on reproducing BH (and other nontrivial) emergent metrics, by suitably engineering the interaction of the electromagnetic field with the appropriate DM. The basis for the study are two kinds of results obtained in earlier investigations and discussed here. On the one hand is the emergence of the Hawking/Unruh effect for specific spatial geometries. On the other hand is the great level of accuracy reached with laser pulses to control spatial and temporal resolution for graphene's electrons dispersion relation [114]. The latter results inspired Ref. [44], where important details are obtained that will pave the way to a full understanding of how to engineer suitable temporal components of the emergent metric, and how to control their dynamics. The two approaches are, of course, tightly related, as one goal will be to rephrase the spatial analysis of previous work into a temporal language, namely by identifying the appropriate transformations among spatial and temporal nontriviality of the emergent metric, and by envisaging the physical setups that could realize those metrics in a laboratory, for instance in the laser–DM interaction.

On the more proper QG side, we expect that lattice effects will play a role, even within the continuous approximation regime [27], but surely at the "very high energy" regimes, where the linear approximation no longer works, these aspects become dominant. In the latter regimes, the (pseudo-)relativistic structure of the Dirac field will be deformed, and the discrete nature of the space(time) becomes so important that the continuum description, in terms of smooth metrics, will no longer be valid. This will become an important point to enforce the analogy with QG scenarios of the discrete spacetime. Indeed, the results in [45], where the natural analog of the Planck length is the lattice spacing ℓ of the material, point in that direction.

4.2. BH Entropy, the Information Paradox and the Xons Model

Having in our hands a suitable emergent gravitational dynamics, along the lines of what is explained in Section 3.4, it surely will be a great advancement and a necessary step toward analog BH thermodynamics. Still it would not be enough, as a suitable and reliable analog of a BH entropy is the key problem to be solved. In this respect, we have two roads in mind, one being easier than the other: **i)** entanglement entropy of the Dirac fields, on the given dynamical emergent BH background; **ii)** computation of the Wald entropy through standard classical calculations based on the experimentally sound geometry/gravity theory that describes the dynamics of the elastic graphene membrane.

The first approach is easier in two respects. First, one does not need an action for the geometry/gravity theory in point. Second, there are many results at our disposal on the entanglement entropy, from the general ones on generic bipartite systems [115], to the specific ones on the BH thermodynamics [84,116,117]. With these in our hands, we can surely attempt various things, and it will be exciting to see how certain issues of the theoretical side are solved here in practice. Given the results on the granular regimes beyond the linear theory [45,118] (see also [49]), we are also in the position to compute QG corrections to the formula, and compare theoretical predictions on the QG side, as well as experiments on the condensed matter side. The second approach is more difficult; nonetheless, we plan to also move steps in that direction because of its more direct link with purely gravitational scenarios. An exciting perspective is that these two approaches are complementary. It will be illuminating to compare the two, Bekenstein/Wald and entanglement.

To have these aspects under control, it is clearly necessary to face, within this approach, the long-standing issue of the information loss. In [6], it was investigated the impact on the Page curve of a picture born in analogy with condensed matter, named there the "quasi-particle picture". In this picture, more fundamental entities exist (we might call them Xons, with Feynman), and they make particles and spacetime at once: hence, the (information preserving) unitarity of the BH evaporation of the Page curve is not tenable. In [40], it was shown how entanglement, Bekenstein and thermodynamic entropies all stem from the same operator, whose structure is the one typical of Takahashi and Umezawa's TFD [86]. We expect that the several interesting new insights gained from this work will substantially help to reach the goals.

Finally, in [46], taking advantage of the peculiarities of the BTZ BH [59], the extremal $M = 0$ case was identified as furnishing an alternative way for the emergence in DMs of a maximal resolution/minimal length, given by the lattice length ℓ, and related to the (negative) cosmological constant as $\ell = 1/\sqrt{-\Lambda}$. Noticeably, a similar independent proposal emerged in the discussion of the entanglement entropy of the BTZ; see [119]. There, the AdS length is promoted to the typical length, below which spatial quantum correlation is traced out. Clearly, this road has the potential to produce very interesting results.

4.3. Other Hep-Th Scenarios on DMs

Many other aspects that will contribute toward the main goal of testing QG scenarios with DMs but that are important on their own right are in sight. Let us mention one.

The action of graphene can be recast [28] in a form very similar to the action of USUSY, for an external non-abelian $SU(2)$ gauge field and a fixed curved background [52]. Indeed, if the geometric background is fixed and the non-abelian gauge field is external (there is no dynamics for the phonons and gauge fields), then the only difference between such actions is the coefficient in front of the torsion term. Interestingly enough, the vacuum sector is defined by configurations that are locally Lorentz flat as is the case of BTZ BHs [120], and $SU(2)$ connections carrying nontrivial global charges [121].

4.4. HELIOS

Let us close by making the case for a laboratory where QG and other fundamental theories of nature are tested with analogs. Bearing in mind what we discussed at length in this review, and that these are the days of the AdS/CFT correspondence (see, for example, [122]), relating gravity and matter, we believe that the times are mature for a dedicated laboratory, entirely devoted to test fundamental theories by using analogs [123–125]. A laboratory built with the same spirit of CERN will unify, systematize and organize those efforts, but will also raise the status of the analog enterprise to the quest to reach beyond the known. The other side of the story is that analogs are often important materials for technological applications, like the case of graphene discussed in this review. Such a laboratory would then be an invaluable think-tank, where unconventional thinking would be routinely applied to create new technology, and to solve fundamental problems. Within our research group in Prague, we call this future facility *HELIOS*, for High Energy Lab for Indirect Observations.

Funding: This research was funded by Charles University Research Center (UNCE/SCI/013), by Fondo Nacional de Desarrollo Científico y Tecnológico–Chile (Fondecyt Grant No. 3200725) and by the Polish National Science Center grant 2018/31/D/ST2/02048.

Institutional Review Board Statement: Not applicable.

Informed Consent Statement: Not applicable.

Data Availability Statement: Not applicable.

Acknowledgments: We gladly thank Raffaele Agostino, Francisco Correa, Arundhati Dasgupta, Gaston Giribet, Paco Guinea, Vit Jakubsky, Siddhartha Sen, Guillermo Silva, Maria Vozmediano and Jorge Zanelli, for the many stimulating and informative discussions on the experimental and theoretical physics of this "graphene analog enterprize". L.S. acknowledges the kind hospitality of the Institute of Particle and Nuclear Physics of Charles University.

Conflicts of Interest: The authors declare no conflict of interest.

Notes

1. Interestingly, there is a proposal called "atoms of spacetime matter" that could be closely related with this concept of Xons [38,39].
2. As the π electrons do not participate in the stronger covalent σ bonds, these electrons are not so attached to the carbon nuclei and are freer to "hop" from an atom to a neighbor one.
3. Here our notations: $\mu, \nu = 0, 1, \ldots, n-1$ are Einstein indices, responding to diffeomorphisms, $a, b = 0, 1, \ldots, n-1$ are flat indices, responding to local Lorentz transformations, while α, β are spin indices. The covariant derivative is

$$\nabla_\mu \psi_\alpha = \partial_\mu \psi_\alpha + \Omega_{\mu\alpha}{}^\beta \psi_\beta,$$

with

$$\Omega_{\mu\alpha}{}^\beta = \frac{1}{2}\omega_\mu^{ab}(J_{ab})_\alpha{}^\beta,$$

where $(J_{ab})_\alpha{}^\beta$ are the Lorentz generators in spinor space, and

$$\omega_\mu{}^a{}_b = e_\lambda^a(\delta_\nu^\lambda \partial_\mu + \Gamma_{\mu\nu}^\lambda)E_b^\nu,$$

is the spin connection, whose relation to the Christoffel connection comes from the full metricity condition $\nabla_\mu e_\nu^a = \partial_\mu e_\nu^a - \Gamma_{\mu\nu}^\lambda e_\lambda^a + \omega_\mu{}^a{}_b e_\nu^b = 0$. We also introduced the Vielbein e_μ^a (and its inverse E_a^μ), satisfying $\eta_{ab} e_\mu^a e_\nu^b = g_{\mu\nu}$, $e_\mu^a E_a^\nu = \delta_\mu^\nu$, $e_\mu^a E_b^\mu = \delta_b^a$, where $\eta_{ab} = \text{diag}(1, -1, \ldots)$. The Weyl dimension of the Dirac field ψ in n dimensions is $d_\psi = (1-n)/2$. Here $n = 3$, and we can move one dimension up (embedding), or down (boundary). More notations can be found in [13].
4. Recall that in three dimensions $\omega_{\mu\,ab} = \epsilon_{abc}\,\omega_\mu{}^c$.
5. Roughly speaking, torsion is the surface density of the Burgers vector $\vec{b}$. For technicalities, see [50,74].
6. Let us explain why we use the word *phase* in quotation marks. Given the general vacuum of a QFT, one can identify several vacua that cannot be obtained one from the other through a smooth unitary transformation. Starting from each of these "sub vacua", and acting with the appropriate creation operators, one builds several (infinite) sectors, sometimes called super-selection sectors. Not all of them correspond to a phase of the system, in the proper statistical mechanical/thermodynamical sense. On the other hand, all such phases need be described by a super-selection sector or by a set of them. On this, see, for example, [88].
7. As it is impossible to distinguish the space corresponding to different topologies of the lattice, the expected value of the entanglement between the fields and the geometrical degrees of freedom is $\langle S \rangle = \sum_i p_{(i)} S_{(i)}$.

References

1. Feynman, R.; Leighton, R.; Sands, M. *The Feynman Lectures on Physics*; Addison-Wesley World Student Series; Addison-Wesley: Boston, MA, USA, 1963; Volume 2.
2. Iorio, A. Two arguments for more fundamental building blocks. *J. Phys. Conf. Ser.* **2019**, *1275*, 012013. [CrossRef]
3. Bekenstein, J.D. Universal upper bound on the entropy-to-energy ratio for bounded systems. *Phys. Rev. D* **1981**, *23*, 287–298. [CrossRef]
4. Bekenstein, J.D. Optimizing entropy bounds for macroscopic systems. *Phys. Rev. E* **2014**, *89*, 052137.

5. Bekenstein, J.D. Information in the holographic universe. *Sci. Am.* **2003**, *289*, 58–65. [CrossRef] [PubMed]
6. Acquaviva, G.; Iorio, A.; Scholtz, M. On the implications of the Bekenstein bound for black hole evaporation. *Ann. Phys.* **2017**, *387*, 317–333. [CrossRef]
7. Bao, N.; Carroll, S.M.; Singh, A. The Hilbert space of quantum gravity is locally finite-dimensional. *Int. J. Mod. Phys. D* **2017**, *26*, 1743013. [CrossRef]
8. Barceló, C.; Liberati, S.; Visser, M. Analogue Gravity. *Living Rev. Relativ.* **2005**, *8*, 12. [CrossRef] [PubMed]
9. Unruh, W.G. Experimental Black-Hole Evaporation? *Phys. Rev. Lett.* **1981**, *46*, 1351–1353. [CrossRef]
10. Unruh, W.G. Notes on black-hole evaporation. *Phys. Rev. D* **1976**, *14*, 870–892. [CrossRef]
11. Hawking, S.W. Particle creation by black holes. *Commun. Math. Phys.* **1975**, *43*, 199–220. [CrossRef]
12. Muñoz de Nova, J.R.; Golubkov, K.; Kolobov, V.I.; Steinhauer, J. Observation of thermal Hawking radiation and its temperature in an analogue black hole. *Nature* **2019**, *569*, 688–691. [CrossRef]
13. Iorio, A. Weyl-gauge symmetry of graphene. *Ann. Phys.* **2011**, *326*, 1334–1353. [CrossRef]
14. Iorio, A. Using Weyl symmetry to make graphene a real lab for fundamental physics. *Eur. Phys. J. Plus* **2012**, *127*, 156. [CrossRef]
15. Iorio, A.; Lambiase, G. The Hawking–Unruh phenomenon on graphene. *Phys. Lett. B* **2012**, *716*, 334–337. [CrossRef]
16. Iorio, A.; Lambiase, G. Quantum field theory in curved graphene spacetimes, Lobachevsky geometry, Weyl symmetry, Hawking effect, and all that. *Phys. Rev. D* **2014**, *90*, 025006. [CrossRef]
17. Iorio, A. Curved spacetimes and curved graphene: A status report of the Weyl symmetry approach. *Int. J. Mod. Phys. D* **2015**, *24*, 1530013. [CrossRef]
18. Lloyd, S.; Gooth, J.; Niemann, A.C.; Meng, T.; Grushin, A.G.; Landsteiner, K.; Gotsmann, B.; Menges, F.; Schmidt, M.; Shekhar, C.; et al. Ultimate physical limits to computation. *Nature* **2000**, *406*, 1047–1054. [CrossRef]
19. Parhizkar, A.; Galitski, V. Strained bilayer graphene, emergent energy scales, and moiré gravity. *Phys. Rev. Res.* **2022**, *4*, L022027. [CrossRef]
20. Parhizkar, A.; Galitski, V. Moiré Gravity and Cosmology. *arXiv* **2022**, arXiv:2204.06574.
21. O'Raifeartaigh, L.; Sreedhar, V. The Maximal Kinematical Invariance Group of Fluid Dynamics and Explosion–Implosion Duality. *Ann. Phys.* **2001**, *293*, 215–227. [CrossRef]
22. Sreedhar, V.V.; Virmani, A. Maximal Kinematical Invariance Group of Fluid Dynamics and Applications. *Universe* **2022**, *8*, 319. [CrossRef]
23. Castorina, P.; Kharzeev, D.; Satz, H. Thermal hadronization and Hawking–Unruh radiation in QCD. *Eur. Phys. J. C* **2007**, *52*, 187. [CrossRef]
24. Castorina, P.; Iorio, A.; Satz, H. Hadron freeze-out and Unruh radiation. *Int. J. Mod. Phys. E* **2015**, *24*, 1550056. [CrossRef]
25. Castorina, P.; Iorio, A. Confinement horizon and QCD entropy. *Int. J. Mod. Phys. A* **2018**, *33*, 1850211. [CrossRef]
26. Castorina, P.; Iorio, A.; Satz, H. Hunting Quantum Gravity with Analogs: The case of High Energy Particle Physics. *arXiv* **2022**, arXiv:2207.11935.
27. Iorio, A.; Pais, P. Revisiting the gauge fields of strained graphene. *Phys. Rev. D* **2015**, *92*, 125005. [CrossRef]
28. Iorio, A.; Pais, P. (Anti-)de Sitter, Poincaré, Super symmetries, and the two Dirac points of graphene. *Ann. Phys.* **2018**, *398*, 265–286. [CrossRef]
29. Iorio, A. Graphene and Black Holes: Novel Materials to Reach the Unreachable. *Front. Mater.* **2015**, *1*, 36. [CrossRef]
30. Castro Neto, A.H.; Guinea, F.; Peres, N.M.R.; Novoselov, K.S.; Geim, A.K. The electronic properties of graphene. *Rev. Mod. Phys.* **2009**, *81*, 109–162. [CrossRef]
31. Wehling, T.; Black-Schaffer, A.; Balatsky, A. Dirac materials. *Adv. Phys.* **2014**, *63*, 1–76. [CrossRef]
32. Dardashti, R.; Thébault, K.P.Y.; Winsberg, E. Confirmation via Analogue Simulation: What Dumb Holes Could Tell Us about Gravity. *Br. J. Philos. Sci.* **2017**, *68*, 55–89. [CrossRef]
33. Penrose, R. On Gravity's role in Quantum State Reduction. *Gen. Relativ. Gravit.* **1996**, *28*, 581–600. [CrossRef]
34. Hawking, S. Black holes and the information paradox. In *General Relativity and Gravitation*; World Scientific Publishing: Singapore, 2005; pp. 56–62.
35. Almheiri, A.; Marolf, D.; Polchinski, J.; Sully, J. Black holes: Complementarity or firewalls? *J. High Energy Phys.* **2013**, *2013*, 62. [CrossRef]
36. 't Hooft, G. The Firewall Transformation for Black Holes and Some of Its Implications. *Found. Phys.* **2017**, *47*, 1503–1542. [CrossRef]
37. Mann, R. *Black Holes: Thermodynamics, Information, and Firewalls*; SpringerBriefs in Physics; Springer International Publishing: Cham, Switzerland, 2015.
38. Singh, T.P. Relativistic weak quantum gravity and its significance for the standard model of particle physics. *arXiv* **2021**, arXiv:2110.02062.
39. Kaushik, P.; Vatsalya, V.; Singh, T.P. An $E_8 \otimes E_8$ unification of the standard model with pre-gravitation, on an octonion-valued twistor space. *arXiv* **2022**, arXiv:2206.06911.
40. Acquaviva, G.; Iorio, A.; Smaldone, L. Bekenstein bound from the Pauli principle. *Phys. Rev. D* **2020**, *102*, 106002. [CrossRef]
41. Wallace, P.R. The Band Theory of Graphite. *Phys. Rev.* **1947**, *71*, 622–634. [CrossRef]
42. Semenoff, G.W. Condensed-Matter Simulation of a Three-Dimensional Anomaly. *Phys. Rev. Lett.* **1984**, *53*, 2449–2452. [CrossRef]

43. Novoselov, K.S.; Geim, A.K.; Morozov, S.V.; Jiang, D.; Zhang, Y.; Dubonos, S.V.; Grigorieva, I.V.; Firsov, A.A. Electric Field Effect in Atomically Thin Carbon Films. *Science* **2004**, *306*, 666–669. [CrossRef]
44. Ciappina, M.F.; Iorio, A.; Pais, P.; Zampeli, A. Torsion in quantum field theory through time-loops on Dirac materials. *Phys. Rev. D* **2020**, *101*, 036021. [CrossRef]
45. Iorio, A.; Pais, P.; Elmashad, I.A.; Ali, A.F.; Faizal, M.; Abou-Salem, L.I. Generalized Dirac structure beyond the linear regime in graphene. *Int. J. Mod. Phys. D* **2018**, *27*, 1850080. [CrossRef]
46. Iorio, A.; Lambiase, G.; Pais, P.; Scardigli, F. Generalized uncertainty principle in three-dimensional gravity and the BTZ black hole. *Phys. Rev. D* **2020**, *101*, 105002. [CrossRef]
47. Gusynin, V.P.; Sharapov, S.G.; Carbotte, J.P. AC conductivity of graphene: From light-binding model to 2 + 1-dimensional quantum electrodynamics. *Int. J. Mod. Phys. B* **2007**, *21*, 4611–4658. [CrossRef]
48. Hands, S. Towards critical physics in 2+1d with U(2N)-invariant fermions. *J. High Energy Phys.* **2016**, *2016*, 15. [CrossRef]
49. Iorio, A.; Ivetic, B.; Mignemi, S.; Pais, P. The three "layers" of monolayer graphene and their generalized uncertainty principles. *arXiv* **2022**, arXiv:2208.02237.
50. Kleinert, H. *Gauge Fields in Condensed Matter*; World Scientific Publishing: Singapore, 1989. Available online: https://www.worldscientific.com/worldscibooks/10.1142/0356#t=aboutBook (accessed on 15 January 2020).
51. Witten, E. 2 + 1 dimensional gravity as an exactly soluble system. *Nucl. Phys. B* **1988**, *311*, 46–78. [CrossRef]
52. Alvarez, P.D.; Valenzuela, M.; Zanelli, J. Supersymmetry of a different kind. *J. High Energy Phys.* **2012**, *2012*, 58. [CrossRef]
53. Alvarez, P.D.; Pais, P.; Zanelli, J. Unconventional supersymmetry and its breaking. *Phys. Lett.* **2014**, *B735*, 314. [CrossRef]
54. Andrianopoli, L.; Cerchiai, B.L.; D'Auria, R.; Gallerati, A.; Noris, R.; Trigiante, M.; Zanelli, J. $\mathcal{N}$-extended $D = 4$ supergravity, unconventional SUSY and graphene. *J. High Energy Phys.* **2020**, *2020*, 84. [CrossRef]
55. Alvarez, P.D.; Delage, L.; Valenzuela, M.; Zanelli, J. Unconventional SUSY and Conventional Physics: A Pedagogical Review. *Symmetry* **2021**, *13*, 628. [CrossRef]
56. Iorio, A.; O'Raifeartaigh, L.; Sachs, I.; Wiesendanger, C. Weyl gauging and conformal invariance. *Nucl. Phys. B* **1997**, *495*, 433–450. [CrossRef]
57. Iorio, A.; Kus, P. Vortex solutions of Liouville equation and quasi spherical surfaces. *Int. J. Geom. Methods Mod. Phys.* **2020**, *17*, 2050106. [CrossRef]
58. Taioli, S.; Gabbrielli, R.; Simonucci, S.; Pugno, N.M.; Iorio, A. Lobachevsky crystallography made real through carbon pseudospheres. *J. Phys. Condens. Matter* **2016**, *28*, 13LT01. [CrossRef] [PubMed]
59. Bañados, M.; Teitelboim, C.; Zanelli, J. Black hole in three-dimensional spacetime. *Phys. Rev. Lett.* **1992**, *69*, 1849–1851. [CrossRef]
60. Maggiore, M. A Generalized uncertainty principle in quantum gravity. *Phys. Lett. B* **1993**, *304*, 65–69. [CrossRef]
61. Kempf, A. Uncertainty relation in quantum mechanics with quantum group symmetry. *J. Math. Phys.* **1994**, *35*, 4483–4496. [CrossRef]
62. Scardigli, F. Generalized uncertainty principle in quantum gravity from micro - black hole Gedanken experiment. *Phys. Lett. B* **1999**, *452*, 39–44. [CrossRef]
63. Das, S.; Vagenas, E.C. Universality of Quantum Gravity Corrections. *Phys. Rev. Lett.* **2008**, *101*, 221301. [CrossRef]
64. Ali, A.F.; Das, S.; Vagenas, E.C. Discreteness of space from the generalized uncertainty principle. *Phys. Lett. B* **2009**, *678*, 497–499. [CrossRef]
65. Ali, A.F.; Das, S.; Vagenas, E.C. Proposal for testing quantum gravity in the lab. *Phys. Rev. D* **2011**, *84*, 044013. [CrossRef]
66. Buoninfante, L.; Luciano, G.G.; Petruzziello, L. Generalized Uncertainty Principle and Corpuscular Gravity. *Eur. Phys. J. C* **2019**, *79*, 663. [CrossRef]
67. Petruzziello, L.; Illuminati, F. Quantum gravitational decoherence from fluctuating minimal length and deformation parameter at the Planck scale. *Nat. Commun.* **2021**, *12*, 4449. [CrossRef] [PubMed]
68. Bosso, P.; Luciano, G.G. Generalized uncertainty principle: From the harmonic oscillator to a QFT toy model. *Eur. Phys. J. C* **2021**, *81*, 982. [CrossRef]
69. Shah, N.A.; Contreras-Astorga, A.; Fillion-Gourdeau, F.m.c.; Ahsan, M.A.H.; MacLean, S.; Faizal, M. Effects of discrete topology on quantum transport across a graphene $n-p-n$ junction: A quantum gravity analog. *Phys. Rev. B* **2022**, *105*, L161401. [CrossRef]
70. Deser, S.; Jackiw, R.; 't Hooft, G. Three-dimensional Einstein gravity: Dynamics of flat space. *Ann. Phys.* **1984**, *152*, 220–235. [CrossRef]
71. Bañados, M.; Henneaux, M.; Teitelboim, C.; Zanelli, J. Geometry of the 2+1 black hole. *Phys. Rev. D* **1993**, *48*, 1506–1525. [CrossRef]
72. Donnay, L. Asymptotic dynamics of three-dimensional gravity. *PoS Proc. Sci.* **2016**, *271*, 1. [CrossRef]
73. Schlenker, J.M.; Witten, E. No Ensemble Averaging Below the Black Hole Threshold. *arXiv* **2022**, arXiv:2202.01372.
74. Katanaev, M.; Volovich, I. Theory of defects in solids and three-dimensional gravity. *Ann. Phys.* **1992**, *216*, 1–28. [CrossRef]
75. Guevara, A.; Pais, P.; Zanelli, J. Dynamical Contents of Unconventional Supersymmetry. *J. High Energy Phys.* **2016**, *2016*, 85. [CrossRef]
76. Pais, P.; Iorio, A. Time-loops in Dirac materials, torsion and unconventional Supersymmetry. *PoS Proc. Sci.* **2021**, *390*, 669. [CrossRef]

77. Jackiw, R.; Pi, S.Y. Soliton solutions to the gauged nonlinear Schrödinger equation on the plane. *Phys. Rev. Lett.* **1990**, *64*, 2969–2972. [CrossRef]
78. Jackiw, R.; Pi, S.Y. Classical and quantal nonrelativistic Chern-Simons theory. *Phys. Rev. D* **1990**, *42*, 3500–3513. [CrossRef]
79. Horváthy, P.A.; Yéra, J.C. Vortex Solutions of the Liouville Equation. *Lett. Math. Phys.* **1998**, *46*, 111–120. [CrossRef]
80. Doplicher, S.; Fredenhagen, K.; Roberts, J.E. The quantum structure of spacetime at the Planck scale and quantum fields. *Commun. Math. Phys.* **1995**, *172*, 187–220. [CrossRef]
81. Page, D.N. Average entropy of a subsystem. *Phys. Rev. Lett.* **1993**, *71*, 1291–1294. [CrossRef] [PubMed]
82. Page, D.N. Information in black hole radiation. *Phys. Rev. Lett.* **1993**, *71*, 3743–3746. [CrossRef]
83. Harlow, D. Jerusalem lectures on black holes and quantum information. *Rev. Mod. Phys.* **2016**, *88*, 015002. [CrossRef]
84. Iorio, A.; Lambiase, G.; Vitiello, G. Entangled quantum fields near the event horizon and entropy. *Ann. Phys.* **2004**, *309*, 151–165. [CrossRef]
85. Takahashi, Y.; Umezawa, H. Thermo field dynamics. *Int. J. Mod. Phys. B* **1996**, *10*, 1755–1805. [CrossRef]
86. Umezawa, H.; Matsumoto, H.; Tachiki, M. *Thermo Field Dynamics and Condensed States*; North-Holland: Amsterdam, The Netherlands, 1982.
87. Dirac, P. *Lectures on Quantum Field Theory*; Monographs Series; Belfer Graduate School of Science, Yeshiva University: New York, NY, USA, 1966.
88. Umezawa, H. *Advanced Field Theory: Micro, Macro, and Thermal Physics*; American Institute of Physics: College Park, MA, USA, 1993.
89. Mukhanov, V.F. Are black holes quantized? *Sov. J. Exp. Theor. Phys. Lett.* **1986**, *44*, 63–66.
90. Bekenstein, J.D.; Mukhanov, V. Spectroscopy of the quantum black hole. *Phys. Lett. B* **1995**, *360*, 7–12. [CrossRef]
91. Bekenstein, J.D. Quantum Black Holes as Atoms. Available online: https://arxiv.org/abs/gr-qc/9710076 (accessed on 6 July 2022).
92. Castorina, P.; Iorio, A.; Smaldone, L. Quantum black holes, partition of integers and self-similarity. *Mod. Phys. Lett. A* **2022**, accepted. [CrossRef]
93. Maggiore, M. Physical Interpretation of the Spectrum of Black Hole Quasinormal Modes. *Phys. Rev. Lett.* **2008**, *100*, 141301. [CrossRef]
94. Stone, M.H. Linear Transformations in Hilbert Space. III. Operational Methods and Group Theory. *Proc. Natl. Acad. Sci. USA* **1930**, *16*, 172–175. [CrossRef]
95. Neumann, J.V. Die Eindeutigkeit der Schrödingerschen Operatoren. *Math. Ann.* **1931**, *104*, 570–578. [CrossRef]
96. Umezawa, H.; Vitiello, G. *Quantum Mechanics*; Bibliopolis: Napoli, Italy, 1985.
97. Friedrichs, K. *Mathematical Aspects of the Quantum Theory of Fields*; Interscience Publishers: Geneva, Switzerland, 1953.
98. Barton, G. *Introduction to Advanced Field Theory*; Interscience Tracts on Physics and Astronomy; Wiley: Hoboken, NJ, USA, 1963.
99. Itzykson, C.; Zuber, J. *Quantum Field Theory*; Dover Books on Physics; Dover Publications: Mineola, NY, USA, 2012.
100. Blasone, M.; Jizba, P.; Smaldone, L. Functional integrals and inequivalent representations in Quantum Field Theory. *Ann. Phys.* **2017**, *383*, 207–238. [CrossRef]
101. Acerbi, F.; Morchio, G.; Strocchi, F. Nonregular representations of CCR algebras and algebraic fermion bosonization. *Rep. Math. Phys.* **1993**, *33*, 7–19. [CrossRef]
102. Schulman, L. A Path Integral for Spin. *Phys. Rev.* **1968**, *176*, 1558–1569. [CrossRef]
103. Schulman, L. *Techniques and Applications of Path Integration*; Dover Books on Physics; Dover Publications: Mineola, NY, USA, 2012.
104. Kastrup, H.A. Quantization of the canonically conjugate pair angle and orbital angular momentum. *Phys. Rev. A* **2006**, *73*, 052104. [CrossRef]
105. Ashtekar, A.; Fairhurst, S.; Willis, J.L. Quantum gravity, shadow states and quantum mechanics. *Class. Quant. Grav.* **2003**, *20*, 1031–1061. [CrossRef]
106. Corichi, A.; Vukašinac, T.; Zapata, J.A. Polymer quantum mechanics and its continuum limit. *Phys. Rev. D* **2007**, *76*, 044016. [CrossRef]
107. Acquaviva, G.; Iorio, A.; Smaldone, L. Quantum groups and polymer quantum mechanics. *Mod. Phys. Lett. A* **2021**, *36*, 2150229. [CrossRef]
108. Acquaviva, G.; Iorio, A.; Smaldone, L. Topologically inequivalent quantizations. *Ann. Phys.* **2021**, *434*, 168641. [CrossRef]
109. Matsumoto, H.; Papastamatiou, N.; Umezawa, H. The boson transformation and the vortex solutions. *Nucl. Phys. B* **1975**, *97*, 90–124. [CrossRef]
110. Matsumoto, H.; Umezawa, H.; Umezawa, M. Extended Objects in Field Theory with Non-Abelian Group Symmetry. *Fortschritte Phys.* **1981**, *29*, 441–462. [CrossRef]
111. Blasone, M.; Jizba, P. Topological Defects as Inhomogeneous Condensates in Quantum Field Theory. *Ann. Phys.* **2002**, *295*, 230–260. [CrossRef]
112. Iorio, A.; Smaldone, L. Quantum Black Holes as Classical Space Factories. Available online: https://www.researchgate.net/publication/362175062_Quantum_black_holes_as_classical_space_factories?channel=doi&linkId=62da79a1aa5823729ed68b24&showFulltext=true (accessed on 6 July 2022)
113. Wald, R.M. Black hole entropy is the Noether charge. *Phys. Rev. D* **1993**, *48*, R3427–R3431. [CrossRef]

114. Heide, C.; Higuchi, T.; Weber, H.B.; Hommelhoff, P. Coherent Electron Trajectory Control in Graphene. *Phys. Rev. Lett.* **2018**, *121*, 207401. [CrossRef]
115. Nielsen, M.; Chuang, I. *Quantum Computation and Quantum Information*; Cambridge Series on Information and the Natural Sciences; Cambridge University Press: Cambridge, UK, 2000.
116. Terashima, H. Entanglement entropy of the black hole horizon. *Phys. Rev. D* **2000**, *61*, 104016. [CrossRef]
117. Maldacena, J.M. Eternal black holes in anti-de Sitter. *J. High Energy Phys.* **2003**, *2003*, 21. [CrossRef]
118. Iorio, A.; Pais, P. Generalized uncertainty principle in graphene. *J. Phys. Conf. Ser.* **2019**, *1275*, 012061. [CrossRef]
119. Cadoni, M.; Melis, M. Holographic Entanglement Entropy of the BTZ Black Hole. *Found. Phys.* **2010**, *40*, 638–657. [CrossRef]
120. Alvarez, P.D.; Pais, P.; Rodríguez, E.; Salgado-Rebolledo, P.; Zanelli, J. The BTZ black hole as a Lorentz-flat geometry. *Phys. Lett. B* **2014**, *738*, 134–135. [CrossRef]
121. Alvarez, P.D.; Pais, P.; Rodríguez, E.; Salgado-Rebolledo, P.; Zanelli, J. Supersymmetric 3D model for gravity with $SU(2)$ gauge symmetry, mass generation and effective cosmological constant. *Class. Quant. Grav.* **2015**, *32*, 175014. [CrossRef]
122. Susskind, L.; Lindesay, J. *An Introduction to Black Holes, Information and the String Theory Revolution*; World Scientific Publishing: Singapore, 2004. Available online: https://www.worldscientific.com/worldscibooks/10.1142/5689#t=aboutBook (accessed on 20 February 2020).
123. Iorio, A. What after CERN? Opportunities from co-responding systems. *J. Phys. Conf. Ser. Iop Publ.* **2015**, *626*, 012035. [CrossRef]
124. Iorio, A. Analog hep-th, on Dirac materials and in general. *PoS Proc. Sci.* **2020**, *376*, 203. [CrossRef]
125. Iorio, A. Making the case for a "CERN for analogs". *PoS Proc. Sci.* **2020**, *390*, 688. [CrossRef]

Article

On the Hilbert Space in Quantum Gravity

Ednardo Paulo Spaniol [1,2,*], Ronni Geraldo Gomes Amorim [3,4] and Sergio Costa Ulhoa [4,5]

1. UDF Centro Universitário, Brasília 70390-045, Brazil
2. Faculdade de Tecnologia e Ciências Sociais Aplicadas, Centro de Ensino Unificado de Brasília (CEUB), Brasília 70790-075, Brazil
3. Faculdade Gama, Centro de Ensino Unificado de Brasília (CEUB), Brasília 70910-900, Brazil
4. Canadian Quantum Research Center, 204-3002 32 Ave, Vernon, BC V1T 2L7, Canada
5. International Center of Physics, Instituto de Física, Centro de Ensino Unificado de Brasília (CEUB), Brasília 70910-900, Brazil
* Correspondence: spaniol.ep@gmail.com

Abstract: This article deals with the fractional problem of Sturm–Liouville and the Hilbert space associated with the solutions of this differential equation. We apply a quantization procedure to Schwarzschild space–time and obtain a fractional differential equation. The Hilbert space for these solutions is established. We used equations arising from quantization for the FRW and Reissner–Nordstron metrics to build the respective Hilbert spaces.

Keywords: quantum gravity; teleparallelism; non-commutative gravity

Citation: Spaniol, E.P.; Amorim, R.G.G.; Ulhoa, S.C. On the Hilbert Space in Quantum Gravity. *Universe* 2022, 8, 413. https://doi.org/10.3390/universe8080413

Academic Editors: Arundhati Dasgupta and Alfredo Iorio

Received: 17 June 2022
Accepted: 3 August 2022
Published: 5 August 2022

Publisher's Note: MDPI stays neutral with regard to jurisdictional claims in published maps and institutional affiliations.

Copyright: © 2022 by the authors. Licensee MDPI, Basel, Switzerland. This article is an open access article distributed under the terms and conditions of the Creative Commons Attribution (CC BY) license (https://creativecommons.org/licenses/by/4.0/).

1. Introduction

The search for a quantum theory of gravitation has been attempted as a stepping stone towards a great unified field theory. Despite this, there is no consensus on how to achieve this goal. The two most widespread approaches are loop quantum gravity [1] and string theory [2]. The latter, despite the mathematical consistency and simple basic principles, does not present experimental verification. On the other hand, loop quantum gravity is based on quantization processes applied to the Hamiltonian formulation of general relativity. This generates a discretization in the space–time itself instead of one in the gravitation source. The lack of gravitational energy in this context leads to a difficult physical interpretation of the result.

In fact, theories, which have a well-defined energy-momentum tensor for the field in question, are less refractory to known quantization techniques. Among the best known of these techniques, we have the canonical quantization that requires the formulation of the field in the phase space [3]. None of these conditions are met by the standard approach to gravitation. On the other hand, an alternative theory equivalent to general relativity allows the conception of a very well-defined gravitational energy, the so-called teleparallel gravity [4,5]. Teleparallelism equivalent to general relativity (TEGR) is constructed out of the tetrad field and was introduced by Einstein himself as part of the same effort to find a unified theory. TEGR is not a priori formulated in the phase space, which makes canonical quantization difficult to implement. The Hamiltonian formulation can certainly overcome this difficulty [6], but we are interested in applying a quantization technique that acts on functions dependent only on coordinates. We refer to Weyl's quantization [7].

Weyl's quantization was introduced in the early days of quantum mechanics and has the property of transforming a coordinate function into an operator [8–10]. On the other hand, it shares with the canonical quantization the arbitrariness in the representation of the operators constructed from the coordinates. Given n variables denoted by $z_1, z_2, \ldots, z_n$, then the prescription

$$(z_1, z_2, \ldots, z_n) \to (\widehat{z_1}, \widehat{z_2}, \ldots, \widehat{z_n}),$$

immediately quantizes a function f dependent on the classical variables. This quantization of the functions f is achieved by the following Weyl's map, $\mathcal{W}: f \to \widehat{f} = \mathcal{W}[f]$,

$$\mathcal{W}[f](z_1, z_2, \ldots, z_n) := \frac{1}{(2\pi)^n} \int d^n k d^n z f(z_1, z_2, \ldots, z_n) \exp\left(i \sum_{l=1}^{n} k_l (\widehat{z_l} - z_l)\right). \quad (1)$$

The operators constructed out of the classical variables obey the following relation

$$[\widehat{z_i}, \widehat{z_j}] = i\beta_{ij},$$

which means that the Weyl's quantization is essentially a non-commutative prescription. It is worth noting that such non-commutativity can be established for all coordinates one pair at a time, by choosing the beta components. On the other hand, non-commutativity involving the temporal coordinate is problematic, so we usually restrict ourselves to spatial coordinates or apply the technique to stationary systems. In addition, the very dependence of the function to be quantized influences this choice. Particularly when energy is used, its dependence on coordinates becomes the natural choice to use non-commutative coordinates. Weyl's quantization combined with TEGR creates a powerful approach to quantum gravity that has achieved consistent results such as the discretization of the charge-mass ratio [11], as well as the achievement of discrete levels of energy in a primordial universe [12]. On the other hand, improvements must be made such as the construction of the Hilbert space associated with the functions on which the resulting operators act. It is interesting to note that the three systems discussed in this article have different possible interpretations. Certainly until the advent of an experimental measurement, we may have a somewhat speculative interpretation. In Schwarzschild's case, a discrete mass spectrum points to a deeper quantum construction of matter, perhaps linked to a geometric structure of particles. Incidentally, such an idea applied to a charged black hole explains why the charge-mass ratio is discrete. It was even possible to associate the non-commutative parameter to the fundamental charge. As for the FLRW metric, its quantization provides a simple and ab initio explanation of the inflation process of the early universe.

This article is divided as follows. In Section 2, we introduce the basic ideas of TEGR, and we show the expression of gravitational energy that will be object of quantization. In Section 3, we approach the fractional problem of Sturm–Liouville and how to build the Hilbert space associated with the solutions of the differential equation. In Section 4, we obtain a fractional equation as a result of the quantization of Schwarzschild space–time. In Section 5, we show that the quantization of Reisner–Nordstrom space–time, obtained in a previous article, has a well-defined Hilbert space. In Section 6, we used the quantization of the FRW metric, already obtained in another article, to establish the associated Hilbert space. Finally, in Section 7 we present our final points. We adopt a unity system such that $G = c = 1$.

2. Teleparallelism Equivalent to General Relativity (TEGR)

TEGR is an alternative theory of gravitation whose dynamic variables are the tetrads. They also determine the state of the observer. That is, there is a single solution of the field equations for each reference system. This arbitrariness in the choice of the observer is a physical feature absent from the standard theory of gravitation, general relativity. In this sense, an expression of gravitational energy must be dependent on the reference system but invariant by coordinate transformations. The tetrad field relates two symmetries. On the one hand, the Greek indices denote coordinates transformation. On the other hand, the Latin indices denote Lorentz transformations.

Let's consider a manifold endowed with the following connection:

$$\Gamma_{\mu\lambda\nu} = e^a{}_\mu \partial_\lambda e_{a\nu},$$

which is the Weitzenböck connection. It is curvature free, but has a torsion given by

$$T^a{}_{\lambda\nu} = \partial_\lambda e^a{}_\nu - \partial_\nu e^a{}_\lambda. \tag{2}$$

It is important to note that the tetrad is related to the metric tensor through $g_{\mu\nu} = e^a{}_\mu e_{a\nu}$. The metric tensor is the foundation of a Riemannian geometry in which Christoffel symbols, ${}^0\Gamma_{\mu\lambda\nu}$, are defined. The scalar curvature obtained from this connection is the invariant in the Hilbert–Einstein's action. Thus, a relationship between Christoffel symbols and Weitzenböck connection determines in itself an equivalence between TEGR and general relativity. Such a relationship is a mathematical identity given by

$$\Gamma_{\mu\lambda\nu} = {}^0\Gamma_{\mu\lambda\nu} + K_{\mu\lambda\nu}, \tag{3}$$

where $K_{\mu\lambda\nu}$, that is defined as

$$K_{\mu\lambda\nu} = \frac{1}{2}(T_{\lambda\mu\nu} + T_{\nu\lambda\mu} + T_{\mu\lambda\nu}), \tag{4}$$

with $T_{\mu\lambda\nu} = e_{a\mu} T^a{}_{\lambda\nu}$, is the contortion tensor.

Due to the identity (3), the following equation holds

$$eR(e) \equiv -e\left(\frac{1}{4}T^{abc}T_{abc} + \frac{1}{2}T^{abc}T_{bac} - T^a T_a\right) + 2\partial_\mu(eT^\mu). \tag{5}$$

It should be noted that the scalar curvature calculated out of the Weitzenböck connection vanishes identically. Hence, the Lagrangian density of TEGR may be written as

$$\begin{aligned}\mathfrak{L}(e_{a\mu}) &= -\kappa e\left(\frac{1}{4}T^{abc}T_{abc} + \frac{1}{2}T^{abc}T_{bac} - T^a T_a\right) - \mathfrak{L}_M \\ &\equiv -\kappa e \Sigma^{abc} T_{abc} - \mathfrak{L}_M,\end{aligned} \tag{6}$$

where $\kappa = 1/(16\pi)$, $\mathfrak{L}_M$ is the Lagrangian density of matter fields and Σ^{abc} is given by

$$\Sigma^{abc} = \frac{1}{4}(T^{abc} + T^{bac} - T^{cab}) + \frac{1}{2}(\eta^{ac}T^b - \eta^{ab}T^c), \tag{7}$$

with $T^a = e^a{}_\mu T^\mu$. The Lagrangian density above yields the Einstein equation but doesn't share the same symmetries of the general relativity Lagrangian density because the total divergence in (5) was dropped out. Thus, the field equations read

$$\partial_\nu\left(e\Sigma^{a\lambda\nu}\right) = \frac{1}{4\kappa} e e^a{}_\mu (t^{\lambda\mu} + T^{\lambda\mu}), \tag{8}$$

where

$$t^{\lambda\mu} = \kappa\left[4\Sigma^{bc\lambda} T_{bc}{}^\mu - g^{\lambda\mu} \Sigma^{abc} T_{abc}\right] \tag{9}$$

is the gravitational energy-momentum tensor. Such a tensor is conserved due to

$$\partial_\lambda \partial_\nu\left(e\Sigma^{a\lambda\nu}\right) \equiv 0. \tag{10}$$

This allows one to define the total energy-momentum vector as

$$P^a = \int_V d^3x\, e e^a{}_\mu (t^{0\mu} + T^{0\mu}), \tag{11}$$

which can be expressed in view of the field equations as

$$P^a = 4k \int_V d^3x\, \partial_\nu\left(e\Sigma^{a0\nu}\right). \tag{12}$$

It should be noted that P^a is a vector under Lorentz transformations, but it is invariant under coordinate transformations, as expected for a energy-momentum vector.

3. The Fractional Sturm–Liouville Theory and Hilbert Space

In this section, we present the fractional theory of Sturm–Liouville. This subject is the background to analyze fractional ordinary differential equations (ODEs) by self-adjunct procedure and to establish a Hilbert space. The content of this section is based on references [13–18].

3.1. Preliminaries

In this subsection, we recall definitions of fractional integrals and derivatives. We focus on the Caputo's approach due to the smoothing of Riemann–Liouville regarding physical interpretations.

Definition 1. *Let's assume $\alpha > 0$, with $n - 1 < \alpha < n$ and $n \in \mathbb{N}$, $[a, b] \subset \mathbb{R}$, in addition let f be a suitable real function. The Caputo fractional derivative is*

$$({}_cD_{a+}^\alpha f)(x) = (I_{a+}^{n-\alpha} D^n f)(x), \quad (x > a), \tag{13}$$

$$({}_cD_{b-}^\alpha f)(x) = (I_{b-}^{n-\alpha} D^n f)(x), \quad (x < b), \tag{14}$$

where

$$(I_{a+}^\alpha f)(x) = \frac{1}{\Gamma(\alpha)} \int_a^x (x-t)^{\alpha-1} f(t) dt, \quad (x > a), \tag{15}$$

$$(I_{b-}^\alpha f)(x) = \frac{1}{\Gamma(\alpha)} \int_x^b (x-t)^{\alpha-1} f(t) dt, \quad (x < b), \tag{16}$$

$\Gamma(\alpha)$ denotes the Euler gamma function, and D^n represents the usual derivative operator $D^n = \frac{\partial^n}{\partial x^n}$.

As an example, let's calculate the Caputo fractional derivative of function $f(x) = x^2$ of order $1/2$. First, we apply the differential operator $D = \frac{\partial}{\partial x}$ in $f(x) = x^2$, obtaining $D(x^2) = 2x$. Now, we calculate the integral

$$({}_cD_0^{1/2} x^2) = \frac{1}{\Gamma(1/2)} \int_0^x (x-t)^{1/2} 2x dt.$$

Performing this integral and using $\Gamma(1/2) = \sqrt{\pi}$, we obtain

$$({}_cD_0^{1/2} x^2) = \frac{2}{\sqrt{\pi}} x^{1/2}. \tag{17}$$

The Caputo fractional derivative satisfies several properties which can be find in references.

Next we present the definition of a relevant function to Caputo differential calculus which is called Mittag–Leffer function.

Definition 2. *The two parameters Mittag–Leffer function, $E_{a,b}(x)$, where $\mathfrak{Re}(a) > 0$, $\mathfrak{Re}(b) > 0$, is defined by*

$$E_{a,b}(x) = \sum_{k=0}^{\infty} \frac{x^k}{\Gamma(ak+b)}, \tag{18}$$

where we notice that $E_{1,1}(x) = e^x$ and $E_{a,1}(x) = E_a(x)$.

It is easy to show that the Caputo derivative of Mittag–Leffer function is given by

$$_cD^\gamma E_\alpha(x^\alpha) = E_\alpha(x^\alpha). \tag{19}$$

The Mittag–Leffer function is a kind of generalization of exponential function to fractional calculus.

3.2. Fractional Sturm–Liouville Theory

In this section, we introduce fractional version to the Sturm–Liouville theory. For this proposal, let's begin with the following definition.

Definition 3. *A Caputo fractional Sturm–Liouville problem is a fractional problem with boundary conditions in the form*

$$-_cD^\alpha_{b-}(_cuD^\alpha_{a+}y)(x) + v(x)y(x) = \lambda r(x)y(x), \tag{20}$$

$$a < x < b, \qquad 1/2 < \alpha < 1 \tag{21}$$

$$\alpha_1 y(a) + \alpha_2 I^{1-\alpha}_{b-}(_cuD^\alpha_{a+}y)(x)|_{x=a} = 0, \tag{22}$$

$$\beta_1 y(b) + \beta_2 I^{1-\alpha}_{b-}(_cuD^\alpha_{a+}y)(x)|_{x=b} = 0, \tag{23}$$

where $L_a = -_cD^\alpha_{b-}(_cuD^\alpha_{a+}) + v$ is a self-adjoint operator and the constants α_i, β_i satisfied in the boundary conditions verify $\alpha_1^2 + \alpha_2^2 \neq 0$, $\beta_1^2 + \beta_2^2 \neq 0$ and the functions u, v, r are continuous, such that $u > 0$ and $r > o$ in $x \in [a,b]$. The function r is called a weight function, and the values of λ for which there exist non-trivial solutions are called eigenvalues of the boundary value problem.

In this sense, the Caputo fractional Sturm–Liouville problem satisfies the following properties.

Caputo Fractional Sturm–Liouville problem properties:

1. All of the eigenvalues of the fractional Sturm–Liouville problem are real.
2. If y_1 and y_2 are two eigenvalues of the fractional Sturm–Liouville problem corresponding to eigenvalues λ_1 and λ_2, respectively, with $\lambda_1 \neq \lambda_2$, then

$$\int_a^b r(x)y_1(x)y_2(x)dx = 0. \tag{24}$$

That is, the eigenvalues corresponding to different eigenvalues have the property of orthogonality with respect to the weight function r.

3. For each eigenvalue, there is only one eigenfunction (except for multiples non zeros).
4. The eigenfunction corresponding to different eigenvalues are linearly independent.

If $y_n(x)$ are complex functions, orthogonality condition, Equation (24) becomes

$$\int_a^b r(x)y_n^*(x)y_m(x)dx = 0; \qquad m \neq n, \tag{25}$$

where $y_n^*(x)$ is the conjugate complex of $y_n(x)$.

Due to the hermiticity of operator $\mathcal{L}_\dagger$, i.e., $\overline{\mathcal{L}} = \mathcal{L}$, their eigenfunctions $y_n(x)$ form a complete set. This completeness means that any well-behaved function $f(x)$ can be approximated by a series

$$f(x) = \sum_{n=0}^\infty a_n y_n(x), \tag{26}$$

where the coefficient a_n are given by

$$a_n = \int_a^b p(x)f(x)y_m^*(x)dx. \tag{27}$$

Equation (26) is unique for a given set of $y_n(x)$. The functions $y_n(x)$ are a basis vector in an infinite dimensional Hilbert space $\mathcal{H}$. In this sense, we can define an Hilbert space $\mathcal{H}$ form the linear space with inner product defined as

$$\langle f(x), g(x) \rangle = \int_a^b r(x) f^*(x) g(x) dx. \tag{28}$$

Those functions $f(x)$ defined in $\mathcal{H}$ are integrable square functions, i.e.,

$$\langle f(x), f(x) \rangle = \int_a^b r(x) f^*(x) f(x) dx < \infty. \tag{29}$$

This framework of fractional Sturm–Liouville theory and Hilbert space will be useful in our discussion about quantum gravity in the next sections. It is worth mentioning that the fractional Sturm–Liouville problem reduces to the usual case when $\alpha = 1$.

4. Quantum Schwarzschild Equation

In this section, we study the eigenvalue equation related to the quantum Schwarzschild system. First, we consider the following metric,

$$ds^2 = -f(r) dt^2 + \frac{dr^2}{f(r)} + r^2 d\Omega^2, \tag{30}$$

where $f(r) = 1 - \frac{2M}{r}$ and $d\Omega$ is a solid angle element. M represent the mass of the system and r is the radial coordinate. From Equation (30), we obtain the density of energy

$$\mathfrak{H} = 4kr \sin\theta (1 - f^{1/2}). \qquad - \tag{31}$$

It should be noted that the energy density is obtained from expression (11) or equivalently from (12), that is $\mathfrak{H} = \frac{\partial P^{(0)}}{\partial V}$. This function can be quantized by using the symmetrization rule and the operators

$$\widehat{\omega} = \omega, \quad \widehat{r} = i\beta \frac{\partial}{\partial \omega}, \tag{32}$$

where $\omega = \sin\theta$ and $\beta = \beta_{12}$ is the non-commutativity parameter. A particularly problematic feature of the energy density is its dependence on the radial coordinate with a fractional exponent. In previous articles [19], this difficulty was overcome through a power series expansion because the non-commutativity parameter must be very small. Here another possibility will be explored, namely the use of the fractional derivative that can be directly used in the energy operator. Then, using the condition $\frac{\beta}{M} \ll 1$ and the result $D^{1/2}(\omega) = \frac{2}{\sqrt{\pi}} \omega^{1/2}$, we obtain the following fractional differential equation,

$$\frac{\epsilon}{2k} \psi(\omega) = \left[i\beta + 2i\beta\omega \frac{\partial}{\partial \omega} - i^{3/2} \beta^{1/2} \sqrt{2M} \omega D^{1/2} - \frac{2i^{3/2} \beta^{1/2} \sqrt{2M}}{\sqrt{\pi}} \omega^{1/2} \right] \psi(\omega), \tag{33}$$

where ϵ is the eigenvalue of operator $\widehat{\mathfrak{H}}$. From now on, we represent the Caputo's fractional derivative $_c D^\alpha$ just by D^α. Equation (33) can be written as

$$c_1 \omega D^1 \psi(\omega) + c_2 \omega D^{1/2} \psi(\omega) + c_3 \omega^{1/2} \psi(\omega) + c_4 \psi(\omega) = 0, \tag{34}$$

where $c_1 = 2i\beta$, $c_2 = -i^{3/2}\beta^{1/2}\sqrt{2M}$, $c_3 = -\dfrac{2i^{3/2}\beta^{1/2}\sqrt{2M}}{\sqrt{\pi}}$, $c_4 = i\beta - \epsilon$. If we multiply Equation (34) by integrator factor

$$\mu(\omega) = E_{1/2}(\frac{c_2}{c_1}\omega^{1/2}),$$

and use the following property of Mittag–Leffer function

$$D^\alpha[E_\alpha(x^\alpha)] = E_\alpha(x^\alpha).$$

Equation (34) becomes

$$D^{1/2}\left[E_{1/2}(\frac{c_2}{c_1}\omega^{1/2})D^{1/2}\psi(\omega)\right] + \frac{c_3}{c_1\omega^{1/2}}E_{1/2}(\frac{c_2}{c_1}\omega^{1/2})\psi(\omega) + \frac{c_4}{c_1\omega}E_{1/2}(\frac{c_2}{c_1}\omega^{1/2})\psi(\omega) = 0. \tag{35}$$

Equation (35) is a Caputo's fractional Sturm–Liouville problem with weight function given by

$$g(\omega) = \frac{c_4}{c_1\omega}E_{1/2}(\frac{c_2}{c_1}\omega^{1/2}). \tag{36}$$

In this sense, the eigenvalues of this problem are real and the solutions $\psi_n(\omega)$ satisfy the following orthogonality condition,

$$\int_a^b \frac{c_4}{c_1\omega}E_{1/2}(\frac{c_2}{c_1}\omega^{1/2})\psi_n\omega\psi_m\omega d\omega = 0, \quad m \neq n. \tag{37}$$

Then, we can define an Hilbert space in such functions where $\psi_n(\omega)$ are square integrable.

5. Quantum Reissner–Nordstrom System

In reference [11], the possibility of a charged particle being described by a quantized version of the Reissner–Nordstrom metric was explored. Here we deal with the Hilbert space as a theoretical advance of the quantum description. The metric is described by the following line element,

$$ds^2 = -f(r)dt^2 + f(r)^{-1}dr^2 + r^2d\theta^2 + r^2\sin^2\theta d\phi^2,$$

with $f(r) = 1 - \dfrac{2M}{r} + \dfrac{Q^2}{r^2}$, which leads to the following energy density

$$\mathfrak{H} = 4r\sin\theta\left[1 - \left(1 - \frac{2M}{r} + \frac{Q^2}{r^2}\right)^{1/2}\right].$$

Thus, by using the following quantization rule, $\sin\theta \to \hat{\sin}\theta = \omega$ and $r \to \hat{r} = \beta\dfrac{\partial}{\partial \omega}$, we have the equation

$$\left\{-\frac{4\beta^2\omega}{2Q}\frac{\partial^2}{\partial\omega^2} + 4\beta\left[\omega\left(1 + \frac{M}{Q}\right) - \frac{\beta}{2Q}\right]\frac{\partial}{\partial\omega} + \left[2\beta\left(1 + \frac{M}{Q}\right) - \epsilon - 4Q\omega\right]\right\}\psi(\omega) = 0. \tag{38}$$

Equation (38) can be written as

$$\left[b_1\omega\frac{\partial^2}{\partial\omega^2} + (b_2\omega - b_3)\frac{\partial}{\partial\omega} + (b_4 + b_5\omega)\right]\psi(\omega) = 0, \tag{39}$$

where $b_1 = -\dfrac{4\beta^2}{2Q}$, $b_2 = 4\beta\left(1 + \dfrac{M}{Q}\right)$, $b_3 = \dfrac{2\beta^2}{Q}$, $b_4 = 2\beta\left(1 + \dfrac{M}{Q}\right) - \epsilon$ and $b_5 = -4Q$.

If we multiply Equation (39) by the integrator factor $\mu(\omega) = \dfrac{b_1}{b_3}e^{\frac{b_3}{b_1}\omega}$, we obtain

$$\frac{\partial}{\partial \omega}\left(\frac{b_1}{b_2}e^{\frac{b_3}{b_1}\omega}\frac{\partial \psi(\omega)}{\partial \omega}\right) + \frac{1}{b_3\omega}e^{\frac{b_3}{b_1}\omega}(b_4 + b_5\omega)\psi(\omega) = 0. \tag{40}$$

Equation (40) is the adjoint form of Equation (39). In this way, Equation (40) represents a usual Sturm–Liouville problem with weight function given by

$$g(\omega) = \frac{b_4}{b_3\omega}e^{\frac{b_3}{b_1}\omega}. \tag{41}$$

In this case, the eigenvalues of operator $\widehat{\mathfrak{H}}$ are real and quantized. The functions $\psi_n(\omega)$ are orthogonal with respect to weight function r, that is

$$\int_0^\infty \frac{b_4}{b_3\omega}e^{\frac{b_3}{b_1}\omega}\psi_n(\omega)\psi_m(\omega)d\omega = 0, \qquad n \neq m. \tag{42}$$

Then, the set $\{\psi_n(\omega)\}$ is complete and an general state $f(\omega)$ can be expressed in terms of this basis,

$$f(\omega) = \sum_{n=0}^{\infty} c_n \psi_n(\omega), \tag{43}$$

with

$$c_n = \int_0^\infty \frac{b_4}{b_3\omega}e^{\frac{b_3}{b_1}\omega}f(\omega)\psi_n(\omega)d\omega. \tag{44}$$

This shows that we have a Hilbert space $\mathcal{H}$ with square integrable functions. Due to $\widehat{\mathfrak{H}}$ being self-adjointed, their eigenvalues are real. In this way, ϵ_n can be represent a measure of a physical quantity.

6. Quantum FLRW System

In reference [12], the quantization of a homogeneous and isotropic universe is explored. The FLRW metric is

$$ds^2 = -dt^2 + a^2\left[\frac{dr^2}{1-kr^2} + r^2\left(d\theta^2 + \sin^2\theta\, d\phi^2\right)\right].$$

Then the energy density is

$$k\mathfrak{H} = \frac{a^2(3\dot{a}^2 - k)\sqrt{\dot{a}^2 + k(1-a^2)}}{4(\dot{a}^2 + k)} - \frac{a(3\dot{a}^2 + k)}{4\sqrt{k}}\arctan\left[\frac{\sqrt{k}\,a}{\sqrt{\dot{a}^2 + k(1-a^2)}}\right].$$

It should be noted that we can choose any representation for the operators in the quantization process. Here we use $a \to \hat{a} = a$ and $\dot{a} \to \hat{\dot{a}} = \beta\dfrac{\partial}{\partial a}$. This leads to the equation

$$\left[15a^2 + \frac{66}{8} + \left(6a^2 + \frac{76}{8}\right)a\frac{d}{da} + \left(\frac{a^2}{2} + \frac{31}{8}\right)a^2\frac{d^2}{da^2}\right]\psi(a) = \epsilon\psi(a), \tag{45}$$

where $\epsilon = \dfrac{k^{3/2}E}{\beta^2}$ and E is the observable energy. Equation (45) can be written as

$$\left[-\epsilon + h_3(a) + h_2(a)a\frac{d}{da} + h_1(a)a^2\frac{d^2}{da^2}\right]\psi(a) = 0, \tag{46}$$

where $h_1 = \frac{1}{2}\left(a^2 + \frac{31}{4}\right)a^2$, $h_2(a) = 2\left(3a^2 + \frac{19}{4}\right)a$, $h_3(a) = 15a^2 + \frac{66}{8}$. Multiplying Equation (46) by integrator factor

$$\mu(a) = \frac{2\left(a^2 + \frac{31}{4}\right)^{127/31}}{a^{45/31}}, \tag{47}$$

we obtain

$$\frac{d}{da}\left(\frac{\left(a^2 + \frac{31}{4}\right)^{158/31}}{a^{76/31}}\frac{d\psi(a)}{da}\right) + h_3(a)\mu(a)\psi(a) - \epsilon\mu(a)\psi(a) = 0. \tag{48}$$

Equation (48) is the self-adjoint of Equation (47). Equation (48) represents a usual Sturm–Liouville problem with weight function given by

$$r(a) = \frac{2\left(a^2 + \frac{31}{4}\right)^{127/31}}{a^{45/31}}. \tag{49}$$

In this case, the eigenvalues of operator $\widehat{\mathfrak{H}}$ are real and quantized. The functions $\psi_n(a)$ are orthogonal with respect to weight function r, that is

$$\int_0^b \frac{2\left(a^2 + \frac{31}{4}\right)^{127/31}}{a^{45/31}} \psi_n(a)\psi_m(a)da = 0, \qquad n \neq m. \tag{50}$$

Then, the set $\{\psi_n(a)\}$ is complete and an general state $f(a)$ can be expressed in terms of this basis,

$$f(a) = \sum_{n=0}^{\infty} c_n \psi_n(a), \tag{51}$$

with

$$c_n = \int_0^{\infty} \frac{2\left(a^2 + \frac{31}{4}\right)^{127/31}}{a^{45/31}} f(a)\psi_n(a)da. \tag{52}$$

This shows that we have a Hilbert space $\mathcal{H}$ with square integrable functions. Because $\widehat{\mathfrak{H}}$ is self-adjointed, their eigenvalues are real. In this way, ϵ_n can represent a measure of a physical quantity.

7. Conclusions

In this article, we apply a quantization procedure to Schwarzschild space–time and obtain a differential equation in terms of the Caputo fractional derivative. With this, we were able to establish Hilbert's space for this configuration. In previous articles, quantization procedures were applied to Reissner–Nordstron space–time, as well as to the FRW metric. The first describes a charged black hole, and the second describes an isotropic and homogeneous Universe. Thus, we show how the respective functions of Hilbert space obey certain orthogonal conditions. It is interesting to note that once Hilbert's space has been defined, the eigenvalues of the equations resulting from the quantization process have real values and represent experimentally verifiable quantities. It is worth clarifying that the quantization process is done in the space of one of the coordinates, that is, one of the operators resulting from the process is always multiplicative. The passage of the gravitational energy function to the quantum operator can be problematic due to the semi-integer powers of the coordinates on which the function depends. This is circumvented with the use of the fractional derivative of Caputo. In particular, the conditions which establish a Hilbert space for the FLRW metric strengthen the interpretation of a multiverse. Each discrete function may describe a specific universe. The orthogonality relationship guarantees the physical independence of each solution. This establishes a very well-defined procedure for the construction of the respective Hilbert space and therefore for a promising quantum gravitational theory. The quantization procedure used here depends only on the existence of a gravitational energy, although we defend the concept of energy obtained in the scope of the TEGR; the method extends to any definition of energy in the literature. In particular, a very well-accepted approach to quantum gravity, the loop quantum gravity, could benefit from this quantization process. As a future perspective we hope to understand the limitations or advantages of Weyl quantization in the 3 + 1 decomposition of the Hamiltonian formulation of gravitation. The greatest difficulty would be in the choice of operators since they are defined from an anti-commutation relation of two (or more) coordinates. In ADM decomposition, the dynamic field is generic. This problem is analogous to the Schroedinger equation in which we need to define the potential so that the equation can be solved; otherwise we have a generic Hamiltonian.

Author Contributions: Conceptualization, E.P.S., R.G.G.A. and S.C.U.; methodology, E.P.S., R.G.G.A. and S.C.U.; formal analysis, E.P.S., R.G.G.A. and S.C.U.; investigation, E.P.S., R.G.G.A. and S.C.U.; writing—original draft preparation, E.P.S. and R.G.G.A.; writing—review and editing, E.P.S., R.G.G.A. and S.C.U.; visualization, S.C.U.; supervision, S.C.U. All authors have read and agreed to the published version of the manuscript.

Funding: This research received no external funding.

Institutional Review Board Statement: The study did not require ethical approval.

Informed Consent Statement: Not applicable.

Data Availability Statement: This manuscript will appear on arXiv.

Conflicts of Interest: The authors declare no conflict of interest.

References

1. Ashtekar, A.; Bianchi, E. A short review of loop quantum gravity. *Rep. Prog. Phys.* **2021**, *84*, 042001. [CrossRef] [PubMed]
2. Mukhi, S. String theory: A perspective over the last 25 years. *Class. Quant. Grav.* **2011**, *28*, 153001. [CrossRef]
3. Ali, S.T.; Englis, M. Quantization methods: A guide for physicists and analysts. *Rev. Math. Phys.* **2005**, *17*, 391–490. [CrossRef]
4. Maluf, J.W. The teleparallel equivalent of general relativity. *Ann. Phys.* **2013**, *525*, 339–357. [CrossRef]
5. Maluf, J.W.; Ulhoa, S.C.; Faria, F.F.; da Rocha-Neto, J.F. The angular momentum of the gravitational field and the Poincaré group. *Class. Quantum Grav.* **2006**, *23*, 6245. [CrossRef]
6. Maluf, J.W.; da Rocha-Neto, J.F. Hamiltonian formulation of general relativity in the teleparallel geometry. *Phys. Rev. D* **2001**, *64*, 084014. [CrossRef]
7. Weyl, H. Quantenmechanik und gruppentheorie. *Z. Phys.* **1927**, *46*, 1–46. [CrossRef]
8. Daubechies, I. On the distributions corresponding to bounded operators in the Weyl quantization. *Commun. Math. Phys.* **1980**, *75*, 229–238. [CrossRef]
9. Daubechies, I. Continuity statements and counterintuitive examples in connection with Weyl quantization. *J. Math. Phys.* **1983**, *24*, 1453–1461. [CrossRef]
10. Dubin, D.A.; Hennings, M.A.; Smith, T.B. *Mathematical Aspects of Weyl Quantization and Phase*; World Scientific: London, UK, 2000.
11. Ulhoa, S.C. On the quantization of the charge—Mass ratio. *Gen. Relativ. Gravit.* **2017**, *49*, 3. [CrossRef]
12. Fernandes, A.S.; Amorim, R.G.G.; Capistrano, A.J.S.; Ulhoa, S.C. Towards energy discretization in quantum cosmology. *Heliyon* **2019**, *5*, 5. [CrossRef] [PubMed]
13. Klimek, M.; Agrawal, O.P. Fractional Sturm–Liouville problem. *Comput. Math. Appl.* **2013**, *66*, 795. [CrossRef]
14. Hammad, M.A.; Khalil, R. Legendre fractional differential equation and Legender fractional polynomials. *Int. J. App. Math. Res.* **2014**, *3*, 214.
15. Bas, E. Fundamental spectral theory of fractional singular Sturm-Liouville operator. *J. Funct. Spaces Appl.* **2013**, *1*, 915830. [CrossRef]
16. Oldham, K.B.; Spanier, J. *The Fractional Calculus: Theory and Applications of Differentiation and Integration to Arbitrary Order*; Elsevier: Amsterdam, The Netherlands, 1974; pp. 111, iii–iv, ix–xiii, 1–234.
17. Herrmann, R. *Fractional Calculus: An Introduction to Physicists*; World Scientific: Singapore, 2011.
18. Torvik, P.J.; Bagley, R.L. On the appearance of the fractional derivative in the behavior of real materials. *J. Appl. Mech.* **1984**, *51*, 294–298. [CrossRef]
19. Ulhoa, S.C.; Amorim, R.G.G. On teleparallel quantum gravity in schwarzschild space-time. *Adv. High Energy Phys.* **2014**, *2014*, 812691. [CrossRef]

Article

Lorentzian Vacuum Transitions in Hořava–Lifshitz Gravity

Hugo García-Compeán * and Daniel Mata-Pacheco

Departamento de Física, Centro de Investigación y de Estudios Avanzados del IPN, P.O. Box 14-740, Ciudad de México 07000, Mexico; dmata@fis.cinvestav.mx
* Correspondence: compean@fis.cinvestav.mx

Abstract: The vacuum transition probabilities for a Friedmann–Lemaître–Robertson–Walker universe with positive curvature in Hořava–Lifshitz gravity in the presence of a scalar field potential in the Wentzel–Kramers–Brillouin approximation are studied. We use a general procedure to compute such transition probabilities using a Hamiltonian approach to the Wheeler–DeWitt equation presented in a previous work. We consider two situations of scalar fields, one in which the scalar field depends on all the spacetime variables and another in which the scalar field depends only on the time variable. In both cases, analytic expressions for the vacuum transition probabilities are obtained, and the infrared and ultraviolet limits are discussed for comparison with the result obtained by using general relativity. For the case in which the scalar field depends on all spacetime variables, we observe that in the infrared limit it is possible to obtain a similar behavior as in general relativity, however, in the ultraviolet limit the behavior found is completely opposite. Some few comments about possible phenomenological implications of our results are given. One of them is a plausible resolution of the initial singularity. On the other hand, for the case in which the scalar field depends only on the time variable, the behavior coincides with that of general relativity in both limits, although in the intermediate region the probability is slightly altered.

Keywords: quantum gravity; Hořava–Lifshitz theory; early universe; vacuum transitions

Citation: García-Compeán, H.;
Mata-Pacheco, D. Lorentzian Vacuum
Transitions in Hořava–Lifshitz
Gravity. *Universe* **2022**, *8*, 237.
https://doi.org/10.3390/
universe8040237

Academic Editors: Arundhati
Dasgupta and Alfredo Iorio

Received: 21 March 2022
Accepted: 8 April 2022
Published: 12 April 2022

Publisher's Note: MDPI stays neutral with regard to jurisdictional claims in published maps and institutional affiliations.

Copyright: © 2022 by the authors. Licensee MDPI, Basel, Switzerland. This article is an open access article distributed under the terms and conditions of the Creative Commons Attribution (CC BY) license (https://creativecommons.org/licenses/by/4.0/).

1. Introduction

The quantum theory of the gravitational phenomena, or quantum gravity, is a theory in construction, which is necessary in order to shed light on the quantum effects of gravitational systems. Among the problems that require the uses of quantum gravity is the study of the microscopic origin of thermodynamic properties of black holes and those describing some cosmological phenomena in the very early universe. Another important problem is the study of the vacuum decay and the transition between vacua at early stages of the evolution of the universe. Euclidean methods have been proposed in order to compute this transition probability by using the path integral approach [1–3]. One of the salient features of this approach is the prediction of transitions between open universes [3]. Later an alternative procedure to compute these transitions using the Hamiltonian approach was developed [4,5]. This method incorporates the Arnowitt, Deser and Misner (ADM) Hamiltonian formalism of general relativity (GR) [6–8]. The vacuum is implemented through a cosmological constant, which is interpreted as the vacuum energy, and the transitions are carried out through a bubble nucleation [9]. In this approach, the transitions between Minkowski and de Sitter spaces are allowed. Very recently, an approach [4,5] was further developed by Cespedes et al. [10] where the vacuum is implemented by the minima of a potential of a scalar field in the curved space. In this reference, it was computed the general vacuum decay transitions in the Hamiltonian formalism in Wheeler's superspace and some examples were implemented in the minisuperspace formalism for the Friedmann–Lemaître–Robertson–Walker (FLRW) cosmology. In this kind of model, it was shown that the transitions between closed universes are allowed, contrary to the Euclidean approximation of Coleman and De Luccia [3]. Later, the formalism of [10] was extended and used to obtain the vacuum decay transition probabilities for some examples of transitions between anisotropic universes [11].

On the other hand, it is well known that GR is not a renormalizable theory. Thus, its application to very small distances such as those associated with the early universe is expected to fail. Instead of that, an important proposal to describe the quantum effects of gravity is the Hořava–Lifshitz (HL) theory [12] (for some recent reviews, see [13–16] and references therein). Hořava–Lifshitz theory is a theory with an anisotropic scaling of spacetime and, consequently, it is not Lorentz invariant at high energies (ultraviolet (UV)). However, it is a well-behaved description at small distances due to the incorporation of higher order derivative terms in the spatial components of the curvature to the usual Einstein–Hilbert action, giving rise to a ghost-free theory. Thus, this theory is more appropriate to describe the quantum effects of the gravitational field, such as the vacuum decay processes in the early stages of the universe evolution.

It is important to remark that HL theory is a theory whose low energy limit which connects with GR is troublesome. The parameters of the theory are the critical exponent z and the foliation parameter λ. This last parameter is associated with a restricted foliation compatible with the Lifshitz scaling. In the low energy limit $z \to 1$, the Lorentz invariance is recovered. In the infrared (IR) limit, the $z \to 1$ limit is accompanied by the limit $\lambda \to 1$, where the full diffeomorphisms symmetry is recovered and, consequently, the usual foliation of the ADM formalism is regained. In addition, the higher order derivative terms in the action have to be properly neglected in order to obtain the correct limit. As we mentioned before, the GR limit is problematic since it remains an additional degree of freedom (in some cases interpreted as dark matter) which leads to a perturbative IR instability [14,16–18]. The non-projectable version of the HL theory has the possibility to remove this unphysical degree of freedom. Thus, it represents an advantage over the projective theory. However, in the case in which one is concerned with the Wheeler–DeWitt (WDW) equation, both approaches give the same result. In consequence, we will work with the projectable version.

Since HL theory represents an improvement over GR in the high energy regime, it is natural that quantum gravity aspects of the theory are of great interest. Indeed, canonical quantization of the theory has been extensively studied. For example, some of the papers describing solutions for Hořava–Lifshitz's gravity in quantum cosmology in the minisuperspace are [19–25].

As we mentioned before, HL gravity is a UV completion of GR, thus, it is a more suitable arena to study the vacuum transitions in the presence of a scalar field potential. This proposal will be carried out in the present article. In order to do that, we use the Hamiltonian formalism of the HL theory, in particular the WDW equation will be discussed in this context following [10,11]. We will particularly focus on the closed FLRW universe and study two types of scalar fields. First, since the anisotropic scaling of spacetime variables is a key ingredient of HL theory, we will consider a scalar field which is allowed to depend on all spacetime coordinates. Lastly, we will also consider a scalar field which only depends on the time variable as it is more usual on the cosmological models.

This work is organized as follows. In Section 2 we give a brief review of the general procedure presented in [11] to study vacuum transition probabilities between two minima of a scalar field potential in the minisuperspace following the formalism of [10]. We will show that this formalism implemented for GR in [10,11] is sufficient to study vacuum transitions in a more general theory, such as HL theory. Section 3 is devoted to obtaining the WDW equation in the context of gravity coupled with matter. In Section 4, we study the vacuum transitions in HL gravity for the scalar field depending on all spacetime variables. The IR and UV limits for the transition probabilities are discussed and compared to the GR result. Then, in Section 5, we study the transition probabilities for the scalar field depending only on the time variable and we also compare the result to the GR one. Finally, in Section 6, we give our conclusions and final remarks.

2. Vacuum Transitions for a Scalar Field

In this section, we review the procedure to obtain a general expression for the transition probability between two minima of the potential of a scalar field by obtaining a semi-classical solution to the WDW equation using a WKB ansatz described in Ref. [11]. We follow closely the notation and conventions given in that reference. It is a remarkable point to see that this procedure is enough to implement theories more general than GR such as the HL gravity.

We start by using the well-known ADM-formulation of GR [6–8] and consider the Hamiltonian constraint expressed in the general form:

$$\mathcal{H} = \frac{1}{2}G^{MN}(\Phi)\pi_M\pi_N + f[\Phi] \approx 0, \tag{1}$$

where we take the coordinates in Wheeler's superspace to be Φ^M with $M, N = 1, \ldots, n$ (which has in general an infinite number of dimensions). These variables are the components of the three-dimensional metric, the matter field variables, etc. and are denoted collectively as Φ. Their corresponding canonical momenta are π_M and the inverse metric in such space is G^{MN}. Finally, $f[\Phi]$ is a function that represents all other additional terms, such as the ^3R term and the potential terms of scalar field in the WDW equation. The general WDW equation that we are going to consider is obtained after carrying out the standard canonical quantization procedure of the Hamiltonian constraint. Thus performing this procedure we obtain:

$$\mathcal{H}\Psi(\Phi) = \left[-\frac{\hbar^2}{2}G^{MN}(\Phi)\frac{\delta}{\delta\Phi^M}\frac{\delta}{\delta\Phi^N} + f[\Phi]\right]\Psi[\Phi] = 0, \tag{2}$$

where $\Psi[\Phi]$ represents the wave functionial which depends on all fields of the theory.

We are interested in obtaining a semi-classical result, therefore, following [10,11] we consider an ansatz of the following WKB form $\Psi[\Phi] = \exp\{\frac{i}{\hbar}S[\Phi]\}$, where S has an expansion in $\hbar$ in the usual form:

$$S[\Phi] = S_0[\Phi] + \hbar S_1[\Phi] + \mathcal{O}(\hbar^2). \tag{3}$$

Inserting Equation (3) into Equation (2) and focusing only on the term at the lowest order in $\hbar$ we obtain:

$$\frac{1}{2}G^{MN}\frac{\delta S_0}{\delta\Phi^M}\frac{\delta S_0}{\delta\Phi^N} + f[\Phi] = 0. \tag{4}$$

On a certain slice of the space of fields a set of integral curves can be specified in the form:

$$C(s)\frac{d\Phi^M}{ds} = G^{MN}\frac{\delta S_0}{\delta\Phi^N}, \tag{5}$$

where s is the parameter of these curves. The classical action appearing in the previous equation has the form:

$$S_0[\Phi_s] = -2\int^s \frac{ds'}{C(s')}\int_X f[\Phi_{s'}]. \tag{6}$$

It is easy to see that Equations (4) and (5) lead to:

$$G_{MN}\frac{d\Phi^M}{ds}\frac{d\Phi^N}{ds} = -\frac{2f[\Phi_s]}{C^2(s)}, \tag{7}$$

where G_{MN} satisfies the standard relation $G_{PM}G^{MN} = \delta_P^N$.

We note that we have a system of equations for the $n+1$ variables: $\left(\frac{d\Phi^M}{ds}, C^2(s)\right)$ defined by (5) and (7). Thus, we can obtain a solution for such a system and then substitute the results back into Equation (6) to obtain the classical action. Therefore, in principle, we have enough information to compute the classical action, and consequently, the wave functional to first order in $\hbar$ regardless of the number of fields in superspace.

Under the ansatz that all the fields Φ^M on the superspace depend only on the time variable, we can obtain a general solution to the system in terms of the volume $\mathrm{Vol}(X)$ of the spatial slice X of the form:

$$C^2(s) = -\frac{2\mathrm{Vol}^2(X)}{f[\Phi]} G^{MN} \frac{\partial f}{\partial \Phi^M} \frac{\partial f}{\partial \Phi^N}, \tag{8}$$

$$\frac{d\Phi^M}{ds} = \frac{f[\Phi]}{\mathrm{Vol}(X)} \frac{G^{MN} \frac{\partial f}{\partial \Phi^N}}{G^{LO} \frac{\partial f}{\partial \Phi^L} \frac{\partial f}{\partial \Phi^O}}. \tag{9}$$

In this article, we will consider gravity coupled to a scalar field provided by a potential which has at least a false and a true minima. Moreover, we will study wave functionals such that the scalar field produces a transition between two minima of the potential.

One can use these wave functionals in order to compute the transition probability with the standard interpretation that these transitions are due to a tunneling between the two minima of the potential involved in the transition. In order to be more precise, in the semi-classical approximation, the probability to produce a transition between two vacua at ϕ_A and ϕ_B is the decay rate which can be written as:

$$P(A \to B) = \left| \frac{\Psi(\varphi_0^I, \phi_B; \varphi_m^I, \phi_A)}{\Psi(\varphi_0^I, \phi_A; \varphi_m^I, \phi_A)} \right|^2 = \left| \frac{\beta e^{\frac{i}{\hbar} S_0(\varphi_0^I, \phi_B; \varphi_m^I, \phi_A)} + \chi e^{-\frac{i}{\hbar} S_0(\varphi_0^I, \phi_B; \varphi_m^I, \phi_A)}}{\beta e^{\frac{i}{\hbar} S_0(\varphi_0^I, \phi_A; \varphi_m^I, \phi_A)} + \chi e^{-\frac{i}{\hbar} S_0(\varphi_0^I, \phi_A; \varphi_m^I, \phi_A)}} \right|^2 = \left| e^{-\Gamma} \right|^2, \tag{10}$$

where φ^I denotes all other fields defined on the superspace except the scalar field, $\Psi(\varphi_0^I, \phi_B; \varphi_m^I, \phi_A)$ is the wave functional associated to the path which starts in $\varphi^I(s=0) = \varphi_0^I$, and where the scalar field takes the value ϕ_B. Moreover, the path ends in $\varphi^I(s=s_M) = \varphi_m^I$, where the scalar field is denoted by ϕ_A. Furthermore, β and χ are the constants of the linear superposition. In the previous equation, we will consider just the dominant contribution of the exponential terms. Then, in the WKB approximation at first order Γ yields:

$$\pm \Gamma = \frac{i}{\hbar} S_0(\varphi_0^I, \phi_B; \varphi_m^I, \phi_A) - \frac{i}{\hbar} S_0(\varphi_0^I, \phi_A; \varphi_m^I, \phi_A), \tag{11}$$

where the choice of the signs $\pm$ indicates the dominant terms in the expression (10). Thus, we finally arrive at the transition probability given by:

$$P(A \to B) = \exp[-2\mathrm{Re}(\Gamma)] = \exp\left\{ \pm 2\mathrm{Re}\left[\frac{i}{\hbar} S_0(\varphi_0^I, \phi_B; \varphi_m^I, \phi_A) - \frac{i}{\hbar} S_0(\varphi_0^I, \phi_A; \varphi_m^I, \phi_A) \right] \right\}. \tag{12}$$

It is worth mentioning that the formalism developed in [10,11], originally for GR, is general enough to include other gravitational theories, since it only depends on a Hamiltonian constraint written in the general form (1). In the next sections, we will show that it can perfectly include higher derivative generalizations of GR as the HL theory. It would be interesting to study to what extent this formalism can be used for more general theories.

3. Wheeler–DeWitt Equation for Hořava–Lifshitz Gravity Coupled to Matter

In this section, we will discuss the action in HL theory, as well as the action that considers the coupling to a scalar field. We will consider a metric describing an FLRW universe and a scalar field depending on the time variable, as well as the spatial variables, and we will obtain the WDW equation for such a system. Although in the context of cosmology it is usual to use a time-dependent field only, in this case, we will allow the scalar field to also depend on the spatial variables since the anisotropic scaling of both sets of variables is a key ingredient for the theory. This type of dependence has been used previously in the context of cosmology for HL. For example, it was used in Ref. [26] to study perturbations coming from a scalar field. However, for completeness and correspondence with the cosmological models, we will also study the case when the field will only depend on the time coordinate in Section 5.

Let us begin by considering the gravitational part of the general action in projectable HL gravity without a cosmological constant and without detailed balance. This action can be written as [19,27,28]:

$$S_{HL} = \frac{M_p^2}{2} \int dt d^3x N\sqrt{h}\left[K^{ij}K_{ij} - \lambda K^2 + R - \frac{1}{M_p^2}\left(g_2 R^2 + g_3 R_{ij}R^{ij}\right) - \frac{1}{M_p^4}\left(g_4 R^3 + g_5 R(R_{ij}R^{ij}) + g_6 R_j^i R_k^j R_i^k + g_7 R D^2 R + g_8 D_i R_{jk} D^i R^{jk}\right)\right], \quad (13)$$

where N is the lapse function, R_{ij} the Ricci tensor with $i,j = 1,2,3$ the spatial indices, R the Ricci scalar, K_{ij} the extrinsic curvature, M_p the Planck mass, D denotes covariant derivative with respect to the three-metric h_{ij} and h denotes its determinant, all g_n ($n = 2,\ldots,8$) are positive dimensionless running coupling constants and the parameter λ runs under the renormalization group flow. GR is, in principle, obtained in the limit $\lambda \to 1$ and $g_n \to 0$. However, this is not actually fulfilled because of the perturbative IR instability and the presence of an unphysical degree of freedom as mentioned in the introduction.

Let us take the FLRW metric with positive curvature that describes a closed homogeneous and isotropic universe. This metric is written as:

$$ds^2 = -N^2(t)dt^2 + a^2(t)\left[dr^2 + \sin^2 r\left(d\theta^2 + \sin^2\theta d\psi^2\right)\right], \quad (14)$$

where, as usual, $0 \leq r \leq \pi$, $0 \leq \theta \leq \pi$ and $0 \leq \psi \leq 2\pi$. In the context of the ADM formalism, we note that for this metric:

$$N = N(t), \quad N_i = 0, \quad (h_{ij}) = a^2(t)\text{diag}\left(1, \sin^2 r, \sin^2 r \sin^2\theta\right), \quad (15)$$

and therefore, we will work with a projectable version of HL gravity. Substituting these values in the action (13) we obtain that for this metric the gravitational action reads:

$$S_{HL} = 2\pi^2 \int dt N\left[-\frac{3(3\lambda-1)M_p^2 a}{2N^2}\dot{a}^2 + 3M_p^2 a - \frac{6}{a}(3g_2 + g_3) - \frac{12}{a^3 M_p^2}(9g_4 + 3g_5 + g_6)\right], \quad (16)$$

where a dot stands for the derivative with respect to the time variable. The integral for the spatial slice has been performed, that is:

$$\text{Vol}(X) = \int_{r=0}^{\pi}\int_{\theta=0}^{\pi}\int_{\psi=0}^{2\pi}\sin^2 r \sin\theta dr d\theta d\psi = 2\pi^2. \quad (17)$$

In order to couple a scalar field $\phi(t, x^i)$ (where x^i denotes collectively the three spatial variables) to this theory we need to consider actions that are compatible with the anisotropic scaling symmetries of the theory and UV renormalizability. In fact, the general scalar action in HL gravity is found to contain up to six order derivatives. This action is written in the form [29]:

$$S_m = \frac{1}{2}\int dt d^3x\sqrt{h}N\left[\frac{(3\lambda-1)}{2N^2}\left(\dot{\phi} - N^i\partial_i\phi\right)^2 + F(\phi)\right], \quad (18)$$

where the function $F(\phi)$ is given by:

$$F(\phi) = \phi\left(c_1\Delta\phi - c_2\Delta^2\phi + c_3\Delta^3\phi\right) - V(\phi), \quad (19)$$

with Δ denoting the three-metric laplacian and $V(\phi)$ is the potential for the scalar field. The constant c_1 is the velocity of light in the IR limit, whereas the two other constants are related to the energy scale M as:

$$c_2 = \frac{1}{M^2}, \quad c_3 = \frac{1}{M^4}. \quad (20)$$

There are three more possible terms that can be part of (19) constructed as products of derivatives, but we restrict ourselves to the terms just described.

Since the three-metric derived from the FLRW metric is just a scale factor times the metric of the three-sphere $\mathbf{S}^3$, we can use the spherical harmonic functions defined in this space to expand our scalar functions [30,31]. These functions can be defined in our spatial three-metric as eigenfunctions of the laplacian of the form:

$$\Delta Y_{nlm}(x^i) = -\frac{n(n+2)}{a^2} Y_{nlm}(x^i), \qquad (21)$$

where n is an integer. They obey the orthonormality condition:

$$\frac{1}{a^3} \int \sqrt{h} Y_{nlm}(x^i) Y^*_{n'l'm'}(x^i) d^3x = \delta_{nn'}\delta_{ll'}\delta_{mm'}. \qquad (22)$$

Since these functions form a complete basis, we can expand any scalar function defined on the sphere in terms of them as:

$$f(x^i) = \sum_{n=0}^{\infty} \sum_{l=0}^{n} \sum_{m=-l}^{l} \alpha_{nlm} Y_{nlm}(x^i) = \sum_{\{n,l,m\}} \alpha_{nlm} Y_{nlm}(x^i). \qquad (23)$$

Therefore, using this basis, we can expand the scalar field as:

$$\phi(t, x^i) = \sum_{\{n,l,m\}} \phi_{nlm}(t) Y_{nlm}(x^i), \qquad (24)$$

where the fields $\phi_{nlm}(t)$ are real functions depending only on the time variable. We can also expand the scalar field potential as:

$$V(\phi) = \sum_{\{n,l,m\}} V_{nlm}(t) Y_{nlm}(x^i), \qquad (25)$$

where the function $V_{nlm}(t)$ depends on all the functions $\phi_{nlm}(t)$ in general. Substituting (24) and (25) back into the action (18), we observe that for the FLRW metric the field part of the action is written as:

$$S_m = \sum_{\{n,l,m\}} \frac{1}{2} \int dt \left\{ \frac{3\lambda-1}{2N} a^3 \dot{\phi}_{nlm}^2 - Na \left[c_1 \beta_n + \frac{c_2 \beta_n^2}{a^2} + \frac{c_3 \beta_n^3}{a^4} \right] \phi_{nlm}^2 - Na^3 \gamma_{nlm} V_{nlm} \right\}, \qquad (26)$$

where $\beta_n = n(n+2)$ and

$$\gamma_{nlm} = \int_{r=0}^{\pi} \int_{\psi=0}^{2\pi} \int_{\theta=0}^{\pi} \sin^2 r \sin\theta Y_{nlm}(r,\theta,\psi) dr d\psi d\theta, \qquad (27)$$

are constants. Finally, by considering both actions (16) and (26), we observe that the full lagrangian describing HL gravity coupled to a scalar field is:

$$\mathcal{L} = 2\pi^2 N \left[-\frac{3M_p^2 \dot{a}^2 a}{2N^2}(3\lambda-1) + 3M_p^2 a - \frac{6}{a}(3g_2 + g_3) - \frac{12}{a^3 M_p^2}(9g_4 + 3g_5 + g_6) \right] \\ + \sum_{\{n,l,m\}} \left\{ \frac{3\lambda-1}{4N} a^3 \dot{\phi}_{nlm}^2 - \frac{Na}{2}\left[c_1 \beta_n + \frac{c_2 \beta_n^2}{a^2} + \frac{c_3 \beta_n^3}{a^4}\right]\phi_{nlm}^2 - \frac{Na^3}{2} \gamma_{nlm} V_{nlm} \right\}. \qquad (28)$$

We observe, in this case, that the degrees of freedom are the fields $\{a, \phi_{nlm}\}$. Their canonical momenta turn out to be:

$$\pi_a = -\frac{6\pi^2 M_p^2 (3\lambda-1)}{N} a\dot{a}, \qquad \pi_{\phi_{nlm}} = \frac{3\lambda-1}{2N} a^3 \dot{\phi}_{nlm}, \qquad (29)$$

and, as it is usual, the lapse function is non-dynamical since $\pi_N = 0$. Therefore, we obtain that the Hamiltonian constraint takes the form:

$$H = N\left\{-\frac{\pi_a^2}{12\pi^2 M_p^2(3\lambda-1)a} + \sum_{\{n,l,m\}} \frac{\pi_{\phi_{nlm}}^2}{(3\lambda-1)a^3}\right.$$
$$+2\pi^2\left[-3M_p^2 a + \frac{6}{a}(3g_2+g_3) + \frac{12}{M_p^2 a^3}(9g_4+3g_5+g_6)\right]$$
$$\left.+\frac{1}{2}\sum_{\{n,l,m\}}\left[\left(c_1\beta_n + \frac{c_2}{a^2}\beta_n^2 + \frac{c_3}{a^4}\beta_n^3\right)a\phi_{nlm}^2 + a^3\gamma_{nlm}V_{nlm}\right]\right\} \simeq 0. \tag{30}$$

4. Vacuum Transitions in Hořava–Lifshitz Gravity

Now that we have obtained the Hamiltonian constraint of the HL gravity coupled to a scalar field depending on all spacetime variables, let us study the probability transition between two vacua of the scalar field potential. We note that the form of the Hamiltonian constraint (30) as obtained in the previous section is of the same general form as the one considered in (1) taking the coordinates on superspace to be $\{a, \phi_{nlm}\}$. The inverse metric is given by:

$$(G^{MN}) = \text{diag}\left(-\frac{1}{6\pi^2 M_p^2(3\lambda-1)a}, \frac{2}{(3\lambda-1)a^3}\mathbf{1}_{nlm}\right), \tag{31}$$

where $\mathbf{1}_{nlm}$ denotes a vector with length equal to all the possible values that the set $\{n, l, m\}$ can have and with 1 in all its entries. In this case then we also have:

$$f(a,\phi_{nlm},V_{nlm}) = 2\pi^2\left[-3M_p^2 a + \frac{6}{a}(3g_2+g_3) + \frac{12}{M_p^2 a^3}(9g_4+3g_5+g_6)\right]$$
$$+\frac{1}{2}\sum_{\{n,l,m\}}\left[\left(c_1\beta_n + \frac{c_2}{a^2}\beta_n^2 + \frac{c_3}{a^4}\beta_n^3\right)a\phi_{nlm}^2 + a^3\gamma_{nlm}V_{nlm}\right]. \tag{32}$$

Therefore, the general procedure to obtain a solution of the WDW equation presented in Section 2 is applicable to the WDW equation obtained after quantizing the Hamiltonian constraint (30) in HL gravity. In order to study transitions between two vacua of a scalar field potential, we consider that all fields V_{nlm} appearing in the expansion of the potential (25) have the same minima, namely, one false minimum at ϕ_{nlm}^A and one true minimum at ϕ_{nlm}^B, and therefore, the two minima of the scalar field $\phi(t, x^i)$ comes only from its time dependence. Therefore, the transition probability in the semi-classical approach between these two minima is given by Equation (12).

Following [11], we can choose the parameter s such that for the interval $[0, \bar{s} - \delta s]$, where $s = 0$ is the initial value, the field remains close to its value at the true minimum ϕ_B, and for the interval $[\bar{s} + \delta s, s_m]$ the field remains very close to its value at the false minimum ϕ_A, that is, we choose the parameter s such that:

$$\phi(s) \approx \begin{cases} \phi_B, & 0 < s < \bar{s} - \delta s, \\ \phi_A, & \bar{s} + \delta s < s < s_M. \end{cases} \tag{33}$$

However, in this case, taking the expansion (24) and since the spherical harmonics are an orthonormal set, the latter implies that:

$$\phi_{nlm}(s) \approx \begin{cases} \phi_{nlm}^B, & 0 < s < \bar{s} - \delta s, \\ \phi_{nlm}^A, & \bar{s} + \delta s < s < s_M, \end{cases} \tag{34}$$

and similarly for the potentials

$$V_{nlm}(s) \approx \begin{cases} V_{nlm}^B, & 0 < s < \bar{s} - \delta s, \\ V_{nlm}^A, & \bar{s} + \delta s < s < s_M. \end{cases} \tag{35}$$

Therefore, using the general form of the action (6) we obtain in this case:

$$S_0\left(a_0, \phi_{nlm}^B; a_m, \phi_{nlm}^A\right) = -4\pi^2\left[\int_0^{\bar{s}-\delta s} \frac{ds}{C(s)}f\Big|_{\phi_{nlm}=\phi_{nlm}^B} + \int_{\bar{s}-\delta s}^{\bar{s}+\delta s}\frac{ds}{C(s)}f + \int_{\bar{s}+\delta s}^{s_m}\frac{ds}{C(s)}f\Big|_{\phi_{nlm}=\phi_{nlm}^A}\right], \qquad (36)$$

and

$$S_0\left(a_0, \phi_{nlm}^A; a_m, \phi_{nlm}^A\right) = -4\pi^2\int_0^{s_m}\frac{ds}{C(s)}f\Big|_{\phi_{nlm}=\phi_{nlm}^A}. \qquad (37)$$

Consequently, the logarithm of the probability (11) is given in this case by:

$$\pm\Gamma = \frac{i}{\hbar}\Big[-4\pi^2\int_0^{\bar{s}-\delta s}\frac{ds}{C(s)}f\Big|_{\phi_{nlm}=\phi_{nlm}^B} + 4\pi^2\int_0^{\bar{s}-\delta s}\frac{ds}{C(s)}f\Big|_{\phi_{nlm}=\phi_{nlm}^A} \\ -4\pi^2\int_{\bar{s}-\delta s}^{\bar{s}+\delta s}\frac{ds}{C(s)}\left\{\frac{1}{2}\sum_{\{n,l,m\}}\left[a\left(c_1\beta_n + \frac{c_2}{a^2}\beta_n^2 + \frac{c_3}{a^4}\beta_n^3\right)\left(\phi_{nlm}^2 - (\phi_{nlm}^A)^2\right)\right.\right. \\ \left.\left. + a^3\gamma_{nlm}(V_{nlm} - V_{nlm}^A)\right]\right\}\Big]. \qquad (38)$$

We note that the last term of Equation (38) can be written as:

$$-4\pi^2 i \int_{\bar{s}-\delta s}^{\bar{s}+\delta s}\frac{ds}{C(s)}\left[\frac{1}{2}\sum_{\{n,l,m\}}a^3\gamma_{n,l,m}(V_{n,l,m} - V_{nlm}^A)\right] = -4\pi^2 i\int_{\bar{s}-\delta s}^{\bar{s}+\delta s}\frac{ds}{C(s)}a^3\left[V_0 - V_0^A\right], \qquad (39)$$

with a potential defined by

$$V_0 = \frac{1}{2}\sum_{\{n,l,m\}}\gamma_{nlm}V_{nlm}. \qquad (40)$$

We note that this term has the same form as the one considered in Refs. [10,11] regarding the portion of the integral in which the scalar field can vary, therefore, we can also interpret this term as a tension term taking:

$$2\pi^2 \bar{a}^3 T_0 = -4\pi^2 i\int_{\bar{s}-\delta s}^{\bar{s}+\delta s}\frac{ds}{C(s)}a^3\left[V_0 - V_0^A\right]. \qquad (41)$$

Moreover, we note that the term that contains c_1 in (38) can be written as:

$$-4\pi^2 i\int_{\bar{s}-\delta s}^{\bar{s}+\delta s}\frac{ds}{C(s)}\left[\frac{1}{2}\sum_{\{n,l,m\}}ac_1\beta_n\left(\phi_{nlm}^2 - (\phi_{nlm}^A)^2\right)\right] = -4\pi^2 i\int_{\bar{s}-\delta s}^{\bar{s}+\delta s}\frac{ds}{C(s)}c_1 a\left[V_1 - V_1^A\right], \qquad (42)$$

with

$$V_1 = \frac{1}{2}\sum_{\{n,l,m\}}\beta_n\phi_{nlm}^2. \qquad (43)$$

Although this function has no minima in the points considered, it is a function of the scalar fields that can be interpreted as a new effective potential with the form of a mass term. Therefore, applying the same logic used in [11] to such terms, we can define a new contribution for the tension term as:

$$2\pi^2 c_1 \bar{a} T_1 = -4\pi^2 i\int_{\bar{s}-\delta s}^{\bar{s}+\delta s}\frac{ds}{C(s)}c_1 a\left[V_1 - V_1^A\right]. \qquad (44)$$

Similarly, we define for the two remaining terms:

$$V_2 = \frac{1}{2}\sum_{\{n,l,m\}}\beta_n^2\phi_{nlm}^2, \qquad V_3 = \frac{1}{2}\sum_{\{n,l,m\}}\beta_n^3\phi_{nlm}^2, \qquad (45)$$

and the two contributions to the tension terms as:

$$2\pi^2\frac{c_2}{\bar{a}}T_2 = -4\pi^2 i\int_{\bar{s}-\delta s}^{\bar{s}+\delta s}\frac{ds}{C(s)}\frac{c_2}{a}\left[V_2 - V_2^A\right], \qquad (46)$$

$$2\pi^2 \frac{c_3}{\bar{a}^3} T_3 = -4\pi^2 i \int_{\bar{s}-\delta s}^{\bar{s}+\delta s} \frac{ds}{C(s)} \frac{c_3}{a^3} \left[V_3 - V_3^A \right]. \tag{47}$$

On the other hand, in order to do the two first integrals in (38) where all scalar fields are constants, we use the general solutions (8) and (9), then, after changing the integration variables from s to a according to $ds = \left(\frac{da}{ds}\right)^{-1} da$ we obtain:

$$-4\pi^2 \int_0^{\bar{s}-\delta s} \frac{ds}{C(s)} f \Big|_{\phi_{nlm}=\phi_{nlm}^{A,B}} = \pm 4\pi^3 M_P \sqrt{3(3\lambda-1)} \int_{a_0}^{\bar{a}-\delta a} \sqrt{-\alpha_1^{A,B} a^2 + \alpha_2^{A,B} + \frac{\alpha_3^{A,B}}{a^2} + V_0^{A,B} a^4} \, da, \tag{48}$$

where

$$\begin{aligned}
\alpha_1^{A,B} &= 6\pi^2 M_p^2 - c_1 V_1^{A,B}, \\
\alpha_2^{A,B} &= 12\pi^2 (3g_2 + g_3) + c_2 V_2^{A,B}, \\
\alpha_3^{A,B} &= \frac{24\pi^2}{M_p^2} (9g_4 + 3g_5 + g_6) + c_3 V_3^{A,B}.
\end{aligned} \tag{49}$$

Therefore, substituting Equations (41), (44), (46), (47) and (48) back into (38) we obtain:

$$\pm \Gamma = \pm \frac{4\pi^3 M_p \sqrt{3(3\lambda-1)}}{\hbar} \left[\int_{a_0}^{\bar{a}-\delta a} F(\alpha_1^B, \alpha_2^B, \alpha_3^B, V_0^B, a) da - \int_{a_0}^{\bar{a}-\delta a} F(\alpha_1^A, \alpha_2^A, \alpha_3^A, V_0^A, a) da \right] + \frac{2\pi^2}{\hbar} \left[\bar{a}^3 T_0 + c_1 \bar{a} T_1 + \frac{c_2}{\bar{a}} T_2 + \frac{c_3}{\bar{a}^3} T_3 \right]. \tag{50}$$

where we have defined the function:

$$F(a, b, c, e, x) = \sqrt{ax^2 - b - \frac{c}{x^2} - ex^4}, \tag{51}$$

and it is worth noting that in the above result, the sign ambiguity on the left-hand side comes from the arguments leading to Equation (11), whereas the one on the right comes from the fact that the general solution (8) and (9) for the system of equations gives a solution for $C^2(s)$, which produces a sign ambiguity in Equation (48). Therefore, both ambiguities are independent to each other.

As it is well known, the IR limit of HL gravity for an FLRW metric is achieved in the limit $\lambda \to 1$ and $a >> 1$, and corresponds to GR with an extra degree of freedom albeit with the instability problems mentioned in the introduction section. We note that the kinetic term for the scale factor in (28) is:

$$-2\pi^2 \left[\frac{3M_p^2 a \dot{a}^2}{2N} (3\lambda - 1) \right]. \tag{52}$$

In the GR case considered in [11], it is given by:

$$-\frac{3a\dot{a}^2}{N}, \tag{53}$$

because the $2\pi^2$ term in that case is a global multiplicative factor to the full lagrangian, it can therefore be ignored. Thus, in order to obtain the same kinetic term in both cases in the limit $\lambda \to 1$, we consider units such as $2\pi^2 M_p^2 = 1$. This choice of units will allow us to directly compare the transition probability to the one obtained in GR.

Then, we finally obtain for the logarithm of the transition probability:

$$\pm \Gamma = \pm \frac{2\pi^2 \sqrt{6(3\lambda-1)}}{\hbar} \left[\int_{a_0}^{\bar{a}-\delta a} F(\alpha_1^B, \alpha_2^B, \alpha_3^B, V_0^B, a) da - \int_{a_0}^{\bar{a}-\delta a} F(\alpha_1^A, \alpha_2^A, \alpha_3^A, V_0^A, a) da \right] + \frac{2\pi^2}{\hbar} \left[\bar{a}^3 T_0 + c_1 \bar{a} T_1 + \frac{c_2}{\bar{a}} T_2 + \frac{c_3}{\bar{a}^3} T_3 \right] \tag{54}$$

with

$$\begin{aligned}
\alpha_1^{A,B} &= 3 - c_1 V_1^{A,B}, \\
\alpha_2^{A,B} &= 12\pi^2(3g_2 + g_3) + c_2 V_2^{A,B}, \\
\alpha_3^{A,B} &= 48\pi^3(9g_4 + 3g_5 + g_6) + c_3 V_3^{A,B}.
\end{aligned} \qquad (55)$$

We note that in contrast to the results obtained using GR for all the types of metrics considered in [11], this transition probability is described by five parameters, and by extremizing the latter with respect to $\bar{a}$ we can at most reduce them by one. It is also important to note that the above integrals cannot be performed explicitly for any values of the α_i constants. Nonetheless, it is an expression valid for any value of the potentials and, interestingly, it is a general expression that does not depend on having to consider the different modes contributing to the expansion in Equation (24) separately.

As it is explained in [10,11], the choice of s as in (33) is useful to obtain exact solutions for the transition probabilities that lead to the same solutions as the ones obtained using euclidean methods. However, there is also room to consider s in different ways. For example, we can also choose s as the distance in field space. This choice allows us to show that we can have classical transitions just because the metric in superspace for the WDW equation considered here coming from the Hamiltonian constraint (30) is non-positive definite, as is the case for all the metrics considered in [11].

Now that we have computed the transition probability for HL gravity in general, let us consider its two limits of importance, namely, the infrared and the ultraviolet limit. The first enables us to directly compare with the result found by using GR and the latter allows us to highlight the contributions for high energies that marks the importance of HL gravity.

Taking the IR limit of (54), that is, taking $\lambda \to 1$ and $a \gg 1$, we obtain:

$$\pm \Gamma_{IR} = \mp \frac{4\pi^2}{\hbar} \sqrt{\frac{1}{3}} \left[\frac{(\alpha_1^B)^{3/2}}{V_0^B} \left(1 - \frac{V_0^B}{\alpha_1^B} a^2\right)^{3/2} \bigg|_{a_0}^{\bar{a}-\delta a} - \frac{(\alpha_1^A)^{3/2}}{V_0^A} \left(1 - \frac{V_0^A}{\alpha_1^A} a^2\right)^{3/2} \bigg|_{a_0}^{\bar{a}-\delta a} \right] + \frac{2\pi^2}{\hbar} \left[\bar{a}^3 T_0 + c_1 \bar{a} T_1 \right]. \qquad (56)$$

Therefore, we find in the infrared an expression quite similar to the GR result plus one degree of freedom extra coming from the c_1 term in the action for the scalar field (26) as expected. In order to directly compare our result to the result obtain for GR, we can, for the moment, set $c_1 = 0$, then the last result simplifies to:

$$\pm \Gamma_{IR} = \mp \frac{12\pi^2}{\hbar} \left[\frac{1}{V_0^B} \left(1 - \frac{V_0^B}{3} a^2\right)^{3/2} \bigg|_{a_0}^{\bar{a}-\delta a} - \frac{1}{V_0^A} \left(1 - \frac{V_0^A}{3} a^2\right)^{3/2} \bigg|_{a_0}^{\bar{a}-\delta a} \right] + \frac{2\pi^2}{\hbar} \bar{a}^3 T_0, \qquad (57)$$

that is the same result obtained for GR in [10,11]. The only difference comes in the choice of a_0, since for the consistency of the integral approximation we have here that $a_0 \gg 1$, therefore, it cannot be chosen to be zero. Thus, the difference between this result and the GR one are only constants. We also note that the potential appearing in this expression is not the potential found originally in the scalar field action, rather it is an effective potential appearing after integration of the harmonic functions. Considering the thin wall limit $\delta a \to 0$ and extremizing the above result with respect to $\bar{a}$ we obtain:

$$T_0 = \pm 2 \left(\sqrt{\frac{1}{\bar{a}^2} - \frac{V_0^A}{3}} - \sqrt{\frac{1}{\bar{a}^2} - \frac{V_0^B}{3}} \right). \qquad (58)$$

Then substituting it back into Equation (57) and choosing the plus sign in the right-hand side we obtain:

$$\pm\Gamma_{IR} = \frac{12\pi^2}{\hbar}\left\{\frac{1}{V_0^B}\left[\left(1-\frac{V_0^B}{3}\bar{a}^2\right)^{3/2}-\left(1-\frac{V_0^B}{3}a_0^2\right)^{3/2}\right]\right.$$
$$\left.-\frac{1}{V_0^A}\left[\left(1-\frac{V_0^A}{3}\bar{a}^2\right)^{3/2}-\left(1-\frac{V_0^A}{3}a_0^2\right)^{3/2}\right]-\frac{\bar{a}^2}{3}\left(\sqrt{1-\frac{V_0^A}{3}\bar{a}^2}-\sqrt{1-\frac{V_0^B}{3}\bar{a}^2}\right)\right\}. \tag{59}$$

Therefore, in this limit, the transition probability is finally described in terms of just one parameter (considering a_0 as a constant).

If we take $a_0 = 0$ in (57), consider the thin wall limit and rename $V_0 \to V$ and $T_0 \to T$, we obtain the result found in [10,11] for GR:

$$\pm\Gamma_{GR} = \mp\frac{12\pi^2}{\hbar}\left[\frac{1}{V_B}\left(1-\frac{V_B}{3}a^2\right)^{3/2}\bigg|_{a_0}^{\bar{a}} - \frac{1}{V_A}\left(1-\frac{V_A}{3}a^2\right)^{3/2}\bigg|_{a_0}^{\bar{a}}\right] + \frac{2\pi^2}{\hbar}\bar{a}^3 T, \tag{60}$$

then, extremizing we obtain:

$$T = \pm 2\left(\sqrt{\frac{1}{\bar{a}^2}-\frac{V_A}{3}} - \sqrt{\frac{1}{\bar{a}^2}-\frac{V_B}{3}}\right). \tag{61}$$

Thus, finally the logarithm of the transition probability in GR is written in terms of just one parameter as:[1]

$$\pm\Gamma_{GR} = \frac{12\pi^2}{\hbar}\left\{\frac{1}{V_B}\left[\left(1-\frac{V_B}{3}\bar{a}^2\right)^{3/2}-1\right] - \frac{1}{V_A}\left[\left(1-\frac{V_A}{3}\bar{a}^2\right)^{3/2}-1\right] - \frac{\bar{a}^2}{3}\left(\sqrt{1-\frac{V_A}{3}\bar{a}^2}-\sqrt{1-\frac{V_B}{3}\bar{a}^2}\right)\right\}. \tag{62}$$

We can see from (61) that since $V_A > V_B$ choosing the plus sign on the right-hand side of (60) implies that $T > 0$ always. The same is true regarding T_0. As we know from [10,11], this choice of sign allows us to obtain the results found using the euclidean approach in [32]. It can be proven that the right-hand side of (62) is always positive and, therefore, in order to have a well-defined probability defined by (12) we choose the plus sign in the left-hand side as well. We note, however, from (58) and (61) that in order to have a well-defined tension we need the terms inside the square roots to be positive. If both potential minima are negative, we see that this is indeed satisfied for all values of $\bar{a}$. However, if at least one of the potential minima is positive, we see that the tension will only be well defined until $\bar{a}$ is big enough, that is, in this case, the tension term is well defined and consequently the expressions (59) and (62) are valid only in an interval from 0 until an upper bound for $\bar{a}$.

Let us study now the ultraviolet limit, this is found when $a \ll 1$. In this limit Equation (54) simplifies to:

$$\pm\Gamma_{UV} = \pm\frac{2\pi^2\sqrt{6(3\lambda-1)}}{\hbar}\left[\left(\sqrt{-\alpha_3^B}-\sqrt{-\alpha_3^A}\right)\ln a\bigg|_{a_0}^{\bar{a}-\delta a}\right] + \frac{2\pi^2}{\hbar}\left[\frac{c_2}{\bar{a}}T_2 + \frac{c_3}{\bar{a}^3}T_3\right]. \tag{63}$$

We note, however, that since all g_n are positive, and the V_3 function defined in (45) is also positive for both minima, the first term is purely imaginary. Therefore, it does not contribute to the transition probability at all and it can be ignored. Thus, we finally obtain in this limit:

$$\pm\Gamma_{UV} = \frac{2\pi^2}{\hbar}\left[\frac{c_2}{\bar{a}}T_2 + \frac{c_3}{\bar{a}^3}T_3\right]. \tag{64}$$

However, this expression does not have an extremal with respect to $\bar{a}$. Then, in this case, the probability is described by three independent parameters. For consistency with the GR result, we will choose the plus sign in the left-hand side of the latter equation, therefore, we note that in order to have a well-defined probability we need the overall sign of the right-hand side to be positive. Thus, we can choose both T_2 and T_3 to take always positive values.

We can see from the GR result (60), or after extremizing (62), that in any case $P(A \to B) \to 1$ when $\bar{a} \to 0$,. However, for the UV limit of HL presented in (64) we have in the contrary $P(A \to B) \to 0$ when $\bar{a} \to 0$ and then the probability increases as $\bar{a}$ increases. Therefore, in this case the UV behavior is completely different for both theories.

Now that we have studied the two limits of interest. We proceed to compare the full result for the transition probability valid for all $\bar{a}$ (54) to the GR result. As we have said, for consistency with GR we are going to choose the plus sign in the left-hand side and the minus sign in the right, and use the thin wall limit, therefore, we will consider:

$$\Gamma = -\frac{2\pi^2 \sqrt{6(3\lambda-1)}}{\hbar} \left[\int_{a_0}^{\bar{a}} F\left(\alpha_1^B, \alpha_2^B, \alpha_3^B, V_0^B, a\right) da - \int_{a_0}^{\bar{a}} F\left(\alpha_1^A, \alpha_2^A, \alpha_3^A, V_0^A, a\right) da \right] + \frac{2\pi^2}{\hbar} \left[\bar{a}^3 T_0 + c_1 \bar{a} T_1 + \frac{c_2}{\bar{a}} T_2 + \frac{c_3}{\bar{a}^3} T_3 \right]. \tag{65}$$

If we want to vary (65), we will obtain an expression involving the tension terms and the functions $F(\alpha_1^{A,B}, \alpha_2^{A,B}, \alpha_3^{A,B}, V^{A,B}, \bar{a})$. However, since those functions are defined in terms of square roots, we need the terms inside to be non-negative in order to have well defined tension terms, that is we need that:

$$\alpha_1^{A,B} \bar{a}^2 - \alpha_2^{A,B} - \frac{\alpha_3^{A,B}}{\bar{a}^2} - V_0^{A,B} \bar{a}^4 \geq 0. \tag{66}$$

We note that in the best scenario, the latter expression implies only a lower bound on $\bar{a}$ coming from the α_3 term. In the other cases, it could happen that we obtain a lower and an upper bound for $\bar{a}$, that is the tension terms would only be well defined over a specific interval, or it could even happen that the latter expression is not satisfied at all for any value of $\bar{a}$ and in that case (65) would not have an extremal. In any case, we see that the extremizing procedure is very dependent on the many parameters of the theory and does not allow us to obtain well-defined tension terms in general, in particular, we never have access to the UV region. Therefore, in order to avoid these difficulties, we are going to compare the results obtained before the extremizing procedure takes place, that is, we will use for the comparison the GR result (60) choosing the signs already mentioned and the HL result (65). Since the HL expression depends on many independent parameters and the integrals cannot be made for any values of the constants involved, we are going to evaluate numerically this expression. Since in the IR limit we saw that after extremizing T_0 and T are always positive, we are going to take positive values for these parameters. On the other hand, in the UV limit we saw that we can take T_2 and T_3 also to be positive. Finally, in order to obtain a well defined probability, we are also going to choose positive values for the remaining free parameter T_1. Thus, we will take positive values for all the tension terms.

In Figure 1, we show a plot of the transition probabilities coming from the two theories. We choose units such as $\frac{24\pi^2}{\hbar} = 1$. For the GR result (blue line), we choose $V_A = 1$, $V_B = 0.1$ and $T = 2$. We note that in this case, the first term in (60) is negative, therefore, we need to choose a value for T great enough to obtain a well-defined probability. We see the behavior outlined earlier, that is, the probability goes to 1 in the limit $\bar{a} \to 0$ and then it decreases as $\bar{a}$ increases going to zero. For the HL plots, we choose $V_0^A = 1$, $V_0^B = 0.1$, $\alpha_1^A = \alpha_2^A = \alpha_3^A = 5$, $\alpha_1^B = \alpha_2^B = \alpha_3^B = 4$, $T_0 = 2$ and $c_1 T_1 = c_2 T_2 = c_3 T_3 = 1$, we plot the probability for three different values of λ to see how this parameter affects the behavior. In this case, we choose $a_0 = 0.000001$ in order to compute the integral numerically, however, we know from the UV analysis, that $a_0 = 0$ can be chosen without any problem and the general form will be unaltered. In this case, we also have that the first term in (65) is negative and increases with λ, therefore, we also need to make sure that the tensions chosen are big enough to have a well-defined probability. This figure shows the behavior that we described earlier by studying the different limits of interest. That is, in the IR region, the probability falls in the same manner as the GR result. We note that $\bar{a}$ has to be big enough so the first term can be positive and then contribute to the probability, therefore, the different values of λ

only affect the curve in the IR region and as λ increases, the probability increases since this term has the opposite sign than the tension terms. In the UV region, the parameter λ has no impact at all, and then, everything is defined by the tension terms. As we have said earlier, the probability goes to 0 as $\bar{a} \to 0$ and then it increases with $\bar{a}$. We note that this behavior comes from the c_2 and c_3 terms, that is, it comes from the extra terms in the action for the scalar field (18) and it can be interpreted as the fact that HL avoids the singularity and predicts these types of transitions to occur in a UV regime (small $\bar{a}$) but not too close to the singularity. Then, the probability starts to decrease and then it goes into the IR behavior just described. We note that the general form of the plot will be maintained regardless of the values of the parameters, we only have to make sure that they are chosen in a way that the tension terms dominate so we can have a well-defined probability. However, specific things such as the maximum height or the point in which both plots match is completely determined by the parameters and, therefore, we cannot say something about them in general.

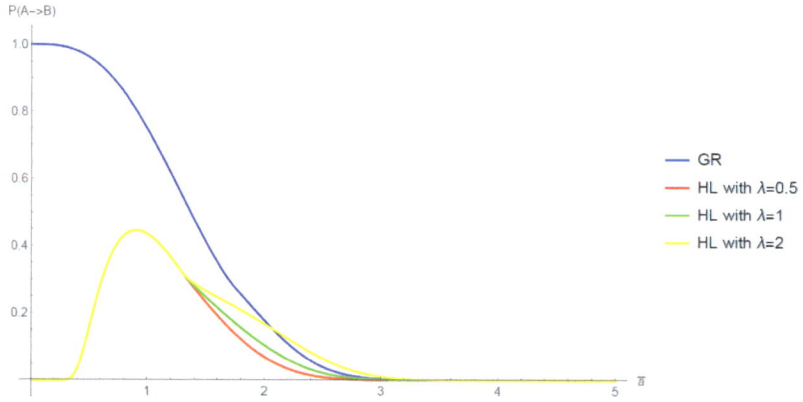

Figure 1. Transition probability in units such that $\frac{24\pi^2}{\hbar} = 1$, with $V_A = V_0^A = 1$, $V_B = V_0^B = 0.1$, $\alpha_1^A = \alpha_2^A = \alpha_3^A = 5$, $\alpha_1^B = \alpha_2^B = \alpha_3^B = 4$, $T = T_0 = 2$, $c_1 T_1 = c_2 T_2 = c_3 T_3 = 1$, for GR (blue line) and HL with a scalar field depending on all spacetime variables with $\lambda = 0.5$ (red line), $\lambda = 1$ (green line) and $\lambda = 2$ (yellow line). For HL we choose $a_0 = 0.000001$ but the same form is expected for $a_0 = 0$.

In the present section, we have obtained a general formula (65) for the transition probability when the scalar field depends on all spacetime variables. The integrals involved cannot be performed in an analytic form, however, a numerical computation was performed and we showed a plot comparing the results with the one coming from GR. As we pointed out earlier, the UV behavior is found to be completely opposite. This result depends on the extra terms in the action for the scalar fields, however, these terms are only present if we take a field that depends on the spatial variables. Therefore, in the next section we are going to study a scalar field depending only on the time variable.

Phenomenological Remarks

Before we end this section, let us discuss some phenomenological aspects regarding our results. The theory of Hořava–Lifshitz has received a lot of attention and much work has been performed since its first proposal from the theoretical, as well as the phenomenological, point of view. For example, in [33–35] the viability of the different versions of the theory have been tested against various experimental (or observational) data sets coming from different sources such as CMB and BAO collaborations. It is found that the theory is in good agreement with such data, and therefore, it supports the importance of considering it as a viable theory. It is also interesting to point out that in these works they always work with an FLRW metric with a non-zero curvature since the flat metric gives the same predictions as in General Relativity. Thus, the importance of studying such metrics as the one we studied in the present article is also supported by these works.

On the other hand, in the vacuum decay process studied by using the euclidean method, discussed in [3], the process is described by the nucleation of true vacuum bubbles and its corresponding expansion. This could lead to phenomenological predictions regarding this kind of phase transitions occurring at some point in the evolution of the universe. However, as it was pointed out in [10] the process studied by using a Hamiltonian approach in the minisuperspace is limited. In fact, in the transition studied, there is no notion of bubble nucleation, we can only compare two configurations of three-metrics and then interpret its ratio as a transition probability. Therefore, it is speculated that this formalism is not describing the same process as the euclidean method. It is believed that it may describe a generalization of the tunneling from nothing scenario, that is, we are obtaining probability distributions of creating universes from a tunneling event between two minima of the scalar potential. If we take seriously this interpretation, then the scale factor $\bar{a}$ appearing in the expression found for the transition probability would correspond to the value that the scale factor of the created universe would have at the time of creation (its corresponding 'size'). Then, the plot in Figure 1 would tell us that in Hořava–Lifshitz gravity, in the case in which the scalar field depends on the spatial variables, the universe would be created with a scalar factor different from zero, and therefore, we would avoid the singularity contrary to GR which predicts a singularity at the beginning of the universe. This, of course, would have potential phenomenological consequences in the physics of the early universe and its corresponding evolution. Therefore, although we are in a speculating phase, these kinds of transitions are worth studying in more detail.

5. Transitions for a Time-Dependent Scalar Field

In the previous sections, we studied a scalar field depending on all coordinates of spacetime and found a transition probability whose behavior differs completely in the UV regime comparing to the GR result. However, in cosmology it is more common to study a scalar field depending only on the time variable as is the case in [29,36,37]. Therefore, in this section we will consider such a dependence for the scalar field and study the vacuum transition probability between two minima of the potential.

In this case, the scalar field action (18) reduces to:

$$S_m = 2\pi^2 \int dt a^3(t) \left[\frac{3\lambda - 1}{4N^2} \dot{\phi}^2 - NV(\phi) \right], \quad (67)$$

where we have redefined the scalar field potential appearing in (19) as $\frac{V}{2} \to V$ so it coincides with the usual scalar potential in the action. Since now we have a global factor of $2\pi^2$ as in the action of the gravitational part (16), we can omit this factor. Then, the lagrangian this time is given by:

$$\mathcal{L} = N \left[-\frac{3M_p^2 \dot{a}^2 a}{2N^2}(3\lambda - 1) + 3M_p^2 a - \frac{6}{a}(3g_2 + g_3) - \frac{12}{a^3 M_p^2}(9g_4 + 3g_5 + g_6) \right] + a^3 \left[\frac{3\lambda - 1}{4N} \dot{\phi}^2 - NV \right]. \quad (68)$$

Therefore, we have only two degrees of freedom a and ϕ, their canonical momenta are:

$$\pi_a = -\frac{3(3\lambda - 1)M_p^2}{N} a\dot{a}, \qquad \pi_\phi = \frac{(3\lambda - 1)a^3}{2N} \dot{\phi}, \quad (69)$$

and the Hamiltonian constraint takes the form:

$$H = N \left[\frac{\pi_\phi^2}{a^3(3\lambda - 1)} - \frac{\pi_a^2}{6(3\lambda - 1)M_p^2 a} - 3M_p^2 a + \frac{6}{a}(3g_2 + g_3) + \frac{12}{M_p^2 a^3}(9g_4 + 3g_5 + g_6) + a^3 V(\phi) \right] \simeq 0. \quad (70)$$

Comparing this last expression to the general form considered in Equation (1), we note that in this case the coordinates in superspace are $\{a, \phi\}$ with inverse metric:

$$G^{\phi\phi} = \frac{2}{(3\lambda - 1)a^3}, \quad G^{aa} = -\frac{1}{3(3\lambda - 1)M_p^2 a}, \tag{71}$$

and we also have

$$f(a, \phi) = -3M_p^2 a + \frac{6}{a}(3g_2 + g_3) + \frac{12}{M_p^2 a^3}(9g_4 + 3g_5 + g_6) + a^3 V(\phi). \tag{72}$$

In order to study transitions between two minima of the potential, we choose the parameter s as in expression (33), then, following a similar procedure as in the previous section, we obtain in this case that choosing units such as $M_p = 1$ (as in the GR case) and in the thin wall limit, the logarithm of the transition probability is written as:

$$\pm \Gamma = \pm \frac{2\pi^2 \sqrt{6(3\lambda - 1)}}{\hbar} \left[\int_{a_0}^{\bar{a}} F(3, \bar{\alpha}_2, \bar{\alpha}_3, V_B, a) da - \int_{a_0}^{\bar{a}} F(3, \bar{\alpha}_2, \bar{\alpha}_3, V_A, a) da \right] + \frac{2\pi^2}{\hbar} \bar{a}^3 T, \tag{73}$$

where

$$\bar{\alpha}_2 = 6(3g_2 + g_3), \quad \bar{\alpha}_3 = 12(9g_4 + 3g_5 + g_6), \tag{74}$$

the function F is defined in (51) and as we have mentioned in Section 4, the sign ambiguities in the last expression are independent.

Now that we have computed the transition probability in general, we move on to study its behavior in the limiting cases considered before. For the IR behavior, we consider $\lambda \to 1$ and $a >> 1$ in the above expression, the result is the same as in (57) with the same subtlety about a_0 as discussed in the previous section. On the other hand, in the UV limit we have $a << 1$. However, in this limit we obtain $\Gamma \to 0$ as $\bar{a} \to 0$. Therefore, we note that the general behavior of these results is the same as the GR result in both extreme cases. In fact, we can variate (73) with respect to $\bar{a}$ to obtain:

$$T = \pm \frac{\sqrt{6(3\lambda - 1)}}{3\bar{a}^2} [F(3, \bar{\alpha}_2, \bar{\alpha}_3, V_A, \bar{a}) - F(3, \bar{\alpha}_2, \bar{\alpha}_3, V_B, \bar{a})]. \tag{75}$$

Substituting it back in (73) we obtain finally:

$$\pm 2 Re[\Gamma] = \pm \frac{4\pi^2 \sqrt{6(3\lambda - 1)}}{\hbar} Re \left[\int_{a_0}^{\bar{a}} F(3, \bar{\alpha}_2, \bar{\alpha}_3, V_B, a) da - \int_{a_0}^{\bar{a}} F(3, \bar{\alpha}_2, \bar{\alpha}_3, V_A, a) da \right. \\ \left. + \frac{\bar{a}}{3} \{ F(3, \bar{\alpha}_2, \bar{\alpha}_3, V_A, \bar{a}) - F(3, \bar{\alpha}_2, \bar{\alpha}_3, V_B, \bar{a}) \} \right]. \tag{76}$$

Thus, the transition probability is also written in terms of just one parameter as in the GR result. Therefore, the only difference between GR and HL in this case is that the transition probability changes by acquiring two more terms in the square root before integration, making the integral not possible to be performed in general and a global factor depending on λ in (76). The qualitative behavior in both the IR and UV limit is unaltered.

In order to compare the result of this section with that of GR in general, not only on the limiting cases, we note that as in the last section, the extremizing procedure leading to Equation (75) gives rise to some restrictions for the validity of (76). In particular, it is never well defined when $\bar{a}$ is small. Therefore, we are going to use the result (73) and choose the minus sign in the right-hand side so in the IR limit it coincides with the GR result and on the left-hand side we will choose the plus sign in accordance with the GR result as well. We will compare it with the GR result (60) with the sign choices made in the above section. In both cases, we will take the tension T as an independent positive parameter and take values big enough so we can have a well-defined probability and compute the integrals numerically. In Figure 2, we show such a comparison. We choose units such as $\frac{24\pi^2}{\hbar} = 1$, with $V_A = -1$, $V_B = -10$, $\bar{\alpha}_2 = \bar{\alpha}_3 = 5$ and $T = 5$. For the HL result we show plots for three values of λ and choose $a_0 = 0.000001$ in order to perform a numerical computation of the integrals, however, doing the UV limit we note that $a_0 = 0$ is possible and it has the same behavior. This figure shows the limiting behavior that we described earlier, that is, in the IR and in the UV limits all curves behave in the same way, it is in the middle region where their behavior is modified. In particular, we note that in the beginning the HL probability is smaller than the one of GR, and the contribution of the λ parameter is not noticeable; however, when the first term in (73) is big enough, the contribution of the first term is big enough to separate the curves, and as in the case considered in the previous section, as λ increases the probability increases. Finally, the three probabilities fall as in the GR case.

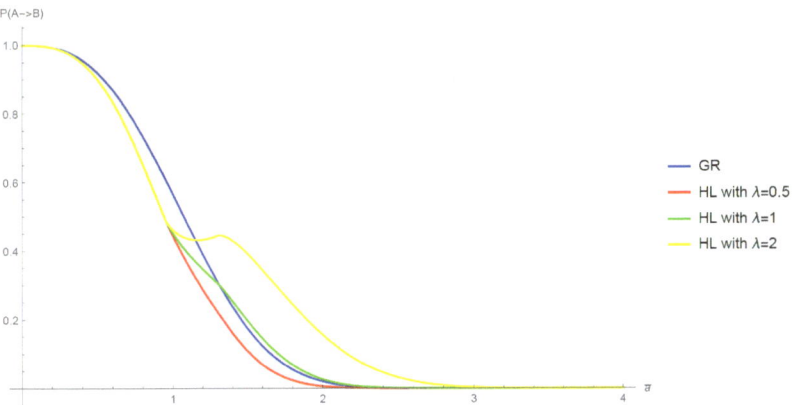

Figure 2. Transition probability in units such that $\frac{24\pi^2}{\hbar} = 1$, with $V_A = -1$, $V_B = -10$, $\bar{\alpha}_2 = \bar{\alpha}_3 = 5$ and $T = 5$, for GR (blue line) and HL with a scalar field depending only on the time variable with $\lambda = 0.5$ (red line), $\lambda = 1$ (green line) and $\lambda = 2$ (yellow line). For HL we choose $a_0 = 0.000001$ but the same form is expected for $a_0 = 0$.

6. Final Remarks

In the present article, we have studied the transition probabilities for an FLRW metric in Hořava–Lifhitz gravity using a WKB approximation to the WDW equation. The general procedure proposed in [10,11] was found to be applicable to this case. We used HL theory without detailed balance and consider an FLRW metric with positive spatial curvature.

We considered two types of scalar fields. First, since the anisotropic scaling between space and time variables is a key ingredient of HL theory, we considered a scalar field which depends on all spacetime variables. This type of dependence is useful to study cosmological perturbations coming from scalar fields [26]. On the other hand, since in cosmology it is customary to propose an ansatz in which the scalar field depends only on the time variable, we studied this kind of dependence as well. For both cases, we found analytic expressions for the logarithm of the transition probabilities in the thin wall limit.

For the scalar field, depending on all spacetime variables, the transition probability (65) was found to depend on five different parameters coming from the new terms present in the action for gravity, as well as the action from the scalar field in HL theory. There is only the possibility to reduce just one of these parameters after an extremizing procedure but such a procedure is not well defined for all values of the scale factor. Taking the IR limit we found that one degree of freedom extra coming from the scalar field action survives, which is a common issue regarding the IR limit of HL theory. However, if we ignore this contribution, we can obtain an expression that differs from the GR result just by constants. In the opposite limit, that is, in the UV limit, we found that the probability is described in terms of three independent parameters and it vanishes in the limit $\bar{a} \to 0$. This is opposite to the GR result in which the probability goes to 1 in that limit. We interpret this result as a way in which HL theory avoids the spatial singularity at $\bar{a} = 0$ and predicts these transitions to occur on the UV regime but away from the singularity. We note that this behavior comes from the terms in the scalar field action with spatial derivatives, and therefore, it is only possible in the case in which the scalar field depends on the spatial variables. In order to visualize these behaviors, we plotted the transition probabilities coming from GR and HL theory. Such plots were presented in Figure 1. For the HL results, the integrals involved were performed numerically and we saw that in all cases the probability begins at zero with $\bar{a} = 0$, then it increases with $\bar{a}$ until at some point it starts to decrease and then it behaves as in the GR case. We noted that the first behavior in the UV region is independent of the λ parameter and it is only on the IR where the dependence on this parameter is noticeable making the probability increase as λ increases.

For the scalar field depending only on the time variable, the logarithm of the transition probability found have the same number of independent parameters as the GR result, that is, after extremizing we only have one parameter left. However, it also has dependence on the many constants g_n appearing in the extra terms in the gravity action for HL theory, as well as in the parameter λ. The behavior of the probability in this case is found to be the same as the GR result in both the IR, as well as the UV limits. In fact, in the IR limit, we obtain the same expression as the one coming from the scalar field with dependence in all spacetime variables when we ignore the degree of freedom that survives this limit and in the UV regime we also found that the probability goes to 1 in the limit $\bar{a} \to 0$. Therefore, with a cosmological ansatz, the behavior in the UV regime found by using GR is unaltered. However, in the intermediate region, the probability is of course modified. In order to visualize the difference in this region, we plotted the transition probabilities coming from GR and HL theory and showed them in Figure 2. In this case, we also carried out a numerical computation of the integrals involved in the HL result. We noted that, at first, the probability of HL is smaller than GR and the contribution from the λ parameter is not noticeable. However, when the scale factor is big enough, this contribution is important and, as in the latter case, the probability increases with λ. It is interesting to note that using HL theory instead of GR for a cosmological ansatz of the scalar field does not have a dramatic change on the transition probability at least at the semi-classical level we used in this article through the WKB approximation.

It is worth pointing out that we have used a WKB approximation and kept only up to first order in the expansion. However, this level of semi-classical approximation is sufficient to obtain the transition probabilities and we can safely explore the UV regime of both GR, as well as HL theory, since the transition probabilities are well-behaved functions in the UV. It was shown that in the case when the scalar field is only dependent on the cosmological time, GR and HL theories give very similar predictions in the WKB approximation. However, the case with a dependence on time and position coordinates for the scalar field, yields very different behavior from the GR case even in the WKB approximation. It would be interesting to work out higher order contributions from the WKB approximation, which presumably will have the contribution of quantum fluctuations.

It is important to remark as well, that we considered closed universes in the HL theory and obtained well-defined transition probabilities. Therefore, one of the important results obtained in Ref. [10] that asserts that these types of transitions can be carried out keeping the closeness of the spatial universe can certainly be extended to include the Hořava–Lifshitz theory of gravity as well.

Author Contributions: All authors contributed equally to this paper. All authors have read and agreed to the published version of the manuscript.

Funding: This research was founded by a Conacyt grant.

Data Availability Statement: Not applicable.

Acknowledgments: D. Mata-Pacheco would like to thank CONACyT for a grant.

Conflicts of Interest: The authors declare no conflict of interest.

Note

1. We note that in [11] we miswrote the sign of the last term in Equation (62).

References

1. Coleman, S.R. The Fate of the False Vacuum. 1. Semiclassical Theory. *Phys. Rev. D* **1977**, *15*, 2929–2936; Erratum in *Phys. Rev. D* **1977**, *16*, 1248. [CrossRef]
2. Callan, C.G., Jr.; Coleman, S.R. The Fate of the False Vacuum. 2. First Quantum Corrections. *Phys. Rev. D* **1977**, *16*, 1762–1768. [CrossRef]
3. Coleman, S.R.; de Luccia, F. Gravitational Effects on and of Vacuum Decay. *Phys. Rev. D* **1980**, *21*, 3305. [CrossRef]
4. Fischler, W.; Morgan, D.; Polchinski, J. Quantum Nucleation of False Vacuum Bubbles. *Phys. Rev. D* **1990**, *41*, 2638. [CrossRef] [PubMed]
5. Fischler, W.; Morgan, D.; Polchinski, J. Quantization of False Vacuum Bubbles: A Hamiltonian Treatment of Gravitational Tunneling. *Phys. Rev. D* **1990**, *42*, 4042–4055. [CrossRef] [PubMed]
6. Arnowitt, R.L.; Deser, S.; Misner, C.W. The Dynamics of general relativity. *Gen. Rel. Grav.* **2008**, *40*, 1997–2027. [CrossRef]
7. Wheeler, J.A. Superspace and the nature of quantum geometrodynamics. In *Topics in Nonlinear Physics*; Zabusky, N.J., Ed.; Springer: New York, NY, USA, 1969; pp. 615–724.
8. DeWitt, B.S. Quantum theory of gravity I, The canonical theory. *Phys. Rev.* **1967**, *160*, 1113. [CrossRef]
9. de Alwis, S.P.; Muia, F.; Pasquarella, V.; Quevedo, F. Quantum Transitions between Minkowski and de Sitter Spacetimes. *Fortsch. Phys.* **2020**, *68*, 2000069. [CrossRef]
10. Cespedes, S.; de Alwis, S.P.; Muia, F.; Quevedo, F. Lorentzian vacuum transitions: Open or closed universes? *Phys. Rev. D* **2021**, *104*, 026013. [CrossRef]
11. García-Compeán, H.; Mata-Pacheco, D. Lorentzian Vacuum Transitions for Anisotropic Universes. *Phys. Rev. D* **2021**, *104*, 106014. [CrossRef]
12. Hořava, P. Quantum Gravity at a Lifshitz Point. *Phys. Rev. D* **2009**, *79*, 084008. [CrossRef]
13. Weinfurtner, S.; Sotiriou, T.P.; Visser, M. Projectable Hořava-Lifshitz gravity in a nutshell. *J. Phys. Conf. Ser.* **2010**, *222*, 012054. [CrossRef]
14. Sotiriou, T.P. Hořava-Lifshitz gravity: A status report. *J. Phys. Conf. Ser.* **2011**, *283*, 012034. [CrossRef]
15. Wang, A. Hořava gravity at a Lifshitz point: A progress report. *Int. J. Mod. Phys. D* **2017**, *26*, 1730014. [CrossRef]
16. Mukohyama, S. Hořava-Lifshitz Cosmology: A Review. *Class. Quant. Grav.* **2010**, *27*, 223101. [CrossRef]
17. Izumi, K.; Mukohyama, S. Nonlinear superhorizon perturbations in Hořava-Lifshitz gravity. *Phys. Rev. D* **2011**, *84*, 064025. [CrossRef]
18. Gumrukcuoglu, A.E.; Mukohyama, S.; Wang, A. General relativity limit of Hořava-Lifshitz gravity with a scalar field in gradient expansion. *Phys. Rev. D* **2012**, *85*, 064042. [CrossRef]
19. Bertolami, O.; Zarro, C.A.D. Hořava-Lifshitz Quantum Cosmology. *Phys. Rev. D* **2011**, *84*, 044042. [CrossRef]
20. Christodoulakis, T.; Dimakis, N. Classical and Quantum Bianchi Type III vacuum Hořava-Lifshitz Cosmology. *J. Geom. Phys.* **2012**, *62*, 2401–2413. [CrossRef]
21. Pitelli, J.P.M.; Saa, A. Quantum Singularities in Hořava-Lifshitz Cosmology. *Phys. Rev. D* **2012**, *86*, 063506. [CrossRef]
22. Vakili, B.; Kord, V. Classical and quantum Hořava-Lifshitz cosmology in a minisuperspace perspective. *Gen. Rel. Grav.* **2013**, *45*, 1313–1331. [CrossRef]
23. Obregon, O.; Preciado, J.A. Quantum cosmology in Hořava-Lifshitz gravity. *Phys. Rev. D* **2012**, *86*, 063502. [CrossRef]
24. Benedetti, D.; Henson, J. Spacetime condensation in (2+1)-dimensional CDT from a Hořava–Lifshitz minisuperspace model. *Class. Quant. Grav.* **2015**, *32*, 215007. [CrossRef]
25. Cordero, R.; García-Compeán, H.; Turrubiates, F.J. A phase space description of the FLRW quantum cosmology in Hořava–Lifshitz type gravity. *Gen. Rel. Grav.* **2019**, *51*, 138. [CrossRef]
26. Mukohyama, S. Scale-invariant cosmological perturbations from Hořava-Lifshitz gravity without inflation. *JCAP* **2009**, *6*, 001. [CrossRef]
27. Sotiriou, T.P.; Visser, M.; Weinfurtner, S. Phenomenologically viable Lorentz-violating quantum gravity. *Phys. Rev. Lett.* **2009**, *102*, 251601. [CrossRef]
28. Sotiriou, T.P.; Visser, M.; Weinfurtner, S. Quantum gravity without Lorentz invariance. *JHEP* **2009**, *10*, 033. [CrossRef]
29. Kiritsis, E.; Kofinas, G. Hořava-Lifshitz Cosmology. *Nucl. Phys. B* **2009**, *821*, 467–480. [CrossRef]
30. Lindblom, L.; Taylor, N.W.; Zhang, F. Scalar, Vector and Tensor Harmonics on the Three-Sphere. *Gen. Rel. Grav.* **2017**, *49*, 139. [CrossRef]
31. Sandberg, V.D. Tensor spherical harmonics on S^2 and S^3 as eigenvalue problems. *J. Math. Phys.* **1978**, *19*, 2441–2446. [CrossRef]
32. Parke, S.J. Gravity, the Decay of the False Vacuum and the New Inflationary Universe Scenario. *Phys. Lett. B* **1983**, *121*, 313–315. [CrossRef]
33. Dutta, S.; Saridakis, E.N. Observational constraints on Hořava-Lifshitz cosmology. *JCAP* **2010**, *1*, 013. [CrossRef]
34. Nilsson, N.A.; Czuchry, E. Hořava–Lifshitz cosmology in light of new data. *Phys. Dark Univ.* **2019**, *23*, 100253. [CrossRef]
35. Nilsson, N.A.; Park, M.I. Tests of Standard Cosmology in Hořava Gravity. *arXiv* **2021**, arXiv:2108.07986.

36. Tavakoli, F.; Vakili, B.; Ardehali, H. Hořava-Lifshitz Scalar Field Cosmology: Classical and Quantum Viewpoints. *Adv. High Energy Phys.* **2021**, *2021*, 6617910. [CrossRef]
37. Tawfik, A.N.; Diab, A.M.; el Dahab, E.A. Friedmann inflation in Hořava-Lifshitz gravity with a scalar field. *Int. J. Mod. Phys. A* **2016**, *31*, 1650042. [CrossRef]

Article

A Unified Quantization of Gravity and Other Fundamental Forces of Nature †

Claus Gerhardt

Institut für Angewandte Mathematik, Ruprecht-Karls-Universität, Im Neuenheimer Feld 205, 69120 Heidelberg, Germany; gerhardt@math.uni-heidelberg.de
† Dedicated to Robert Finn on the occasion of his 100th birthday.

Abstract: We quantized the interaction of gravity with Yang–Mills and spinor fields; hence, offering a quantum theory incorporating all four fundamental forces of nature. Let us abbreviate the spatial Hamilton functions of the standard model by H_{SM} and the Hamilton function of gravity by H_G. Working in a fiber bundle E with base space $S_0 = \mathbb{R}^n$, where the fiber elements are Riemannian metrics, we can express the Hamilton functions in the form $H_G + H_{SM} = H_G + t^{-\frac{2}{3}}\tilde H_{SM}$, if $n = 3$, where $\tilde H_{SM}$ depends on metrics σ_{ij} satisfying $\det \sigma_{ij} = 1$. In the quantization process, we quantize H_G for general σ_{ij} but $\tilde H_{SM}$ only for $\sigma_{ij} = \delta_{ij}$ by the usual methods of QFT. Let v resp. ψ be the spatial eigendistributions of the respective Hamilton operators, then the solutions u of the Wheeler–DeWitt equation are given by $u = wv\psi$, where w satisfies an ODE and u is evaluated at (t, δ_{ij}) in the fibers.

Keywords: quantization of gravity; quantum gravity; standard model; temporal and spatial eigenfunctions; Fourier quantization; symmetric spaces

1. Introduction

General relativity is a Lagrangian theory, i.e., the Einstein equations are derived as the Euler–Lagrange equation of the Einstein–Hilbert functional

$$\int_N (\bar R - 2\Lambda), \tag{1}$$

where $N = N^{n+1}$, $n \geq 3$, is a globally hyperbolic Lorentzian manifold, $\bar R$ is the scalar curvature, and Λ is a cosmological constant. We also omitted the integration density in the integral. In order to apply a Hamiltonian description of general relativity, one usually defines a time function x^0 and considers the foliation of N given by the slices

$$M(t) = \{x^0 = t\}. \tag{2}$$

We may, without loss of generality, assume that the spacetime metric splits

$$d\bar s^2 = -w^2(dx^0)^2 + g_{ij}(x^0, x)dx^i dx^j, \tag{3}$$

cf. [1] (Theorem 3.2). Then, the Einstein equations also split into a tangential part

$$G_{ij} + \Lambda g_{ij} = 0 \tag{4}$$

and a normal part

$$G_{\alpha\beta}\nu^\alpha\nu^\beta - \Lambda = 0, \tag{5}$$

where the naming refers to the given foliation. For the tangential Einstein equations, one can define equivalent Hamilton equations due to the groundbreaking paper by Arnowitt,

Deser, and Misner [2]. The normal Einstein equations can be expressed by the so-called Hamilton condition

$$\mathcal{H} = 0, \tag{6}$$

where $\mathcal{H}$ is the Hamiltonian used in defining the Hamilton equations. In the canonical quantization of gravity, the Hamiltonian is transformed to a partial differential operator of a hyperbolic type $\hat{\mathcal{H}}$ and the possible quantum solutions of gravity are supposed to satisfy the so-called Wheeler–DeWitt equation

$$\hat{\mathcal{H}} u = 0 \tag{7}$$

in an appropriate setting, i.e., only the Hamilton condition, (6) was quantized, or equivalently, the normal Einstein equation, while the tangential Einstein equations were ignored.

In [1], we solved the Equation (7) in a fiber bundle E with the base space $\mathcal{S}_0$,

$$\mathcal{S}_0 = \{x^0 = 0\} \equiv M(0), \tag{8}$$

and fibers $F(x)$, $x \in \mathcal{S}_0$,

$$F(x) \subset T_x^{0,2}(\mathcal{S}_0), \tag{9}$$

the elements of which are the positive definite symmetric tensors of order two, the Riemannian metrics in $\mathcal{S}_0$. The hyperbolic operator $\hat{\mathcal{H}}$ is then expressed in the form

$$\hat{\mathcal{H}} = -\Delta - (R - 2\Lambda)\varphi, \tag{10}$$

where Δ is the Laplacian of the DeWitt metric given in the fibers, R the scalar curvature of the metrics $g_{ij}(x) \in F(x)$, and φ is defined by

$$\varphi^2 = \frac{\det g_{ij}}{\det \rho_{ij}}, \tag{11}$$

where ρ_{ij} is a fixed metric in $\mathcal{S}_0$, such that, instead of densities, we consider functions. The Wheeler–DeWitt equation could be solved in E but only as an abstract hyperbolic equation. The solutions could not be split into corresponding spatial and temporal eigenfunctions.

In a recent paper [3], we overcame this difficulty by quantizing the Hamilton equations instead of the Hamilton condition.

As a result, we obtained the equation

$$-\Delta u = 0 \tag{12}$$

in E, where the Laplacian is the Laplacian in (10). The lower order terms of $\hat{\mathcal{H}}$

$$(R - 2\Lambda)\varphi \tag{13}$$

were eliminated during the quantization process. This result was valid for all dimensions $3 \leq n$, provided $n \neq 4$.

The fibers add additional dimensions to the quantized problem, namely,

$$\dim F = \frac{n(n+1)}{2} \equiv m + 1. \tag{14}$$

The fiber metric, the DeWitt metric, which is responsible for the Laplacian in (12), can be expressed in the form

$$ds^2 = -\frac{16(n-1)}{n} dt^2 + \varphi G_{AB} d\xi^A d\xi^B, \tag{15}$$

where the coordinate system is

$$(\xi^a) = (\xi^0, \xi^A) \equiv (t, \xi^A). \tag{16}$$

The (ξ^A), $1 \leq A \leq m$, are coordinates for the hypersurface

$$M \equiv M(x) = \{(g_{ij}) : t^4 = \det g_{ij}(x) = 1, \forall\, x \in \mathcal{S}_0\}. \tag{17}$$

We also assumed that $\mathcal{S}_0 = \mathbb{R}^n$ and that the metric ρ_{ij} in (11) is the Euclidean metric δ_{ij}. It is well-known that M is a symmetric space

$$M = SL(n, \mathbb{R})/SO(n) \equiv G/K. \tag{18}$$

It is also easily verified that the induced metric of M in E is isometric to the Riemannian metric of the coset space G/K.

Now, we were in a position to use the separation of variables, namely, we wrote a solution of (12) in the form

$$u = w(t)v(\xi^A), \tag{19}$$

where v is a spatial eigenfunction of the induced Laplacian of M

$$-\Delta_M v \equiv -\Delta v = (|\lambda|^2 + |\rho|^2)v \tag{20}$$

and w is a temporal eigenfunction satisfying the ODE

$$\ddot{w} + mt^{-1}\dot{w} + \mu_0 t^{-2} w = 0 \tag{21}$$

with

$$\mu_0 = \frac{16(n-1)}{n}(|\lambda|^2 + |\rho|^2). \tag{22}$$

The eigenfunctions of the Laplacian in G/K are well-known and we chose the kernel of the Fourier transform in G/K in order to define the eigenfunctions. This choice also allowed us to use the Fourier quantization similar to the Euclidean case, such that the eigenfunctions were transformed to Dirac measures and the Laplacian to a multiplication operator in the Fourier space.

In the present paper, we quantize the Einstein–Hilbert functional combined with the functionals of the other fundamental forces of nature, i.e., we look at the Lagrangian functional

$$\begin{aligned} J = {} & \alpha_N^{-1} \int_{\tilde{\Omega}} (\bar{R} - 2\Lambda) - \int_{\tilde{\Omega}} \tfrac{1}{4} \gamma_{\bar{a}\bar{b}} \bar{g}^{\mu\rho_2} \bar{g}^{\lambda\rho_1} F^{\bar{a}}_{\mu\rho_1} F^{\bar{b}}_{\rho_2\lambda} \\ & - \int_{\tilde{\Omega}} \{\tfrac{1}{2} \bar{g}^{\mu\lambda} \gamma_{\bar{a}\bar{b}} \Phi^{\bar{a}}_\mu \bar{\Phi}^{\bar{b}}_\lambda + V(\Phi)\} \\ & + \int_{\tilde{\Omega}} \{\tfrac{1}{2} [\tilde{\psi}_I E^\mu_a \gamma^a (D_\mu \psi)^I + \overline{\tilde{\psi}_I E^\mu_a \gamma^a (D_\mu \psi)^I}] + m\tilde{\psi}_I \psi^I\}, \end{aligned} \tag{23}$$

where α_N is a positive coupling constant, $\tilde{\Omega} \Subset N = N^{n+1}$ and N is a globally hyperbolic spacetime with metric $\bar{g}_{\alpha\beta}$, $0 \leq \alpha, \beta \leq n$, where the metric splits as in (3).

The functional J consists of the Einstein–Hilbert functional, the Yang–Mills and Higgs functional, and a massive Dirac term.

The Yang–Mills field (A_μ)

$$A_\mu = f_{\bar{c}} A^{\bar{c}}_\mu \tag{24}$$

corresponds to the adjoint representation of a compact, semi-simple Lie group $\mathcal{G}$ with Lie algebra $\mathfrak{g}$. The $f_{\bar{c}}$,

$$f_{\bar{c}} = (f^{\bar{a}}_{\bar{c}\bar{b}}) \tag{25}$$

are the structure constants of $\mathfrak{g}$.

We assume the Higgs field $\Phi = (\Phi^{\bar{a}})$ to have complex valued components.

The spinor field $\psi = (\psi_A^I)$ has a spinor index A, $1 \le A \le n_1$, and a color index I, $1 \le I \le n_2$. Here, we suppose that the Lie group has a unitary representation R, such that

$$t_{\tilde{c}} = R(f_{\tilde{c}}) \tag{26}$$

are anti-Hermitian matrices acting on $\mathbb{C}^{n_2}$. The symbol $A_\mu \psi$ is now defined by

$$A_\mu \psi = t_{\tilde{c}} \psi A_\mu^{\tilde{c}}. \tag{27}$$

There are some major difficulties in achieving a quantization of the functional in (23). Quantizing the Hamilton equations, to avoid the problem with the scalar curvature term, runs into technical difficulties, even if the required quantization of the matter fields in the curved spacetimes could be achieved since the resulting operator would no longer be hyperbolic because the elliptic parts of the gravitational resp. matter Hamiltonians would have different signs in the case of $n = 3$. This particular problem would not occur when the Hamilton condition would be quantized. The Hamilton condition has the form

$$H_G + H_{YM} + H_D + H_H = 0, \tag{28}$$

where the subscripts refer to gravity, Yang–Mills, Dirac, and Higgs. On the left-hand side are the Hamilton functions of the respective fields. They depend on the Riemannian metrics g_{ij}, the Yang–Mills connections, and the spinor and Higgs fields. The main part of the quantized gravitational Hamiltonian is a second-order hyperbolic differential operator with respect to the variables g_{ij} while the scalar curvature term R is of zero-order. With this in mind, we also shall apply these categories to the gravitational Hamilton function where the main part, quadratic in the conjugate momenta, is said to be of the second-order and the zero-order terms consist of the scalar curvature and the cosmological constant Λ. Similarly, we consider the matter Hamilton functions to be zero-order terms with respect to the metric g_{ij}, i.e., there is no qualitative difference by assuming g_{ij} to be flat or non-flat, or more precisely, quantizing a matter Hamiltonian in a curved spacetime (when g_{ij} is a given, fixed metric and not a variable) is qualitatively the same as quantizing it for the Euclidean metric, though the task is certainly more difficult.

Thus, the difficulties arising from quantizing the Hamilton condition can best be explained by considering the Wheeler–DeWitt equation

$$\hat{H}_G u = 0 \quad \text{in } E, \tag{29}$$

cf. (7), where we wrote $\hat{\mathcal{H}}$ instead of $\hat{H}_G$. This is a hyperbolic differential equation, which can be expressed by

$$\hat{H}_G u = -\Delta u + \varphi(R - 2\Lambda)u = 0, \tag{30}$$

where the Laplacian is the Laplacian of the fiber metric (15). In the coordinate system (16), we have

$$\hat{H}_G u = t^{-m} \frac{\partial}{\partial t}(t^m \frac{\partial u}{\partial t}) - t^{-2} \Delta_M u + t^2 (R - 2\Lambda) u, \tag{31}$$

where M is the hypersurface (17). Since M is isometric to the symmetric space (18) it is mathematically irresistible to solve (31) by applying separation of variables and using the functions of the Fourier kernel of M as spatial eigenfunctions v, where $v = v(\sigma_{ij})$, σ_{ij} are the elements of M. Since

$$g_{ij}(x) = t^{\frac{4}{n}} \sigma_{ij}(x) \tag{32}$$

the critical term R can be expressed as

$$R(g_{ij}) = t^{-\frac{4}{n}} R(\sigma_{ij}) \tag{33}$$

due to the relation between the scalar curvatures of conformal metrics.

Thus, it is obvious that the ansatz

$$u = wv, \qquad (34)$$

where $w = w(t)$ solves an ODE is only possible if $R(\sigma_{ij})$ is constant

$$R(\sigma_{ij}) = \lambda_0. \qquad (35)$$

The constant is arbitrary but determined by the metrics we consider to be important, e.g., in the case of a black hole, we would choose σ_{ij} to be the limit metric of a converging sequence of Cauchy hypersurfaces of the interior region of the black hole, which converge to the event horizon topologically but the induced metrics of which converge to a Riemannian metric, cf. [4,5] or [6] (Chapters 4 and 5). In the present case, where we want to include the matter fields of the standard model, we could choose $\sigma_{ij} = \delta_{ij}$.

However, this ansatz implies that the Wheeler–DeWitt equation is not solved for all (t, σ_{ij}) but only for the σ_{ij} satisfying (35). Given the simplicity and mathematical beauty of the solution, we are inclined to accept this restriction.

Let us now consider the quantization of the Hamilton condition (28) taking all Hamilton functions into account. In view of the relation (32), let us propose the following model: If we were able to express the non-gravitational Hamiltonians as

$$H_{YM} = t^p \tilde{H}_{YM}, \quad H_D = t^p \tilde{H}_D, \quad H_H = t^p \tilde{H}_H, \qquad (36)$$

where the embellished Hamiltonians depend on σ_{ij}, then, by choosing in addition $n = 3$ and $\sigma_{ij} = \delta_{ij}$, these Hamiltonians could be quantized by the known methods of QFT, if the Lie groups would be chosen appropriately. The Wheeler–DeWitt equation would then not be solved for all (t, σ_{ij}) but only for (t, δ_{ij}). However, the spatial eigendistributions of the Hamilton operator $\hat{H}_G$, i.e., the eigendistributions of the Laplacian of M, cf. (20), would still be used but they would be evaluated at $\sigma_{ij} = \delta_{ij}$.

In Section 4, we prove that the expressions in (36) are indeed valid with $p = -\frac{2}{3}$ provided $n = 3$ and provided that the mass term in the Dirac Lagrangian and the Higgs Lagrangian is slightly modified. The embellished Hamiltonians are then standard Hamiltonians without any modifications, for details, we refer to Section 4. The Hamilton constraint then has the form

$$\begin{aligned} H &= H_G + H_{YM} + H_H + H_D \\ &= H_G + t^{-\frac{2}{3}}(\tilde{H}_{YM} + \tilde{H}_H + \tilde{H}_D) \\ &\equiv H_G + t^{-\frac{2}{3}}\tilde{H}_{SM} = 0, \end{aligned} \qquad (37)$$

where the subscript SM refers to the fields of the standard model or a corresponding subset of fields. The solutions of the Wheeler–DeWitt equation

$$\hat{H}u = 0 \qquad (38)$$

can then be achieved by using the separation of variables. We proved:

Theorem 1. *Let $n = 3$, $v = e_{\lambda, b_0}$ and let ψ be an eigendistribution of $\tilde{H}_{SM}$ when $\sigma_{ij} = \delta_{ij}$ such that*

$$-\Delta_M e_{\lambda, b_0} = (|\lambda|^2 + 1)e_{\lambda, b_0}, \qquad (39)$$

$$\tilde{H}_{SM}\psi = \lambda_1 \psi, \qquad \lambda_1 \geq 0, \qquad (40)$$

and let w be a solution of the ODE

$$t^{-m}\frac{\partial}{\partial t}(t^m \frac{\partial w}{\partial t}) + \frac{32}{3}(|\lambda|^2+1)t^{-2}w + \frac{32}{3}\alpha_N^{-1}\lambda_1 t^{-\frac{2}{3}}w + \frac{64}{3}\alpha_N^{-2}\Lambda t^2 w = 0 \quad (41)$$

then

$$u = w e_{\lambda,b_0} \psi \quad (42)$$

is a solution of the Wheeler–DeWitt equation

$$\hat{H}u = 0, \quad (43)$$

where e_{λ,b_0} is evaluated at $\sigma_{ij} = \delta_{ij}$ and where we note that $m = 5$.

We shall refer to e_{λ,b_0} and ψ as the spatial eigenfunctions and w as the temporal eigenfunction.

Remark 1. *We could also apply the respective Fourier transforms to* $-\tilde{\Delta} e_{\lambda,b_0}$ *resp.* $\tilde{H}_{SM}\psi$ *and consider*

$$w \hat{e}_{\lambda,b_0} \hat{\psi} \quad (44)$$

as the solution in Fourier space, where $\hat{\psi}$ would be expressed with the help of the ladder operators.

The temporal eigenfunctions are analyzed in Section 5. They must satisfy an ODE of the form

$$\ddot{w} + 5t^{-1}\dot{w} + m_1 t^{-2} w + m_2^2 t^{-\frac{2}{3}} w + m_3 t^2 w = 0, \quad (45)$$

where

$$m_1 \geq \frac{32}{3}, \quad m_2 \geq 0, \quad m_3 \in \mathbb{R}. \quad (46)$$

For simplicity, we shall only state the result when $m_3 = 0$, which is tantamount to setting $\Lambda = 0$.

Theorem 2. *Assume $m_3 = 0$ and $m_2 > 0$, then the solutions of the ODE (45) are generated by*

$$J(\tfrac{3}{2}\sqrt{m_1-4}\,i, \tfrac{3}{2}m_2 t^{\frac{2}{3}})t^{-2} \quad (47)$$

and

$$J(-\tfrac{3}{2}\sqrt{m_1-4}\,i, \tfrac{3}{2}m_2 t^{\frac{2}{3}})t^{-2}, \quad (48)$$

where $J(\lambda, t)$ is the Bessel function of the first kind.

Lemma 1. *The solutions in the theorem above diverge to complex infinity if t tends to zero and they converge to zero if t tends to infinity.*

2. Definitions and Notations

Greek indices α, β range from 0 to n, Latin i, j, k from 1 to n, and we stipulate $0 \leq a, b \leq n$ but $1 \leq a', b' \leq n$. Barred indices $\bar{a}$ refer to the Lie algebra $\mathfrak{g}$, $1 \leq \bar{a} \leq n_0 = \dim \mathfrak{g}$.
$\gamma_{\bar{a}\bar{b}}$ is the Cartan–Killing metric.
The Dirac matrices are denoted by γ^a and they satisfy

$$\gamma^a \gamma^b + \gamma^b \gamma^a = 2\eta^{ab} I, \quad (49)$$

where η_{ab} is the Minkowski metric with signature $(-,+,\ldots,+)$. γ^0 is anti-Hermitian and $\gamma^{a'}$ Hermitian.

The indices a, b are always raised or lowered with the help of the Minkowski metric, Greek indices with the help of the spacetime metric $\bar{g}_{\alpha\beta}$.

The γ^a act in

$$\mathbb{C}^{2^{\frac{n+1}{2}}}, \tag{50}$$

if n is odd and in

$$\mathbb{C}^{2^{\frac{n}{2}}} \oplus \mathbb{C}^{2^{\frac{n}{2}}}, \tag{51}$$

if n is even. In both cases, we simply refer to these spaces as

$$\mathbb{C}^{n_1}, \tag{52}$$

i.e., the spinor index A has a range of $1 \leq A \leq n_1$.

The color index I has a range of $1 \leq I \leq n_2$ and, hence, a spinor field ψ_A^I has values in

$$\mathbb{C}^{n_1} \otimes \mathbb{C}^{n_2}. \tag{53}$$

Finally, a Hermitian form $\langle \cdot, \cdot \rangle$ is anti-Hermitian in the first argument.

3. Spinor Fields

The Lagrangian of the spinor field is stated in (23). Here, $\psi = (\psi_A^I)$ is a multiplet of the spinors with spin $\frac{1}{2}$; A is the spinor index, $1 \leq A \leq n_1$, and I, $1 \leq I \leq n_2$, the *color* index. We shall also lower or raise the index I with the help of the Euclidean metric (δ_{IJ}).

Let Γ_μ be the spinor connection

$$\Gamma_\mu = \tfrac{1}{4} \omega_\mu{}^b{}_a \gamma_b \gamma^a, \tag{54}$$

then the covariant derivative $D_\mu \psi$ is defined by

$$D_\mu \psi = \psi_{,\mu} + \Gamma_\mu \psi + A_\mu \psi. \tag{55}$$

Let (e_λ^b) be a n-bein, such that

$$\bar{g}_{\mu\lambda} = \eta_{ab} e_\mu^a e_\lambda^b, \tag{56}$$

where (η_{ab}) is the Minkowski metric, and let (E_a^μ) be its inverse

$$E_a^\mu = \eta_{ab} \bar{g}^{\mu\lambda} e_\lambda^b, \tag{57}$$

cf. [7] (p. 246).

The covariant derivative of E_a^α with respect to $(\bar{g}_{\alpha\beta})$ is then given by

$$E_{a;\mu}^\alpha = E_{a,\mu}^\alpha + \bar{\Gamma}_{\mu\beta}^\alpha E_a^\beta \tag{58}$$

and

$$\omega_\mu{}^b{}_a = E_{a;\mu}^\lambda e_\lambda^b = -E_a^\lambda e_{\lambda;\mu}^b, \tag{59}$$

hence, the spin connection Γ_μ can be expressed as

$$\Gamma_\mu = \tfrac{1}{4} \omega_\mu{}^b{}_a \gamma_b \gamma^a = \tfrac{1}{4} E_{a;\mu}^\lambda e_\lambda^b \gamma_b \gamma^a = -\tfrac{1}{4} E_a^\lambda e_{\lambda;\mu}^b \gamma_b \gamma^a. \tag{60}$$

We shall first show:

Lemma 2. *Let $\bar{g}_{\alpha\beta}$ be a fixed spacetime metric that is split by the time function x^0, then there exists an orthonormal frame (e_λ^a), such that*

$$e_k^0 = 0, \quad 1 \leq k \leq n, \tag{61}$$

and
$$e_{k;0}^{a'} = e_{,0}^{a'} - \bar{\Gamma}_{k0}^{\lambda} e_{\lambda}^{a'} = 0 \tag{62}$$

for all $1 \leq a' \leq n$ and $1 \leq k \leq n$.

Proof. Assume that
$$\bar{g}_{00} = -w^2, \tag{63}$$

then define the conformal metric
$$\tilde{g}_{\alpha\beta} = w^{-2} \bar{g}_{\alpha\beta}. \tag{64}$$

The curves
$$(\gamma^{\alpha}(t,x)) = (t, x^i), \qquad x \in \mathcal{S}_0, \tag{65}$$

are then geodesics with respect to $\tilde{g}_{\alpha\beta}$. Let $(\hat{e}_{\lambda}^{a'})$, $1 \leq a' \leq n$, be an orthonormal frame in $T^{0,1}(\mathcal{S}_0) \hookrightarrow T^{0,1}(N)$, such that
$$\hat{e}_0^{a'} = 0 \qquad \forall\, 1 \leq a' \leq n. \tag{66}$$

The $\hat{e}^{a'}$ depend on $x = (x^i) \in \mathcal{S}_0$. Let $(\tilde{e}_{\lambda}^{a'})(t,x)$ be the solutions of the flow equations
$$\begin{aligned} \frac{D}{dt} \tilde{e}_{\lambda}^{a'} &= 0, \\ \tilde{e}_{\lambda}^{a'}(0, x) &= \hat{e}_{\lambda}^{a'}(x), \end{aligned} \tag{67}$$

i.e., we parallel transport $\hat{e}^{a'}$ along the geodesics. Setting
$$(\tilde{e}_{\lambda}^0) = (1, 0, \ldots, 0) \tag{68}$$

the $(\tilde{e}_{\lambda}^a)$ are then an orthonormal frame of 1-forms in $(N, \tilde{g}_{\alpha\beta})$ such that the $\tilde{e}^a$ satisfy
$$\tilde{e}_{\lambda;0}^a = 0 \qquad \forall\, 0 \leq a \leq n, \tag{69}$$

where we indicate covariant differentiation with respect to $\tilde{g}_{\alpha\beta}$ by a colon.

Define e_{λ}^a by
$$e_{\lambda}^a = w \tilde{e}_{\lambda}^a, \tag{70}$$

then the e_{λ}^a are orthonormal frames in $(N, \bar{g}_{\alpha\beta})$. The Christoffel symbols $\bar{\Gamma}_{\alpha\beta}^{\gamma}$ resp. $\tilde{\Gamma}_{\alpha\beta}^{\gamma}$ are related by the formula
$$\bar{\Gamma}_{\alpha\beta}^{\gamma} = \tilde{\Gamma}_{\alpha\beta}^{\gamma} - w^{-1} w_{\alpha} \delta_{\beta}^{\gamma} + w^{-1} w_{\beta} \delta_{\alpha}^{\gamma} - w^{-1} \check{w}^{\gamma} \tilde{g}_{\alpha\beta}, \tag{71}$$

where
$$\check{w}^{\gamma} = \tilde{g}^{\gamma\lambda} w_{\lambda}. \tag{72}$$

In view of (69), we then infer
$$0 = \tilde{e}_{j;0}^{a'} = \dot{\tilde{e}}_j^{a'} - \tilde{\Gamma}_{0j}^k \tilde{e}_k^{a'} \tag{73}$$

and we deduce further
$$\begin{aligned} e_{j;0}^{a'} &= \dot{w} \tilde{e}_j^{a'} + w \dot{\tilde{e}}_j^{a'} - \bar{\Gamma}_{0j}^k w \tilde{e}_k^{a'} \\ &= \dot{w} \tilde{e}_j^{a'} + \tilde{\Gamma}_{0j}^k w \tilde{e}_k^{a'} - \bar{\Gamma}_{0j}^k w \tilde{e}_k^{a'} \\ &= 0 \end{aligned} \tag{74}$$

because of (71). □

Subsequently, we shall always use these particular orthonormal frames.

We are now able to simplify the expressions for the spin connections

$$\Gamma_\mu = -\tfrac{1}{4} E_a^\lambda e^b_{\lambda;\mu} \gamma_a \gamma^b. \tag{75}$$

We have

$$\begin{aligned}
4\Gamma_0 &= -E_a^\lambda e^b_{\lambda;0} \gamma_b \gamma^a \\
&= -E_a^\lambda e^0_{\lambda;0} \gamma_0 \gamma^a - E_a^\lambda e^{b'}_{\lambda;0} \gamma_{b'} \gamma^a \\
&= -E_0^0 e^0_{0;0} \gamma_0 \gamma^0 - E_{a'}^i e^0_{i;0} \gamma_0 \gamma^{a'} - E_0^0 e^{b'}_{0;0} \gamma_{b'} \gamma^0 - E_{a'}^i e^{b'}_{i;0} \gamma_{b'} \gamma^{a'} \\
&= -E_{a'}^i e^0_{i;0} \gamma_0 \gamma^{a'} - E_0^0 e^{b'}_{0;0} \gamma_{b'} \gamma^0
\end{aligned} \tag{76}$$

in view of Lemma 2 and the fact that

$$e^0_{0;0} = 0. \tag{77}$$

The matrices $\gamma_0 \gamma^{a'}$ and $\gamma_{b'} \gamma^0$ are Hermitian, since γ^0 is anti-Hermitian, $\gamma^{a'}$ Hermitian, and there holds

$$\gamma_0 \gamma^{a'} = -\gamma^{a'} \gamma_0. \tag{78}$$

Hence, the quadratic form

$$\tilde\psi E_a^0 \gamma^a \Gamma_0 \psi = -i E_0^0 \tilde\psi \Gamma_0 \psi \tag{79}$$

is imaginary and will be eliminated by adding its complex conjugate. Γ_0 can therefore be ignored, which we shall indicate by writing

$$\Gamma_0 \simeq 0. \tag{80}$$

A similar notation should apply to other terms that will be canceled when adding the complex conjugates.

Let us consider Γ_k:

$$\begin{aligned}
4\Gamma_k &= -E_a^\lambda e^b_{\lambda;k} \gamma_b \gamma^a \\
&= -E_a^\lambda e^0_{\lambda;k} \gamma_0 \gamma^a - E_a^\lambda e^{b'}_{\lambda;k} \gamma_{b'} \gamma^a \\
&= -E_0^0 e^0_{0;k} \gamma_0 \gamma^0 - E_{a'}^i e^0_{i;k} \gamma_0 \gamma^{a'} - E_0^0 e^{b'}_{0;k} \gamma_{b'} \gamma^0 - E_{a'}^i e^{b'}_{i;k} \gamma_{b'} \gamma^{a'}.
\end{aligned} \tag{81}$$

The first term on the right-hand side vanishes, since

$$e^0_{0;k} = w_k - \bar\Gamma^0_{0k} w = 0. \tag{82}$$

Furthermore, there holds

$$e^0_{i;k} = -\bar\Gamma^0_{ik} w = -\tfrac{1}{2} \dot g_{ik} w^{-1} \tag{83}$$

and

$$e^{b'}_{0;k} = -\bar\Gamma^j_{0k} e^{b'}_j = -\tfrac{1}{2} g^{lj} \dot g_{kl} e^{b'}_j, \tag{84}$$

yielding

$$\begin{aligned}
4\Gamma_k &= \tfrac{1}{2} \dot g_{ik} w^{-1} E_{a'}^i \gamma_0 \gamma^{a'} + \tfrac{1}{2} w^{-1} g^{lj} \dot g_{kl} e^{b'}_i \gamma_{b'} \gamma^0 - E_{a'}^i e^{b'}_{i;k} \gamma_{b'} \gamma^{a'} \\
&= w^{-1} \dot g_{ik} E_{a'}^i \gamma_0 \gamma^{a'} - E_{a'}^i e^{b'}_{i;k} \gamma_{b'} \gamma^{a'},
\end{aligned} \tag{85}$$

since

$$\gamma_0 \gamma^{a'} = -\gamma^{a'} \gamma_0. \tag{86}$$

The first term on the right-hand side of (85) has to be eliminated because of the presence of $\dot g_{ik}$. To achieve this, fix a Riemannian metric $\rho_{ij} = \rho_{ij}(x) \in T^{0,2}(\mathcal{S}_0)$, and define the function φ by

$$\varphi = \sqrt{\frac{\det g_{ij}}{\det \rho_{ij}}} \tag{87}$$

and the spinors $\chi = (\chi_A^i)$ by

$$\chi = \sqrt{\varphi}\psi, \tag{88}$$

then

$$\dot\chi = \sqrt{\varphi}\dot\psi + \tfrac{1}{4}g^{ij}\dot g_{ij}\chi \tag{89}$$

and

$$\chi_{,k} = \tfrac{1}{2}\varphi_k \varphi^{-1/2}\chi + \sqrt{\varphi}\psi_{,k}. \tag{90}$$

Looking at the real part of the quadratic form

$$i\tilde\chi E_{a'}^k \gamma^{a'} \chi_{,k} \tag{91}$$

we deduce that

$$\chi_{,k} \simeq \sqrt{\varphi}\psi_{,k}. \tag{92}$$

Moreover, we infer

$$\begin{aligned}
i\bar\psi E_{c'}^k \gamma^{c'} \Gamma_k \psi &= i\bar\psi E_{c'}^k \gamma^0 \gamma^{c'} \Gamma_k \psi \\
&= \tfrac{1}{4} i\bar\psi E_{c'}^k E_{a'}^j w^{-1} \dot g_{jk} \gamma^0 \gamma^{c'} \gamma_0 \gamma^{a'} \psi \\
&\quad - \tfrac{1}{4} i\bar\psi E_{c'}^k E_{a'}^j e_{j;k}^{b'} \gamma^0 \gamma^{c'} \gamma_{b'} \gamma^{a'} \psi.
\end{aligned} \tag{93}$$

We now observe that

$$\gamma^0 \gamma^{c'} \gamma_0 \gamma^{a'} = -\gamma^0 \gamma_0 \gamma^{c'} \gamma^{a'} = -\gamma^{c'} \gamma^{a'}, \tag{94}$$

hence,

$$E_{c'}^k E_{a'}^j \gamma^0 \gamma^{c'} \gamma_0 \gamma^{a'} = -E_{c'}^k E_{a'}^j \gamma^{c'} \gamma^{a'} = -g^{jk} \tag{95}$$

and we conclude

$$\begin{aligned}
i\bar\psi E_c^\mu \gamma^c D_\mu \psi \varphi &\simeq -i\tilde\chi \dot\chi w^{-1} \\
&\quad + i\tilde\chi E_{c'}^k \gamma^0 \gamma^{c'} \{\chi_{,k} - \tfrac{1}{4} E_{a'}^j e_{j;k}^{b'} \gamma_{b'} \gamma^{a'} \chi + A_k \chi\}
\end{aligned} \tag{96}$$

Remark 2. *The term in the braces is the covariant derivative of χ with respect to the spin connection $\tilde\Gamma_k$*

$$\tilde\Gamma_{ka'}^{b'} = \tfrac{1}{4}\tilde\omega_{ka'}^{b'} = -\tfrac{1}{4} E_{a'}^j e_{j;k}^{b'} \gamma_{b'} \gamma^{a'} \tag{97}$$

and the Yang–Mills connection (A_μ) satisfying $A_0 = 0$, such that

$$\tilde D_k \chi = \chi_{,k} + \tilde\Gamma_k \chi + A_k \chi. \tag{98}$$

The gauge transformations for both the Yang–Mills connection as well as for the spin connection do not depend on x^0 but only on $x \in \mathcal{S}_0$. In case of the Yang–Mills connection, this has already been proved in [8] (Lemma 2.6), while the proof for the spin connection $\tilde\Gamma_k$ follows from (97) and (85) if we only consider Lorentzian metrics of the form

$$d\tilde s^2 = -dt^2 + g_{ij}(x)dx^i dx^j \tag{99}$$

in a product manifold $N = I \times \mathcal{S}_0$, as will be the case after the quantization of the Dirac field.

Summarizing the preceding results, we obtain:

Lemma 3. *The Dirac Lagrangian can be expressed in the form*

$$L_D = \tfrac{i}{2}(\bar{\chi}_I \dot{\chi}^I - \dot{\bar{\chi}}^I \chi_I) w^{-1} \varphi^{-1} + m i \bar{\chi}_I \gamma^0 \chi^I \varphi^{-1} \\ - \tfrac{i}{2}\{\bar{\chi}_I \gamma^0 E^k_{a'} \gamma^{a'} \tilde{D}_k \chi^I - \overline{\bar{\chi}_I \gamma^0 E^k_{a'} \gamma^{a'} \tilde{D}_k \chi^I}\} \varphi^{-1}, \tag{100}$$

where χ and $\tilde{D}_k$ are defined in (88) resp. (98).

4. Quantization of the Lagrangian

We consider the functional

$$J = \alpha_N^{-1} \int_{\tilde{\Omega}} (\bar{R} - 2\Lambda) - \int_{\tilde{\Omega}} \tfrac{1}{4} \gamma_{\bar{a}\bar{b}} \bar{g}^{\mu\rho_2} \bar{g}^{\lambda\rho_1} F^{\bar{a}}_{\mu\rho_1} F^{\bar{b}}_{\rho_2 \lambda} \\ - \int_{\tilde{\Omega}} \{\tfrac{1}{2} \bar{g}^{\mu\lambda} \gamma_{\bar{a}\bar{b}} \Phi^{\bar{a}}_\mu \Phi^{\bar{b}}_\lambda + V(\Phi)\} \\ + \int_{\tilde{\Omega}} \{\tfrac{1}{2} [\bar{\psi}_I E^\mu_a \gamma^a (D_\mu \psi)^I + \overline{\bar{\psi}_I E^\mu_a \gamma^a (D_\mu \psi)^I}] + m \bar{\psi}_I \psi^I\}, \tag{101}$$

where α_N is a positive coupling constant and $\tilde{\Omega} \Subset N$.

We use the action principle that, for an arbitrary $\tilde{\Omega}$ as above, a solution $(A, \Phi, \psi, \bar{g})$ should be a stationary point of the functional with respect to compact variations. This principle requires no additional surface terms for the functional.

As we proved in [1], we may only consider metrics $\bar{g}_{\alpha\beta}$ that split with respect to some fixed globally defined time function x^0, such that

$$d\bar{s}^2 = -w^2(dx^0)^2 + g_{ij} dx^i dx^j \tag{102}$$

where $g(x^0, \cdot)$ are Riemannian metrics in $\mathcal{S}_0$,

$$\mathcal{S}_0 = \{x^0 = 0\}. \tag{103}$$

The first functional on the right-hand side of (101) can be written in the form

$$\alpha_N^{-1} \int_a^b \int_\Omega \{\tfrac{1}{4} G^{ij,kl} \dot{g}_{ij} \dot{g}_{kl} w^{-2} + R - 2\Lambda\} w \varphi, \tag{104}$$

where

$$G^{ij,kl} = \tfrac{1}{2}\{g^{ik} g^{jl} + g^{il} g^{jk}\} - g^{ij} g^{kl} \tag{105}$$

is the DeWitt metric,

$$(g^{ij}) = (g_{ij})^{-1}, \tag{106}$$

R the scalar curvature of the slices

$$\{x^0 = t\} \tag{107}$$

with respect to the metric $g_{ij}(t, \cdot)$, and where we also assumed that $\tilde{\Omega}$ is a cylinder

$$\tilde{\Omega} = (a, b) \times \Omega, \quad \Omega \Subset \mathcal{S}_0, \tag{108}$$

such that $\tilde{\Omega} \subset U_k$ for some $k \in \mathbb{N}$, where the U_k are special coordinate patches of N, such that there exists a local trivialization in U_k with the properties that there is a fixed Yang–Mills connection

$$\bar{A} = (\bar{A}^{\bar{a}}_\mu) = f_{\bar{a}} \bar{A}^{\bar{a}}_\mu dx^\mu \tag{109}$$

satisfying

$$\bar{A}^{\bar{a}}_0 = 0 \quad \text{in } U_k, \tag{110}$$

cf. [8] (Lemma 2.5). We may then assume that the Yang–Mills connections $A = (A_\mu^{\bar{a}})$ are of the form

$$A_\mu^{\bar{a}}(t,x) = \bar{A}_\mu^{\bar{a}}(0,x) + \tilde{A}_\mu^{\bar{a}}(t,x), \tag{111}$$

where $(\tilde{A}_\mu^{\bar{a}})$ is a tensor, see [8] (Section 2).

The Riemannian metrics $g_{ij}(t,\cdot)$ are elements of the bundle $T^{0,2}(\mathcal{S}_0)$. Denote by E the fiber bundle with base $\mathcal{S}_0$ where the fibers $F(x)$ consist of the Riemannian metrics (g_{ij}). We shall consider each fiber to be a Lorentzian manifold equipped with the DeWitt metric. Each fiber F has the dimension

$$\dim F = \frac{n(n+1)}{2} \equiv m+1. \tag{112}$$

Let (ξ^r), $0 \leq r \leq m$, be coordinates for a local trivialization, such that

$$g_{ij}(x, \xi^r) \tag{113}$$

is a local embedding. The DeWitt metric is then expressed as

$$G_{rs} = G^{ij,kl} g_{ij,r} g_{kl,s}, \tag{114}$$

where a comma indicates partial differentiation. In the new coordinate system, the curves

$$t \to g_{ij}(t,x) \tag{115}$$

can be written in the form

$$t \to \xi^r(t,x) \tag{116}$$

and we infer

$$G^{ij,kl} \dot{g}_{ij} \dot{g}_{kl} = G_{rs} \dot{\xi}^r \dot{\xi}^s. \tag{117}$$

Hence, we can express (104) as

$$J = \int_a^b \int_\Omega \alpha_n^{-1} \{\tfrac{1}{4} G_{rs} \dot{\xi}^r \dot{\xi}^s w^{-1} \varphi + (R - 2\Lambda) w \varphi\}, \tag{118}$$

where we now refrain from writing down the density $\sqrt{\rho}$ explicitly, since it does not depend on (g_{ij}) and, therefore, should not be part of the Legendre transformation. Here, we follow Mackey's advice in [9] (p. 94) to always consider rectangular coordinates when applying canonical quantization, which can be rephrased that the Hamiltonian has to be a coordinate invariant, hence no densities are allowed.

Denoting the Lagrangian function in (118) by L, we define

$$\pi_r = \frac{\partial L}{\partial \dot{\xi}^r} = \varphi G_{rs} \frac{1}{2\alpha_N} \dot{\xi}^s w^{-1} \tag{119}$$

and we obtain for the Hamiltonian function $\hat{H}_G$

$$\begin{aligned}
\hat{H}_G &= \dot{\xi}^r \frac{\partial L}{\partial \dot{\xi}^r} - L \\
&= \varphi G_{rs} \left(\frac{1}{2\alpha_N} \dot{\xi}^r w^{-1}\right) \left(\frac{1}{2\alpha_N} \dot{\xi}^s w^{-1}\right) w \alpha_N - \alpha_N^{-1}(R - 2\Lambda) \varphi w \\
&= \varphi^{-1} G^{rs} \pi_r \pi_s w \alpha_N - \alpha_N^{-1}(R - 2\Lambda) \varphi w \\
&\equiv H_G w,
\end{aligned} \tag{120}$$

where G^{rs} is the inverse metric. Hence,

$$H_G = \alpha_N \varphi^{-1} G^{rs} \pi_r \pi_s - \alpha_N^{-1}(R - 2\Lambda) \varphi \tag{121}$$

is the Hamiltonian that will enter the Hamilton constraint, for details see [6] (Chapter 1.4).

Let us recall that the fibers F can be considered Lorentzian manifolds, even globally hyperbolic manifolds, equipped with the DeWitt metric ($\varphi G^{ij,kl}$), where φ is a time function, cf. [6] (Theorem 1.4.2). In the fibers, we can introduce new coordinates, $(\xi^a) = (\xi^0, \xi^A) \equiv (t, \xi^A)$, $0 \leq a \leq m$, and $1 \leq A \leq m$, such that

$$t = \sqrt{\varphi} \tag{122}$$

and (ξ^A) are coordinates for the hypersurface

$$M = \{\varphi = 1\} = \{\xi^0 = 1\}. \tag{123}$$

The Lorentzian metric in the fibers can then be expressed in the form

$$ds^2 = -\frac{16(n-1)}{n}dt^2 + t^2 G_{AB} d\xi^A d\xi^B, \tag{124}$$

where (G_{AB}) is a Riemannian metric on M, which is independent of t. When we work in a local trivialization of the bundle E, the coordinates (ξ^A) are independent of x. The time coordinate t is also independent of x, cf. [1] (Lemma 1.8). Moreover, the fiber elements (g_{ij}) can be expressed in the form

$$g_{ij} = t^{\frac{4}{n}} \sigma_{ij}, \tag{125}$$

where (σ_{ij}) is an element of M, i.e.,

$$t(\sigma_{ij}) = 1, \tag{126}$$

or equivalently,

$$\det \sigma_{ij} = \det \rho_{ij}. \tag{127}$$

Next, let us look at the Yang–Mills Lagrangian, which can be expressed as

$$L_{YM} = \tfrac{1}{2}\gamma_{\bar{a}\bar{b}} g^{ij} \tilde{A}^{\bar{a}}_{i,0} \tilde{A}^{\bar{b}}_{j,0} w^{-1} \varphi - \tfrac{1}{4} F_{ij} F^{ij} w \varphi. \tag{128}$$

Let E_0 be the adjoint bundle

$$E_0 = (\mathcal{S}_0, \mathfrak{g}, \pi, \mathrm{Ad}(\mathcal{G})) \tag{129}$$

with base space $\mathcal{S}_0$, where the gauge transformations only depend on the spatial variables $x = (x^i)$. Then the mappings $t \to \tilde{A}^{\bar{a}}_i(t, \cdot)$ can be looked at as curves in $T^{1,0}(E_0) \otimes T^{0,1}(\mathcal{S}_0)$, where the fibers of $T^{1,0}(E_0) \otimes T^{0,1}(\mathcal{S}_0)$ are the tensor products

$$\mathfrak{g} \otimes T^{0,1}_x(\mathcal{S}_0), \quad x \in \mathcal{S}_0, \tag{130}$$

which are vector spaces equipped with the metric

$$\gamma_{\bar{a}\bar{b}} \otimes g^{ij}. \tag{131}$$

For our purposes, it is more convenient to consider the fibers to be Riemannian manifolds endowed with the above metric. Let (ζ^p), $1 \leq p \leq n_1 n$, where $n_0 = \dim \mathfrak{g}$, be local coordinates and

$$(\zeta^p) \to \tilde{A}^{\bar{a}}_i(\zeta^p) \equiv \tilde{A}(\zeta) \tag{132}$$

be a local embedding, then the metric has the coefficients

$$G_{pq} = \langle \tilde{A}_p, \tilde{A}_q \rangle = \gamma_{\bar{a}\bar{b}} g^{ij} \tilde{A}^{\bar{a}}_{i,p} \tilde{A}^{\bar{b}}_{j,q}. \tag{133}$$

Hence, the Lagrangian L_{YM} in (128) can be expressed in the form

$$L_{YM} = \tfrac{1}{2} G_{pq} \dot{\zeta}^p \dot{\zeta}^q w^{-1} \varphi - \tfrac{1}{4} F_{ij} F^{ij} w \varphi \tag{134}$$

and we deduce

$$\tilde{\pi}_p = \frac{\partial L_{YM}}{\partial \dot{\zeta}^p} = G_{pq}\dot{\zeta}^q w^{-1}\varphi \qquad (135)$$

yielding the Hamilton function

$$\begin{aligned}\hat{H}_{YM} &= \pi_p \dot{\zeta}^p - L_{YM} \\ &= \tfrac{1}{2}G_{pq}(\dot{\zeta}^p w^{-1}\varphi)(\dot{\zeta}^q w^{-1}\varphi)w\varphi^{-1} + \tfrac{1}{4}F_{ij}F^{ij}w\varphi \\ &= \tfrac{1}{2}G^{pq}\tilde{\pi}_p\tilde{\pi}_q w\varphi^{-1} + \tfrac{1}{4}F_{ij}F^{ij}w\varphi \\ &\equiv H_{YM}w.\end{aligned} \qquad (136)$$

Thus, after introducing a normal Gaussian coordinate system, such that $w = 1$, the Hamiltonian that will enter the Hamilton constraint equation is

$$H_{YM} = \tfrac{1}{2}\varphi^{-1}G^{pq}\tilde{\pi}_p\tilde{\pi}_q + \tfrac{1}{4}F_{ij}F^{ij}\varphi. \qquad (137)$$

Combining, now, (122), (125) and (133) we infer that the Yang–Mills Hamiltonian can be expressed as

$$H_{YM} = t^{\frac{4}{n}-2}\tilde{G}^{pq}\tilde{\pi}_p\tilde{\pi}_q + \tfrac{1}{4}F_{ij}F^{ij}t^{2-\frac{8}{n}}, \qquad (138)$$

where the indices in the last term are raised with respect to the metric σ_{ij}, i.e.,

$$F^{ij} = \sigma^{ik}\sigma^{jl}F_{kl}. \qquad (139)$$

In the case of $n = 3$, the exponents of t in (138) are equal

$$\frac{4}{3} - 2 = 2 - \frac{8}{3} = -\frac{2}{3} \qquad (140)$$

and we can write

$$\begin{aligned}H_{YM} &= t^{-\frac{2}{3}}\{\tilde{G}^{pq}\tilde{\pi}_p\tilde{\pi}_q + \tfrac{1}{4}F_{ij}F^{ij}\} \\ &\equiv t^{-\frac{2}{3}}\tilde{H}_{YM}.\end{aligned} \qquad (141)$$

Moreover, if (σ_{ij}) as well as (ρ_{ij}) are equal to the Euclidean metric (δ_{ij}), then the quantization of $\tilde{H}_{YM}$ would be achieved by known methods of QFT.

Hence, we shall attempt to express the Hamiltonians of the other physical forces, such as the Dirac and Higgs Hamiltonians, when evaluated for

$$\sigma_{ij} = \rho_{ij} = \delta_{ij} \qquad (142)$$

and in the case of $n = 3$ in the form

$$H_D = t^{-\frac{2}{3}}\tilde{H}_D \qquad (143)$$

resp.

$$H_H = t^{-\frac{2}{3}}\tilde{H}_H \qquad (144)$$

such that the quantization of the spatial Hamiltonian

$$\tilde{H}_{YM} + \tilde{H}_D + \tilde{H}_H \qquad (145)$$

would be well known, and in the end, all spatial Hamiltonians of the standard model could be incorporated.

Let us first consider the Dirac Hamiltonian. In the Dirac Lagrangian L_D, defined in Equation (100) on page 11, the volume density $\sqrt{g}$ is missing, i.e., in order to define the Hamiltonian, we have to multiply the Lagrangian with $\sqrt{g}$, or, since we work with functions instead of densities, we have to multiply the Lagrangian with φ.

In addition, we shall also consider—at least locally—a normal Gaussian coordinate system, such that $w = 1$. Then, the final Dirac Lagrangian has the form

$$L_D = \tfrac{i}{2}(\bar{\chi}_I \dot{\chi}^I - \dot{\bar{\chi}}^I \chi_I) + mi\bar{\chi}_I \gamma^0 \chi^I \\ - \tfrac{i}{2}\{\bar{\chi}_I \gamma^0 E^k_{a'} \gamma^{a'} \tilde{D}_k \chi^I - \overline{\bar{\chi}_I \gamma^0 E^k_{a'} \gamma^{a'} \tilde{D}_k \chi^I}\}, \tag{146}$$

The spinorial variables χ^I_A are anti-commuting Grassmann variables. They are elements of a Grassmann algebra with involution, where the involution corresponds to the complex conjugation and will be denoted by a bar.

The χ^I_A are complex variables and we define the real resp. imaginary parts as

$$\xi^I_A = \tfrac{1}{\sqrt{2}}(\chi^I_A + \bar{\chi}^I_A) \tag{147}$$

resp.

$$\eta^I_A = \tfrac{1}{\sqrt{2}i}(\chi^I_A - \bar{\chi}^I_A). \tag{148}$$

Then,

$$\chi^I_A = \tfrac{1}{\sqrt{2}}(\xi^I_A + i\eta^I_A) \tag{149}$$

and

$$\bar{\chi}^I_A = \tfrac{1}{\sqrt{2}}(\xi^I_A - i\eta^I_A). \tag{150}$$

With these definitions, we obtain

$$\tfrac{i}{2}(\bar{\chi}_I \dot{\chi}^I - \dot{\bar{\chi}}^I \chi_I) = \tfrac{i}{2}(\xi^A_I \dot{\xi}^I_A + \eta^A_I \dot{\eta}^I_A). \tag{151}$$

Casalbuoni quantized the Bose–Fermi system in [10] (section 4), the results of which can be applied to spin $\tfrac{1}{2}$ fermions. The Lagrangian in [10] is the same as the main part our Lagrangian in (146) on page 15, and the left derivative is used in that paper; hence, we use left derivatives as well such that the conjugate momenta of the odd variables are, e.g.,

$$\pi^A_I = \frac{\partial L}{\partial \dot{\xi}^I_A} = -\tfrac{i}{2}\xi^A_I, \tag{152}$$

and, thus, the conclusions in [10] can be applied.

The Lagrangian has been expressed in real variables—at least the important part of it—and it follows that the odd variables ξ^I_A, η^I_A satisfy, after introducing anti-commutative Dirac brackets as in [10] (equ. (4.11)),

$$\{\xi^A_I, \xi^J_B\}^*_+ = -i\delta^J_I \delta^A_B, \tag{153}$$

$$\{\eta^A_I, \eta^J_B\}^*_+ = -i\delta^J_I \delta^A_B, \tag{154}$$

and

$$\{\xi^A_I, \eta^J_B\}^*_+ = 0, \tag{155}$$

cf. [10] (equ. (4.19)).

In view of (149), (150) we then derive

$$\{\bar{\chi}^A_I, \chi^J_B\}^*_+ = -i\delta^J_I \delta^A_B, \tag{156}$$

where $\bar{\chi}^A_I$ are the conjugate momenta.

Canonical quantization—with $\hbar = 1$—then requires that the corresponding operators $\hat{\chi}^I_A, \hat{\bar{\chi}}^B_J$ satisfy the anti-commutative rules

$$[\hat{\chi}^I_A, \hat{\bar{\chi}}^B_J]_+ = i\{\chi^I_A, \bar{\chi}^B_J\}^*_+ = \delta^I_J \delta^B_A \tag{157}$$

and
$$[\hat{\bar{\chi}}_I^A, \hat{\bar{\chi}}_J^B]_+ = [\hat{\chi}_A^I, \hat{\chi}_B^J]_+ = 0, \tag{158}$$

cf. [11] (equ. (3.10)) and [10] (equ. (5.17)).

From (146), we then deduce that the spinorial Hamilton function is equal to

$$H_D = \tfrac{i}{2}\{\bar{\chi}_I \gamma^0 E_{a'}^k \gamma^{a'} \tilde{D}_k \chi^I - \overline{\bar{\chi}_I \gamma^0 E_{a'}^k \gamma^{a'} \tilde{D}_k \chi^I}\} \\ - mi\bar{\chi}_I \gamma^0 \chi^I. \tag{159}$$

When we attempt to quantize this Hamilton function, then the vielbein $e_k^{a'}$ and its inverse $E_{a'}^k$ will correspond to a given element $g_{ij}(x)$ in the fiber F, which can be expressed as in (125), and we deduce that the vielbein

$$\tilde{e}_k^{a'} = t^{-\frac{2}{n}} e_k^{a'} \tag{160}$$

and its inverse

$$\tilde{E}_{a'}^k = t^{\frac{2}{n}} E_{a'}^k \tag{161}$$

correspond to the metric σ_{ij}. Furthermore, the covariant derivative $\tilde{D}_k \chi^I$ is independent of t, in view of (97) and (98) on page 10. Thus, the Hamilton function H_D can be expressed as

$$H_D = t^{-\frac{2}{3}}\left(\tfrac{i}{2}\{\bar{\chi}_I \gamma^0 \tilde{E}_{a'}^k \gamma^{a'} \tilde{D}_k \chi^I - \overline{\bar{\chi}_I \gamma^0 \tilde{E}_{a'}^k \gamma^{a'} \tilde{D}_k \chi^I}\}\right) \\ - mi\bar{\chi}_I \gamma^0 \chi^I, \tag{162}$$

i.e., the main part already has the form that we looked for in (143), provided $n = 3$, only the mass term spoils the necessary configuration. To overcome this setback, we either have to omit the mass term or modify it by multiplying the mass term in (23) on page 3 with the factor

$$\varphi^{-\frac{1}{n}}, \tag{163}$$

where φ is defined in (87) on page 10. Note that $\varphi = 1$ if

$$g_{ij} = \rho_{ij} = \delta_{ij} \tag{164}$$

as is the case in QFT. Either by omitting or by modifying the mass term, the Dirac Hamilton function can be expressed in the required form

$$H_D = t^{-\frac{2}{3}} \tilde{H}_D, \tag{165}$$

where the underlying Riemannian metric is σ_{ij}, provided $n = 3$.

The remaining Hamiltonian is the Hamiltonian of the Higgs field. The Higgs Lagrangian is defined by

$$L_H = -\tfrac{1}{2} \tilde{g}^{\alpha\beta} \gamma_{ab} \Phi_\alpha^a \Phi_\beta^b - V(\Phi), \tag{166}$$

where V is a smooth potential. We assume that in a local coordinate system Φ has real coefficients. The covariant derivatives of Φ are defined by a connection $A = (A_\mu^a)$ in E_0

$$\Phi_\mu^a = \Phi_{,\mu}^a + f_{cb}^a A_\mu^c \Phi^b. \tag{167}$$

As in the preceding section, we work in a local trivialization of E_0 using the temporal gauge, i.e.,

$$A_0^a = 0, \tag{168}$$

hence, we conclude

$$\Phi_0^a = \Phi_{,0}^a. \tag{169}$$

Expressing the density g as in (87) on page 10, we obtain Lagrangian

$$L_H = \tfrac{1}{2}\gamma_{ab}\Phi^a_{,0}\Phi^b_{,0}\varphi - \tfrac{1}{2}g^{ij}\gamma_{ab}\Phi^a_i\Phi^b_j\varphi - V(\Phi)\varphi, \tag{170}$$

where, again, we use local coordinates, such that $w=1$. In order to apply our approach, outlined in (144), we have to modify the Lagrangian. Instead of the above Lagrangian, we have to consider

$$L_{Hmod} = \{\tfrac{1}{2}\gamma_{ab}\Phi^a_{,0}\Phi^b_{,0} - \tfrac{1}{2}g^{ij}\gamma_{ab}\Phi^a_i\Phi^b_j\}\varphi^{1+\gamma_1} - V(\Phi)\varphi^{1+\gamma_2}. \tag{171}$$

Let us define

$$p_a = \frac{\partial L_H}{\partial \dot\Phi^a}, \qquad \dot\Phi^a = \Phi^a_{,0}, \tag{172}$$

then we obtain the Hamilton function

$$\begin{aligned}H_{Hmod} &= p_a\dot\Phi^a - L_H \\ &= \tfrac{1}{2}\gamma^{ab}p_a p_b \varphi^{-(1+\gamma_1)} + \tfrac{1}{2}g^{ij}\gamma_{ab}\Phi^a_i\Phi^b_j\varphi^{1+\gamma_1} + V(\Phi)\varphi^{1+\gamma_2}.\end{aligned} \tag{173}$$

After quantization, the g_{ij} are elements of the fiber F, i.e.,

$$g_{ij} = t^{\frac{4}{n}}\sigma_{ij}. \tag{174}$$

If $n=3$, then γ_1 has to be chosen, such that

$$-2(1+\gamma_1) = -\tfrac{4}{3} + 2(1+\gamma_1) = -\tfrac{2}{3} \tag{175}$$

which is the case if

$$\gamma_1 = -\tfrac{2}{3}. \tag{176}$$

For γ_2, we obtain

$$2(1+\gamma_2) = -\tfrac{2}{3} \tag{177}$$

yielding

$$\gamma_2 = -\tfrac{4}{3}. \tag{178}$$

Thus, the Hamilton function of the modified Higgs field has the required form

$$H_{Hmod} = t^{-\frac{2}{3}}\tilde H_{Hmod}, \tag{179}$$

where

$$\tilde H_{Hmod} = \tfrac{1}{2}\gamma^{ab}p_a p_b + \tfrac{1}{2}\sigma^{ij}\gamma_{ab}\Phi^a_i\Phi^b_j + V(\Phi) \tag{180}$$

is a standard Hamiltonian of a Higgs field in QFT by choosing $\sigma_{ij} = \delta_{ij}$ and Φ, $V(\Phi)$ as well as the Yang–Mills connection appropriately.

Combining the four Hamilton functions in (120), (138), (179) and (162), the Hamilton constraint has the form

$$\begin{aligned}H &= H_G + H_{YM} + H_H + H_D \\ &= H_G + t^{-\frac{2}{3}}(\tilde H_{YM} + \tilde H_H + \tilde H_D) \\ &\equiv H_G + t^{-\frac{2}{3}}\tilde H_{SM} = 0,\end{aligned} \tag{181}$$

where we omit the subscript *mod* and where SM refers to the fields of the standard model or to a corresponding subset of fields.

The Hamiltonian

$$H_G = \alpha_N \varphi^{-1} G^{rs} \pi_r \pi_s - \alpha_N^{-1}(R - 2\Lambda)\varphi \tag{182}$$

we quantize, as in our former papers [1,12], to obtain

$$H_G = -\alpha_N \Delta - \alpha_N^{-1} R t^2 + 2\alpha_N^{-1} \Lambda t^2, \tag{183}$$

where the Laplacian is the Laplacian of the metric (124) acting in the fibers F of E. The Laplacian acts on smooth functions u of the form $u = u(g_{ij})$. Choosing the Gaussian coordinate system $(\xi^a) = (t, \xi^A)$, such that the fiber metric has form as in (124), then, the hyperbolic term $-\Delta u$ can be expressed as

$$-\Delta u = \frac{n}{16(n-1)} t^{-m} \frac{\partial}{\partial t}(t^m \frac{\partial u}{\partial t}) - t^{-2}\bar{\Delta} u, \tag{184}$$

where $\bar{\Delta}$ is the Laplacian of the hypersurface

$$M = \{t = 1\}. \tag{185}$$

Using the separation of variables we consider functions u which are products

$$u(t, \xi^A) = w(t)v(\xi^A), \tag{186}$$

where v is a spatial eigenfunction, or eigendistribution, of the Laplacian $\bar{\Delta}$

$$-\bar{\Delta} v = \lambda v. \tag{187}$$

The hypersurface

$$M = \{\varphi = 1\} \tag{188}$$

can be considered a subbundle of E, where each fiber $M(x)$ is a hypersurface in the fiber $F(x)$ of E. We shall use the same notation M for the subbundle as well as for the hypersurface, and in general, we shall omit the reference to the base point $x \in \mathcal{S}_0$. Furthermore, we specify the metric $\rho_{ij} \in T^{0,2}(\mathcal{S}_0)$, which we used to define φ, to be equal to the Euclidean metric, such that in Euclidean coordinates

$$\varphi^2 = \frac{\det g_{ij}}{\det \delta_{ij}} = \det g_{ij}. \tag{189}$$

Then, it is well-known that each $M(x)$ with the induced metric (G_{AB}) is a symmetric space, namely, it is isometric to the coset space

$$G/K = SL(n, \mathbb{R})/SO(n), \tag{190}$$

cf. [13] (equ. (5.17), p. 1123) and [14] (p. 3). The eigenfunctions in symmetric spaces, and especially of the coset space in (190), are well-known; they are the so-called *spherical functions*. One can also define a Fourier transformation for functions in $L^2(G/K)$ and prove a Plancherel formula, similar to the Euclidean case, cf. [15] (Chapter III). Moreover, similar to the Euclidean case, we shall use the Fourier kernel to define the eigenfunctions or eigendistributions, cf. [3] (Section 5).

Let

$$G = NAK \tag{191}$$

be an Iwasawa decomposition of G, where N is the subgroup of unit upper triangle matrices, A is the abelian subgroup of the diagonal matrices with strictly positive diagonal components and $K = SO(n)$. The corresponding Lie algebras are denoted by

$$\mathfrak{g}, \mathfrak{n}, \mathfrak{a} \text{ and } \mathfrak{k}. \tag{192}$$

Here,

$$\begin{aligned} \mathfrak{g} &= \text{real matrices with zero trace} \\ \mathfrak{n} &= \text{subspace of strictly upper triangular matrices with zero diagonal} \\ \mathfrak{a} &= \text{subspace of diagonal matrices with zero trace} \\ \mathfrak{k} &= \text{subspace of skew-symmetric matrices.} \end{aligned} \tag{193}$$

The Iwasawa decomposition is unique, when

$$g = nak \tag{194}$$

we define the maps n, A, k by

$$g = n(g)A(g)k(g). \tag{195}$$

We also use the expression $\log A(g)$, where log is the matrix logarithm. In the case of diagonal matrices,

$$a = \text{diag}(a_1, \ldots, a_n) \tag{196}$$

with positive entries

$$\log a = \text{diag}(\log a_i), \tag{197}$$

hence,

$$A(g) = e^{\log A(g)}. \tag{198}$$

Remark 3. *(i) The Lie algebra $\mathfrak{a}$ is a $(n-1)$-dimensional real algebra, which, as a vector space, is equipped with a natural real, symmetric scalar product, namely, the trace form*

$$\langle H_1, H_2 \rangle = \text{tr}(H_1 H_2), \qquad H_i \in \mathfrak{a}. \tag{199}$$

(ii) Let $\mathfrak{a}^*$ be the dual space of $\mathfrak{a}$. Its elements will be denoted by Greek symbols, some of which have special meanings in the literature. The linear forms are also called additive characters.
(iii) Let $\lambda \in \mathfrak{a}^*$, then there exists a unique matrix $H_\lambda \in \mathfrak{a}$, such that

$$\lambda(H) = \langle H_\lambda, H \rangle \qquad \forall H \in \mathfrak{a}. \tag{200}$$

This definition allows defining a dual trace form in $\mathfrak{a}^*$ by setting for $\lambda, \mu \in \mathfrak{a}^*$

$$\langle \lambda, \mu \rangle = \langle H_\lambda, H_\mu \rangle. \tag{201}$$

The Fourier theory in $X = G/K$, which we summarized in [3] (Section 6), uses the functions

$$e_{\lambda,b}(x) = e^{(i\lambda + \rho) \log A(x,b)}, \qquad (\lambda, b) \in \mathfrak{a}^* \times B, \ x \in X, \tag{202}$$

as the Fourier kernel, where

$$B = K/M. \tag{203}$$

Here, M is the centralizer of A in K and ρ is a special character with the norm

$$\langle \rho, \rho \rangle = \frac{1}{12}(n-1)^2 n, \tag{204}$$

cf. [3] (Lemma 1). If $n = 3$, then

$$|\rho|^2 = 1. \tag{205}$$

For a precise definition of $A(x,b) \in A$, we refer to [3] (p. 19), which also contains references to the corresponding mathematical literature given, especially to Helgason's book [15] (Chapter III).

The Fourier transform for functions $f \in C_c^\infty(X, \mathbb{C})$ is then defined by

$$\hat{f}(\lambda, b) = \int_X f(x) e^{(-i\lambda + \rho) \log A(x,b)} dx \qquad (206)$$

for $\lambda \in \mathfrak{a}^*$ and $b \in B$, or, if we use the definition in (202)

$$e_{\lambda, b}(x) = e^{(i\lambda + \rho) \log A(x,b)}, \qquad (207)$$

by

$$\hat{f}(\lambda, b) = \int_X f(x) \bar{e}_{\lambda, b}(x) dx. \qquad (208)$$

The functions $e_{\lambda, b}$ are real analytics in x and are eigenfunctions of the Laplacian, cf. [15] (Prop. 3.14, p. 99),

$$-\tilde{\Delta} e_{\lambda, b} = (|\lambda|^2 + |\rho|^2) e_{\lambda, b}, \qquad (209)$$

where

$$|\lambda|^2 = \langle \lambda, \lambda \rangle, \qquad (210)$$

cf. (201), and similarly for $|\rho|^2$. We also denote the Fourier transform by $\mathcal{F}$, such that

$$\mathcal{F}(f) = \hat{f}. \qquad (211)$$

In Equation (209), we identified

$$\tilde{\Delta} = \Delta_M = \Delta_X. \qquad (212)$$

In [3], we finally dropped the embellishment and simply wrote Δ when referring to the above Laplacian, but at the moment we refrain from doing so to avoid confusion.

We shall consider the eigenfunctions $e_{\lambda, b}$ as tempered distributions of the Schwartz space $\mathcal{S}(X)$ and shall use their Fourier transforms

$$\hat{e}_{\lambda, b} = \delta_{(\lambda, b)} = \delta_\lambda \otimes \delta_b \qquad (213)$$

as the spatial eigenfunctions of

$$\mathcal{F}(-\Delta) = m(\mu) = (|\mu|^2 + |\rho|^2), \qquad (214)$$

which is a multiplication operator, such that

$$\mathcal{F}(-\Delta) \hat{e}_{\lambda, b} = m(\mu) \hat{e}_{\lambda, b} = (|\lambda|^2 + |\rho|^2) \hat{e}_{\lambda, b}, \qquad (215)$$

cf. [3] (Section 6) for details.

Looking at the Fourier transformed eigenfunctions

$$\hat{e}_{\lambda, b} = \delta_\lambda \otimes \delta_b \qquad (216)$$

it is obvious that the dependence on b has to be eliminated, since there is neither a physical nor a mathematical motivation to distinguish between $e_{\lambda, b}$ and $e_{\lambda, b'}$. We discard the integration over B in [3] (Section 6) and pick instead a special element $b_0 \in B$, namely,

$$b_0 = eM, \qquad e = \mathrm{id} \in K, \qquad (217)$$

and only consider the eigenfunctions e_{λ, b_0} with corresponding Fourier transforms

$$\delta_\lambda \equiv \delta_\lambda \otimes \delta_{b_0} = \hat{e}_{\lambda, b_0}, \qquad \lambda \in \mathfrak{a}^*. \qquad (218)$$

For justification, see [3] (Lemma 4) and the arguments preceding the referenced Lemma.

The eigenfunctions e_{λ,b_0} depend on the characters $\lambda \in \mathfrak{a}^*$ but not all characters are physically relevant. For a definition of the physically relevant characters, let us rephrase [3] (Remark 2, p. 18):

Remark 4. *There are characters α_{ij}, $1 \leq i < j \leq n$, that will represent the elementary gravitons stemming from the degrees of freedom in choosing the coordinates*

$$g_{ij}, \quad 1 \leq i < j \leq n, \tag{219}$$

of a metric tensor. The diagonal elements offer, in general, additional n degrees of freedom, but in our case, where we consider metrics satisfying

$$\det g_{ij} = 1, \tag{220}$$

only $(n-1)$ diagonal components can be freely chosen, and we shall choose the first $(n-1)$ entries, namely,

$$g_{ii}, \quad 1 \leq i \leq n-1. \tag{221}$$

The corresponding additive characters are named α_i, $1 \leq i \leq n-1$.

The characters α_i, $1 \leq i \leq n-1$, and α_{ij} $1 \leq i < j \leq n$ will represent the $\frac{(n+2)(n-1)}{2}$ elementary gravitons at the character level. We shall normalize the characters by defining

$$\tilde{\alpha}_i = \|H_{\alpha_i}\|^{-1} \alpha_i \tag{222}$$

and

$$\tilde{\alpha}_{ij} = \|H_{\alpha_{ij}}\|^{-1} \alpha_{ij} \tag{223}$$

such that the normalized characters have unit norm, cf. (201).

We can now define the corresponding forms in $\mathfrak{a}^*$ with arbitrary energy levels:

Definition 1. *Let $\lambda \in \mathbb{R}_+$ be arbitrary. Then we consider the characters*

$$\lambda \tilde{\alpha}_i \quad \wedge \quad \lambda \tilde{\alpha}_{ij}, \tag{224}$$

where we recall that the terms embellished by a tilde refer to the corresponding unit vectors, Then the eigenfunctions representing the elementary gravitons are $e_{\lambda \tilde{\alpha}_i, b_0}$ and $e_{\lambda \tilde{\alpha}_{ij}, b_0}$.

The corresponding eigenvalue with respect to $-\tilde{\Delta}$ is $|\lambda|^2 + |\rho|^2$, where, by a slight abuse of notation, $|\lambda|^2 = \lambda^2$ and $|\rho|^2 = \langle \rho, \rho \rangle$. Note that $|\rho|^2 = 1$ if $n = 3$, cf. (205).

We define a zero-point energy eigenfunction by choosing $\lambda \in \mathfrak{a}^* = 0$. The corresponding eigenfunction would be e_{0,b_0}, satisfying

$$-\tilde{\Delta} e_{0,b_0} = |\rho|^2 e_{0,b_0} = e_{0,b_0}. \tag{225}$$

if $n = 3$.

We are now able to quantize the Hamiltonian H in (181). For brevity we denote the quantized Hamiltonians, which are operators, by using the same symbols as for the Hamilton functions. For the Hamilton operator H_G, we express as in (183)

$$\begin{aligned} H_G u = &-\alpha_N \frac{n}{16(n-1)} t^{-m} \frac{\partial}{\partial t}(t^m \frac{\partial w}{\partial t})v - \alpha_N t^{-2} w \bar{\Delta} v \\ &- \alpha_N^{-1} t^{2-\frac{4}{n}} R(\sigma_{ij}) wv + 2\alpha_N^{-1} \Lambda t^2 wv, \end{aligned} \tag{226}$$

where we use the separation of variables in (186), the form of the metric in (125), namely,

$$g_{ij} = t^{\frac{4}{n}} \sigma_{ij} \tag{227}$$

and the relation between the scalar curvatures of conformal metrics

$$R(g) = t^{-\frac{4}{n}} R(\sigma). \tag{228}$$

Let us recall that for the quantization of $\tilde{H}_{SM}$ we shall specify $\sigma_{ij} = \delta_{ij}$, such that the spatial eigendistributions, or approximate eigendistributions, ψ, satisfying

$$\tilde{H}_{SM}\psi = \lambda_1 \psi, \qquad \lambda_1 \geq 0 \tag{229}$$

can be derived by applying standard methods of QFT. We then solve the Wheeler–DeWitt equation

$$Hu = 0 \tag{230}$$

not for all $(t, \sigma_{ij}) \in \mathbb{R}^+ \times M$ but only for (t, δ_{ij}), where $t > 0$ is arbitrary. Thus, we shall solve

$$-\tilde{\Delta} v = (|\lambda|^2 + |\rho|^2) v \tag{231}$$

by using

$$v = e_{\lambda, b_0} \tag{232}$$

for arbitrary $\sigma_{ij} \in M$, but we shall evaluate e_{λ, b_0} only at $\sigma_{ij} = \delta_{ij}$. Furthermore, we observe that for $x = gK \in X$ and $b = kM \in B$, we have

$$A(x, b) = A(gK, kM) = A(k^{-1}g), \tag{233}$$

cf. [3] (equ. (202), p. 18), hence, if $b = b_0$, i.e., if $k = e = \text{id}$, then

$$A(x, b_0) = A(g). \tag{234}$$

Moreover, let

$$\pi : G/K \to M \tag{235}$$

be the isometry, then

$$\pi(gK) = gg^*, \tag{236}$$

where g^* is the adjoint. Thus, if $g = (\delta_{ij}) = e$, we infer

$$\sigma_{ij} = \delta_{ij} \in M \implies e_{\lambda, b_0}(\sigma_{ij}) = 1, \tag{237}$$

and we have proved:

Theorem 3. *Let $n = 3$, $v = e_{\lambda, b_0}$, and let ψ be an eigendistribution of $\tilde{H}_{SM}$ when $\sigma_{ij} = \delta_{ij}$ such that*

$$-\tilde{\Delta} e_{\lambda, b_0} = (|\lambda|^2 + 1) e_{\lambda, b_0}, \tag{238}$$

$$\tilde{H}_{SM}\psi = \lambda_1 \psi, \qquad \lambda_1 \geq 0, \tag{239}$$

and let w be a solution of the ODE

$$t^{-m}\frac{\partial}{\partial t}(t^m \frac{\partial w}{\partial t}) + \frac{32}{3}(|\lambda|^2 + 1)t^{-2}w + \frac{32}{3}\alpha_N^{-1}\lambda_1 t^{-\frac{2}{3}}w + \frac{64}{3}\alpha_N^{-2}\Lambda t^2 w = 0 \tag{240}$$

then

$$u = w e_{\lambda, b_0} \psi \tag{241}$$

is a solution of the Wheeler–DeWitt equation

$$Hu = 0, \qquad (242)$$

where e_{λ,b_0} is evaluated at $\sigma_{ij} = \delta_{ij}$ and where we note that $m = 5$.

We shall refer to e_{λ,b_0} and ψ as the spatial eigenfunctions and to w as the temporal eigenfunction.

Remark 5. *We could also apply the respective Fourier transforms to $-\tilde{\Delta} e_{\lambda,b_0}$ resp. $\tilde{H}_{SM}\psi$ and consider*

$$w \hat{e}_{\lambda,b_0} \hat{\psi} \qquad (243)$$

as the solution in the Fourier space, where $\hat{\psi}$ would be expressed with the help of the ladder operators.

In the next section, we shall analyze the temporal eigenfunctions.

5. Temporal Eigenfunctions

The temporal eigenfunctions have to satisfy the ODE (240) or equivalently

$$\ddot{w} + 5t^{-1}\dot{w} + \frac{32}{3}(|\lambda|^2 + 1)t^{-2}w + \frac{32}{3}\alpha_N^{-1}\lambda_1 t^{-\frac{2}{3}}w \\ + \frac{64}{3}\alpha_N^{-2}\Lambda t^2 w = 0, \qquad (244)$$

where we used that $m = 5$, since we assume $n = 3$. Let us denote the other constants in front of the three lower order terms by m_i, m_2^2 resp. m_3, then the ODE appears as

$$\ddot{w} + 5t^{-1}\dot{w} + m_1 t^{-2}w + m_2^2 t^{-\frac{2}{3}}w + m_3 t^2 w = 0, \qquad (245)$$

where

$$m_1 \geq \frac{32}{3}, \quad m_2 \geq 0, \quad m_3 \in \mathbb{R}. \qquad (246)$$

The ODE (245) has two linearly independent solutions that are smooth and defined for all $t > 0$. However, if m_2, as well as m_3 are both different from zero, then the solution cannot be expressed by known functions, such as variants of the Bessel functions. Only if this is not valid, the solutions can be expressed by known functions.

Theorem 4. *Assume $m_3 = 0$ and $m_2 > 0$, then the solutions of the ODE (245) are generated by*

$$J(\tfrac{3}{2}\sqrt{m_1 - 4}\, i, \tfrac{3}{2} m_2 t^{\frac{2}{3}}) t^{-2} \qquad (247)$$

and

$$J(-\tfrac{3}{2}\sqrt{m_1 - 4}\, i, \tfrac{3}{2} m_2 t^{\frac{2}{3}}) t^{-2}, \qquad (248)$$

where $J(\lambda, t)$ is the Bessel function of the first kind.

Proof. We used Mathematica to obtain these solutions. The verification that these functions are indeed solutions is straightforward. □

Lemma 4. *The solutions in the theorem above diverge to complex infinity if t tends to zero and they converge to zero if t tends to infinity.*

Proof. The results can be derived by looking at a series expansion of the corresponding Bessel functions near the origin resp. near infinity. □

Next, let us consider the solutions when $m_2 = 0$ and $m_3 \neq 0$. Then we distinguish two cases $m_3 > 0$ resp. $m_3 < 0$. For a better distinction, we shall express m_3 in the form

$$m_3 = m_4^2, \quad m_4 > 0, \tag{249}$$

in the first case, and as

$$m_3 = -m_4^2, \quad m_4 > 0, \tag{250}$$

in the second case.

Theorem 5. *Assume $m_2 = 0$ and $m_3 > 0$, then the solutions of the ODE (245) are generated by the functions*

$$J(\tfrac{1}{2}\sqrt{m_1 - 4}\, i, \tfrac{1}{2} m_4 t^2) t^{-2} \tag{251}$$

and

$$J(-\tfrac{1}{2}\sqrt{m_1 - 4}\, i, \tfrac{1}{2} m_4 t^2) t^{-2}, \tag{252}$$

where $J(\lambda, t)$ is the Bessel function of the first kind.

Similarly, we obtain in the second case:

Theorem 6. *Assume $m_2 = 0$ and $m_3 < 0$, then the solutions of the ODE (245) are generated by the functions*

$$I(\tfrac{1}{2}\sqrt{m_1 - 4}\, i, \tfrac{1}{2} m_4 t^2) t^{-2} \tag{253}$$

and

$$I(-\tfrac{1}{2}\sqrt{m_1 - 4}\, i, \tfrac{1}{2} m_4 t^2) t^{-2}, \tag{254}$$

where $I(\lambda, t)$ is the modified Bessel function of the first kind. In Mathematica, this function is denoted by BesselI$[\lambda, t]$.

The arguments in the proof of Theorem 4 also apply in the case of Theorems 5 and 6.

Lemma 5. *The solutions in Theorem 5 resp. Theorem 6 diverge to complex infinity if t tends to zero, as well as if t tends to infinity.*

Proof. The same arguments as in the proof of Lemma 4 apply. □

6. Conclusions

The temporal eigenfunctions in the theorems of the previous section all become unbounded if $t \to 0$, which can be described as a big bang on a quantum level. Furthermore, if we consider $t < 0$, then the functions

$$\tilde{w}(t) = w(-t), \quad t < 0, \tag{255}$$

also satisfy the ODE (244) for $t < 0$, if we replace $t^{-\frac{2}{3}}$ by $|t|^{-\frac{2}{3}}$, i.e., they are also temporal eigenfunctions if the light cone in E is flipped.

Thus, we conclude

Theorem 7. *The quantum model we derived for gravity combined with the forces of the standard model can be described by products of spatial and temporal eigenfunctions of corresponding self-adjoint operators with a continuous spectrum.*

We have a zero-point energy state as a spatial eigendistribution of the gravitational Hamiltonian with the smallest eigenvalue $|\rho|^2 = 1$, which could be considered the source of the dark energy.

Furthermore, we have a big bang singularity in $t = 0$. Since the same quantum model is also valid by switching from $t > 0$ to $t < 0$, with appropriate changes to the temporal eigenfunctions, one could argue that at the big bang, two universes with different time orientations could have been created, such that, in view of the CPT theorem, one was filled with matter and the other with anti-matter.

Remark 6. *One of the reviewers raised two questions. First, he wondered about the logic to combine a low energy event, the quantization of the fields of the standard model with a flat metric, with an high-energy event, the quantization of gravity. As we have already pointed out in the introduction, a unified quantization of gravity and matter fields leads to a hyperbolic equation of second order in a fiber space, where the main part of the hyperbolic operator acts in the fibers. The zero-order terms of the operator contain the contributions of the quantized matter Hamiltonian and the interaction of gravity with matter fields occurs with the help of the fiber variables (t, σ_{ij}). The metric σ_{ij} is used in the quantization of the matter fields. Looking at the spatial eigenfunction v of the gravitational Hamiltonian and its eigenvalue, which expresses the energy, then the eigenvalue is independent of the metric σ_{ij} at which v is evaluated and only the evaluation point is relevant for the interaction, i.e., even if a non-flat metric σ_{ij} would have been used in the quantization of the matter fields, the contribution to the unified operator would not have changed qualitatively. Furthermore, as we already mentioned in the introduction, due to the scalar curvature term R, we cannot expect to solve the Wheeler–DeWitt equation for all (t, σ_{ij}) if we use the separation of variables, instead, we have to choose metrics with constant scalar curvatures. Thus, we opted for $\sigma_{ij} = \delta_{ij}$, also out of necessity because we could not quantize the matter field in the curved spacetime.*

The second interaction with respect to the variable t, the quantum time, is realized in the ODE, where the contributions by the spatial gravitational resp. matter eigenfunctions and also by the cosmological constant Λ have a power of t as a multiplicative factor with different exponents. For small t, the gravitational energy dominates because of the factor t^{-2}, for larger t, the matter energy dominates because of the factor $t^{-\frac{2}{3}}$, and if $\Lambda \neq 0$, then the cosmological constant dominates for very large t because of the factor t^2. This is also reflected in the results of Lemmas 4 and 5.

The second question raised concerned the QFT renormalizability in this unified setting.

The quantization of gravity takes place in the fibers of E while the quantization of the matter fields takes place in the base space $S_0 = \mathbb{R}^n$, which we equipped with the Euclidean metric for this task. Hence, the usual renormalization techniques can be used to deal with infinities. The fibers are ignored in this process.

Remark 7. *The Academic Editor of the journal also requested some observational predictions of the theory presented in this paper.*

In Theorem 7, we already offered possible answers to two open questions, namely, the source of the dark energy and why matter dominates anti-matter.

The Big Bang is only predicted by the singularity of the Friedmann model, a classical theory. In this paper, the Big Bang is predicted on a quantum level, which is a more appropriate level because the Big Bang is certainly a quantum event.

Powerful gravitational waves might be caused by quantum gravitational forces, such as the collision of two black holes. If this is the case, then they should satisfy an ODE similar to what we analyzed in Section 5. The patterns produced by the wave detectors should be similar to the plots produced by the solutions of the ODE in (244) on page 23, though the scalar curvature term does not appear in the ODE since $R(\delta_{ij}) = 0$, and in the case of black holes, R would be constant but different from zero, i.e., the ODE should contain a term, probably positive, with the factor $t^{\frac{2}{3}}$, and most likely no contribution by the standard model fields.

Funding: This research received no external funding.

Data Availability Statement: Not applicable.

Conflicts of Interest: The author declares no conflict of interest.

References

1. Gerhardt, C. The quantization of gravity in globally hyperbolic spacetimes. *Adv. Theor. Math. Phys.* **2013**, *17*, 1357–1391. [CrossRef]
2. Arnowitt, R.; Deser, S.; Misner, C.W. The dynamics of general relativity. In *Gravitation: An Introduction to Current Research*; Louis, W., Ed.; John Wiley: New York, NY, USA, 1962; pp. 227–265.
3. Gerhardt, C. The quantization of gravity: Quantization of the Hamilton equations. *Universe* **2021**, *7*, 91. [CrossRef]
4. Gerhardt, C. The quantization of a black hole. *arXiv* **2016**, arXiv:1608.08209.
5. Gerhardt, C. The quantization of a Kerr-AdS black hole. *Adv. Math. Phys.* **2018**, *2018*, 4328312. [CrossRef]
6. Gerhardt, C. *The Quantization of Gravity*, 1st ed.; Fundamental Theories of Physics; Springer: Cham, Switzerland, 2018; Volume 194. [CrossRef]
7. Eguchi, T.; Gilkey, P.B.; Hanson, A.J. Gravitation, gauge theories and differential geometry. *Phys. Rep.* **1980**, *66*, 213. [CrossRef]
8. Gerhardt, C. A unified quantum theory I: Gravity interacting with a Yang-Mills field. *Adv. Theor. Math. Phys.* **2014**, *18*, 1043–1062. [CrossRef]
9. Mackey, G.W. *The Mathematical Foundations of Quantum Mechanics: A Lecture-Note Volume*; W.A. Benjamin, Inc.: New York, NY, USA, 1963.
10. Casalbuoni, R. The classical mechanics for Bose-Fermi systems. *Il Nuovo C. A (1971–1996)* **1976**, *33*, 389–431. [CrossRef]
11. Casalbuoni, R. On the quantization of systems with anticommuting variables. *Il Nuovo C. A (1971–1996)* **1976**, *33*, 115–125. [CrossRef]
12. Gerhardt, C. The quantization of gravity. *Adv. Theor. Math. Phys.* **2018**, *22*, 709–757. [CrossRef]
13. DeWitt, B.S. Quantum Theory of Gravity. I. The Canonical Theory. *Phys. Rev.* **1967**, *160*, 1113–1148. [CrossRef]
14. Jorgenson, J.; Lang, S. *Spherical Inversion on SLn(R)*; Springer: New York, NY, USA, 2001. [CrossRef]
15. Helgason, S. Geometric analysis on symmetric spaces, Mathematical surveys and monographs. *Am. Math. Soc.* **1994**, *39*. [CrossRef]

MDPI AG
Grosspeteranlage 5
4052 Basel
Switzerland
Tel.: +41 61 683 77 34

Universe Editorial Office
E-mail: universe@mdpi.com
www.mdpi.com/journal/universe

Disclaimer/Publisher's Note: The statements, opinions and data contained in all publications are solely those of the individual author(s) and contributor(s) and not of MDPI and/or the editor(s). MDPI and/or the editor(s) disclaim responsibility for any injury to people or property resulting from any ideas, methods, instructions or products referred to in the content.